Advancements in Powder Metallurgy:

Processing, Applications, and Properties

Shashanka Rajendrachari
Bartin University, Turkey

Rayappa Shrinivas Mahale
Jain College of Engineering and Research, India

A volume in the Advances in Chemical and Materials Engineering (ACME) Book Series

IGI Global
PUBLISHER of TIMELY KNOWLEDGE

Published in the United States of America by
IGI Global
Engineering Science Reference (an imprint of IGI Global)
701 E. Chocolate Avenue
Hershey PA, USA 17033
Tel: 717-533-8845
Fax: 717-533-8661
E-mail: cust@igi-global.com
Web site: http://www.igi-global.com

Library of Congress Cataloging-in-Publication Data

Names: Rajendrachari, Shashanka, 1986- editor. | Mahale, Rayappa (Rayappa Shrinivas), 1986- editor.
Title: Advancements in powder metallurgy : processing, applications, and properties / edited by Shashanka Rajendrachari, Rayappa Mahale.
Description: Hershey PA : Engineering Science Reference, [2024] | Includes bibliographical references. | Summary: "This book covers most of the basic concept of mechanical alloying along with recent advancements. Most importantly it covers almost all types of ferrous and non-ferrous alloys, their processing, properties and novel applications"-- Provided by publisher.
Identifiers: LCCN 2023038628 (print) | LCCN 2023038629 (ebook) | ISBN 9781668493854 (hardcover) | ISBN 9781668493861 (paperback) | ISBN 9781668493878 (ebook)
Subjects: LCSH: Powder metallurgy.
Classification: LCC TN695 .A34 2024 (print) | LCC TN695 (ebook) | DDC 671.3/7--dc23/eng/20240226
LC record available at https://lccn.loc.gov/2023038628
LC ebook record available at https://lccn.loc.gov/2023038629

This book is published in the IGI Global book series Advances in Chemical and Materials Engineering (ACME) (ISSN: 2327-5448; eISSN: 2327-5456)

British Cataloguing in Publication Data
A Cataloguing in Publication record for this book is available from the British Library.

Advances in Chemical and Materials Engineering (ACME) Book Series

ISSN:2327-5448
EISSN:2327-5456

Editor-in-Chief: J. Paulo Davim, University of Aveiro, Portugal

MISSION

The cross disciplinary approach of chemical and materials engineering is rapidly growing as it applies to the study of educational, scientific and industrial research activities by solving complex chemical problems using computational techniques and statistical methods.

The **Advances in Chemical and Materials Engineering (ACME) Book Series** provides research on the recent advances throughout computational and statistical methods of analysis and modeling. This series brings together collaboration between chemists, engineers, statisticians, and computer scientists and offers a wealth of knowledge and useful tools to academics, practitioners, and professionals through high quality publications.

COVERAGE

- Sustainable Materials
- Biomaterials
- Industrial Chemistry
- Metallic Alloys
- Wear of Materials
- Nanomaterials
- Heat Treatments
- Artificial Intelligence Methods
- Fracture Mechanics
- Ductility and Crack-Resistance

IGI Global is currently accepting manuscripts for publication within this series. To submit a proposal for a volume in this series, please contact our Acquisition Editors at Acquisitions@igi-global.com or visit: http://www.igi-global.com/publish/.

The Advances in Chemical and Materials Engineering (ACME) Book Series (ISSN 2327-5448) is published by IGI Global, 701 E. Chocolate Avenue, Hershey, PA 17033-1240, USA, www.igi-global.com. This series is composed of titles available for purchase individually; each title is edited to be contextually exclusive from any other title within the series. For pricing and ordering information please visit http://www.igi-global.com/book-series/advances-chemical-materials-engineering/73687. Postmaster: Send all address changes to above address.

Titles in this Series

For a list of additional titles in this series, please visit:
http://www.igi-global.com/book-series/advances-chemical-materials-engineering/73687

Machining Polymer Matrix Composites Tools, Techniques, and Sustainability
Francisco Mata Cabrera (Universidad de Castilla-La Mancha, Spain) and Issam Hanafi (Université Abdelmalek Essaâdi, Morocco)
Engineering Science Reference • © 2024 • 320pp • H/C (ISBN: 9781668499276) • US $300.00

Building a Low-Carbon Society Through Applied Environmental Materials Science
Takaomi Kobayashi (Nagaoka University of Technology, Japan)
Engineering Science Reference • © 2024 • 300pp • H/C (ISBN: 9798369300039) • US $255.00

FlexPDE and Finite Element Method Applications in Thermal Energy Storage and Cavities
Khashayar Hosseinzadeh (Sharif University of Technology, Iran) Mehdi Mahboobtosi (Babol Noshirvani University of Technology, Iran) and Davood Domairry Ganji (Babol Noshirvani University of Technology, Iran)
Engineering Science Reference • © 2024 • 300pp • H/C (ISBN: 9798369309322) • US $225.00

Innovative Multifunctional Nanomaterial for Photocatalysis, Sensing, and Imaging
Azad Kumar (Siddharth University, India)
Engineering Science Reference • © 2023 • 283pp • H/C (ISBN: 9781668487433) • US $270.00

Innovations in Materials Chemistry, Physics, and Engineering Research
Eugene de Silva (Virginia Research Institute, USA) and Pramudi Abeydeera (Polymeric Ltd., USA)
Engineering Science Reference • © 2023 • 446pp • H/C (ISBN: 9781668468302) • US $270.00

Handbook of Research on Advanced Functional Materials for Orthopedic Applications
R. Ranjith (SNS College of Technology, India) and J. Paulo Davim (University of Aveiro, Portugal)
Engineering Science Reference • © 2023 • 371pp • H/C (ISBN: 9781668474129) • US $325.00

IGI Global
PUBLISHER of TIMELY KNOWLEDGE

701 East Chocolate Avenue, Hershey, PA 17033, USA
Tel: 717-533-8845 x100 • Fax: 717-533-8661
E-Mail: cust@igi-global.com • www.igi-global.com

Table of Contents

Detailed Table of Contents

This chapter discusses the distinguishing features between wet and dry ball milling processes used in industries. Wet milling and dry milling are the two categories under which industrial milling are most frequently employed in the manufacturing sector. It is important to comprehend the fundamentals in order to understand the differences between the two and the difficulties associated with the milling process. The process of milling involves applying various mechanical forces, such as grinding media, pegs, rods, pebbles, and screens to break down particles. The components of the mill work on the mixture's solids to separate or crush them, further reducing their size, as the material is pumped through it.

Development of electrochemical sensors and its research extends over a wide range of applications including clinical analysis, food safety, environmental monitoring, and many more. Electrochemical sensing devices are more convenient for the analytical determination of various electroactive analytes due to its simple fabrication procedure and the reliability in the obtained results. The sensitivity of the electrochemical sensors can be improved by using various chemically modified electrodes instead of traditional unmodified electrodes. Mechanically alloyed materials are one of the fascinating materials used for the modification of electrodes in the development of electrochemical sensors. Metal alloys and their nano porous structures provide good conductivity and high surface area making it more convenient for the sensing applications. The chapter summarizes the analytical characteristics of the mechanically synthesized metal alloys modified electrochemical sensors giving a special emphasis to various dyes, neurotransmitters, biological molecules, drugs, etc. as the target analytes.

This chapter introduces the emerging field of high entropy ceramics, which are characterized by their complex and disordered structures resulting from the amalgamation of multiple elements in nearly equal amounts. Unlike traditional ceramics composed of one or a few elements, high entropy ceramics exhibit exceptional properties such as high strength, hardness, and thermal stability. The chapter delves into high-entropy ceramics' synthesis, properties, and potential applications, highlighting their versatility in various fields. Additionally,

applications, with challenges to address for full potential realization. This chapter discusses the different approaches opted for mechanical alloying of magnesium and titanium alloys.

SMA (shape memory alloys) materials are a kind of smart material that undergoes shape changes with respect to temperature variations. In this chapter, the authors focus on the microstructure, classifications, fabrication, and future applications of these materials in brief. They discuss the four main types of SMA processing techniques, which include mechanical, thermomechanical, powder metallurgy, and melting route. Finally, its applications in various fields like transportation, civil structures, robotics, aerospace, automation, and medical fields will be briefly discussed.

Researchers face a significant challenge in designing low-carbon bainite rail steel, which offers better mechanical properties and corrosion resistance than high-carbon pearlitic steel tracks. High-carbon bainitic steel takes a long time to manufacture, so low-carbon bainite steel is preferred for heavy-haul railway tracks due to its shorter production time and cost-effectiveness. The steel is prepared through a specific heat treatment process, and although carbide formation is limited at lower temperatures, precipitation hardening is used to improve its properties. Mechanical alloying with the elements like Si, Ni, and Co is also considered to enhance further the mechanical and chemical properties of the bainitic rail steel. This chapter aims to find an economical solution to reduce accidents caused by the failure of pearlitic steel and improve safety in rail transportation.

The process of sintering mechanically alloyed powders entails subjecting a powder mixture of diverse metals or alloys, which have undergone mechanical milling to elevated temperatures to produce a compacted mass. This procedure is employed to fabricate substances possessing distinctive mechanical, physical, and chemical characteristics. The process of mechanical alloying entails the milling of two or more metals or alloys together in a high-energy ball mill. The procedure yields a powder amalgamation that exhibits a uniform dispersion of its constituent elements, thereby engendering a microstructural morphology characterized by fine-grained features.

In mineral processing, particle size is an important factor to consider since it affects plant design and performance, which later determines the plant's economic return. A certain particle size above and below the mine's optimal size range can adversely influence the recovery of valuable minerals. Particle size analysis is crucial in mechanical alloying controlling alloy composition and homogeneity of the mixture, affecting milling kinetics, reducing contamination, optimizing the milling process, and predicting properties. This chapter includes the understanding of particle size distribution, which helps researchers to tailor the alloying process for specific applications. The trend has been the wet-sieving analysis as a particle size analysis tool for the past 25 years. It provides an overview of different industries where the particle size analysis is used and associated end-users. The chapter also reports on the variables and principles governing factors that affect wet sieving analysis and the challenges that can give misleading and erroneous particle size distribution results.

Preface

Welcome to *Advancements in Powder Metallurgy: Processing, Applications, and Properties*. As editors of this comprehensive reference book, we are thrilled to present a compilation of nine insightful chapters, each authored by distinguished pioneers in the field of powder metallurgy from around the globe. This collective effort aims to provide a thorough exploration of the evolution, methods, and applications of mechanical alloying, spanning from its historical roots to contemporary advancements.

In this book, readers will embark on a journey through the intricate realm of powder metallurgy, specifically focusing on the technique of mechanical alloying. The narrative unfolds with a historical overview of the evolution of mechanical alloying, paying tribute to the innovators who shaped its development.

The subsequent chapters delve into the nuanced landscape of mechanical alloying, encompassing topics such as types of ball milling, optimization of milling parameters, advantages over conventional powder metallurgical methods, achieving maximum density, and characterization of mechanically alloyed powders. The discussions extend to the synthesis of various alloys, including high entropy alloys, oxide dispersed stainless steels, iron-based alloys, tungsten heavy alloys, cermets, titanium, magnesium, aluminum alloys, and shape memory alloys.

ORGANIZATION OF THE BOOK

Chapter 1: This chapter serves as an illuminating exploration of the distinctions between wet and dry ball milling processes, prevalent in industrial applications. Offering a comprehensive understanding of the fundamental principles underlying these milling techniques, the chapter sheds light on the mechanical forces at play—utilizing grinding media, pegs, rods, and screens—to break down particles. Readers gain insights into the intricacies of the milling process, where various components within the mill act on solid mixtures, facilitating the separation and size reduction of particles.

Chapter 2: Delving into the expansive realm of electrochemical sensors, this chapter presents an overview of their development and diverse applications, from clinical analysis to environmental monitoring. A special focus is placed on the utilization of mechanically alloyed materials in modifying electrodes for enhanced sensor performance. Readers will find a detailed exploration of chemically modified electrodes, emphasizing the analytical characteristics of mechanically synthesized metal alloys in electrochemical sensing applications targeting dyes, neurotransmitters, biological molecules, drugs, and more.

Chapter 3: This chapter introduces readers to the emerging field of high entropy ceramics, characterized by complex and disordered structures resulting from the amalgamation of multiple elements. Providing insights into synthesis, properties, and potential applications, the chapter showcases the exceptional attributes of high-entropy ceramics, such as high strength, hardness, and thermal stability. Readers will also gain a comprehensive understanding of the fabrication techniques involved, encompassing material selection, powder preparation, powder compaction, and sintering.

Chapter 4: A concise exploration of mechanical alloying's application to aluminum alloys, this chapter navigates through the principles, advantages, and potential applications of this solid-state powder processing technique. Detailing the process of subjecting aluminum powder and alloying elements to intense mechanical deformation, the chapter elucidates the creation of a homogenous and fine-grained microstructure. Readers will appreciate the unique alloying mechanisms involving solid-state diffusion and deformation-induced mixing, distinguishing mechanical alloying from conventional methods.

Chapter 5: This chapter unveils the realm of advanced Mg-Ti alloys synthesized through controlled microstructural changes induced by high-energy ball milling. Parameters such as duration, ball-to-powder ratio, and speed are dissected for their impact on microstructure, revealing phase changes and element distribution. The analytical tools employed shed light on the enhanced mechanical properties of Mg-Ti alloys, promising applications in aerospace and automotive sectors. Challenges and approaches in realizing the full potential of mechanical alloying for these alloys are thoughtfully discussed.

Chapter 6: Focusing on shape memory alloys (SMA), this chapter offers a brief yet comprehensive exploration of their microstructure, classifications, fabrication techniques, and future applications. The four main types of SMA processing techniques—mechanical, thermomechanical, powder metallurgy, and melting route—are outlined, providing readers with a foundational understanding. The chapter concludes with a glimpse into diverse fields where SMAs find applications, from transportation and civil structures to robotics, aerospace, automation, and the medical domain.

Chapter 7: This chapter addresses the significant challenge of designing low-carbon bainite rail steel, surpassing high-carbon pearlitic steel in mechanical properties and corrosion resistance. The focus on time-efficient production makes low-carbon bainite steel favorable for heavy-haul railway tracks. Specific heat treatment processes and the role of elements like Si, Ni, and Co are discussed, along with the potential benefits of mechanical alloying. The chapter aims to propose an economical solution to enhance safety in rail transportation by reducing accidents associated with pearlitic steel failure.

Chapter 8: Exploring the process of sintering mechanically alloyed powders, this chapter elucidates the steps involved in subjecting a mechanically milled powder mixture to elevated temperatures. Emphasizing the production of substances with distinctive mechanical, physical, and chemical characteristics, the chapter provides insights into the uniform dispersion of constituent elements during mechanical alloying. The resulting microstructural morphology, characterized by fine-grained features, is discussed, highlighting the significance of this procedure in material fabrication.

Chapter 9: Addressing the crucial role of particle size in mineral processing and mechanical alloying, this chapter delves into its impact on plant design, performance, and economic return. Particle size analysis is explored as a vital factor in controlling alloy composition and homogeneity, optimizing the milling process, and predicting properties. The chapter provides an overview of the wet-sieving analysis tool, detailing its trend over the past 25 years, along with variables and principles governing factors affecting results. Various industries benefiting from particle size analysis are highlighted, offering a comprehensive understanding for researchers.

Each chapter in this book contributes unique insights and expertise to the expansive field of powder metallurgy and mechanical alloying, making it a valuable resource for scholars, researchers, and practitioners alike.

What sets this book apart is its inclusive approach, addressing both ferrous and non-ferrous alloys and offering valuable insights into the comparison between them concerning ease of preparation and advantages. Moreover, the book explores novel applications and electrochemical sensor applications of mechanically alloyed materials, expanding the reader's understanding of the diverse and evolving landscape of powder metallurgy.

Our motivation in crafting this book was to bridge the gap in existing literature, providing a resource that not only covers fundamental aspects but also includes advanced topics rarely found in conventional works. The content is presented in a manner accessible to a broad audience, from research scholars and engineering students to industrial professionals seeking to enhance their knowledge of mechanical alloying.

Three noteworthy features distinguish this book:

Unique Comparison of Wet and Dry Milling: The chapters dive into the distinctive comparison of wet and dry milling, exploring their effects on microstructures, properties, and morphology of alloy powders, a topic seldom found in other publications.

Comprehensive Coverage of Alloys: Unlike many existing works, our book encompasses the processing, properties, and applications of a wide array of alloys, including shape memory alloys, high entropy alloys, cermets, and various ferrous and non-ferrous alloys.

Insightful Comparison Studies: The chapters provide valuable insights into the preparation of ferrous vs. non-ferrous alloys through mechanical alloying, aiding readers in identifying which alloys are more amenable to this method.

We recognize the declining interest in metallurgy and mechanical engineering programs and aspire for this book to serve as a catalyst for renewed enthusiasm. By presenting complex concepts in an accessible manner, we hope to empower the next generation of researchers and engineers to explore the full spectrum of possibilities within the realm of mechanical alloying.

Thank you for embarking on this intellectual journey with us.

Shashanka Rajendrachari
Bartin University, Turkey

Rayappa Shrinivas Mahale
Jain College of Engineering and Research, India

Chapter 1
Comparison of Wet and Dry Milling

R. Raghavendra Rao
Malnad College of Engineering, India

B. N. Sharath
https://orcid.org/0000-0001-7521-1759
Malnad College of Engineering, India

K. S. Madhu
Malnad College of Engineering, India

S. Pradeep
Malnad College of Engineering, India

ABSTRACT

This chapter discusses the distinguishing features between wet and dry ball milling processes used in industries. Wet milling and dry milling are the two categories under which industrial milling are most frequently employed in the manufacturing sector. It is important to comprehend the fundamentals in order to understand the differences between the two and the difficulties associated with the milling process. The process of milling involves applying various mechanical forces, such as grinding media, pegs, rods, pebbles, and screens to break down particles. The components of the mill work on the mixture's solids to separate or crush them, further reducing their size, as the material is pumped through it.

1. INTRODUCTION

Milling processes are of utmost importance in diverse industries, serving as essential operations for the reduction of particle size and the processing of materials. Wet

DOI: 10.4018/978-1-6684-9385-4.ch001

milling and dry milling are two widely investigated and utilised milling techniques within the field. Wet milling is characterised by the utilisation of a liquid medium throughout the milling procedure, whereas dry milling is predicated on the absence of a liquid medium (Loh et al., 2015). The techniques provide notable benefits and pose distinct difficulties, contingent upon the application and intended result. The examination of wet and dry milling methods is of considerable importance owing to its implications for the efficiency of particle size reduction, the quality of the resulting product, the consumption of energy, and the impact on the environment (Cañellas, Al-Rifai, Padrela, Tajber, Khamiakova, Otava, & Geertman, 2023). Gaining a comprehensive understanding of the distinctive attributes and efficacy of each approach empowers researchers, engineers, and industry practitioners to make judicious choices when it comes to selecting the optimal milling method that aligns with their particular requirements.Milling is an essential procedure employed across various industries, encompassing pharmaceuticals, food processing, cosmetics, and chemical manufacturing. The process entails the reduction in size of solid materials, including grains, ores, and particles, in order to achieve the desired characteristics of the final product. Wet and dry milling are two distinct methodologies utilised in this process, each possessing its own distinct advantages and limitations (Protonotariou et al., 2021). Wet milling is a size reduction technique that involves the utilisation of a liquid medium, commonly water or a solvent, in the milling process. The substance intended for milling is submerged in a liquid medium, facilitating the creation of a mixture with a semi-liquid consistency. On the other hand, dry milling entails the utilisation of solid-state procedures, wherein the absence of any liquid is observed, and the substance is processed in its dehydrated state. The selection between wet and dry milling is contingent upon various factors, including the desired particle size, attributes of the product, and the inherent properties of the material undergoing processing. Both methods possess distinct advantages and limitations, underscoring the importance of comprehending their disparities and assessing their suitability for particular milling needs (Nykamp et al., 2002).

1.1 Brief Overview of Milling Processes and Their Significance in Various Industries

The history of wet and dry milling can be traced back to ancient times, as both methods have been used for processing various types of materials. Here's a brief overview of the history of wet and dry milling (Figure.1).

The earliest evidence of milling dates back to the Stone Age, where stone tools were used to grind grains and other food materials. In this primitive form of milling, the grains were crushed and ground between two stones, creating a coarse powder or paste. In ancient civilizations like Egypt and Greece, advancements in milling

Figure 1. Overview of milling processes and their significance in various industries

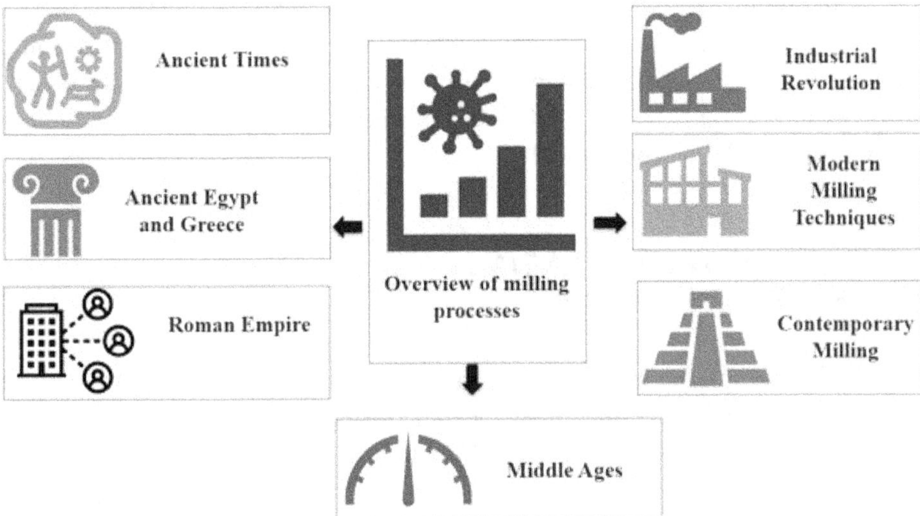

techniques were made. Hand-operated quern stones, which were flat stones with a concave lower stone and a convex upper stone, were used to grind grains. This method was labour-intensive and required a lot of physical effort. During the Roman Empire, the water mill was invented, which revolutionized milling. Water mills were powered by flowing water, which turned a waterwheel that in turn powered the millstones. This allowed for a more efficient and continuous milling process. Wet milling, where grains were soaked in water before milling, became more common during this time. In the Middle Ages, windmills were introduced as another source of power for milling. Windmills used wind energy to turn the millstones and grind various materials. Both wet and dry milling methods continued to be used during this period. The Industrial Revolution in the 18th and 19th centuries brought significant advancements in milling technology. Steam engines were introduced, replacing water and wind power, and allowing mills to be located in more diverse areas. This led to the development of larger-scale milling operations. In the 20th century, milling techniques further evolved with the introduction of electricity and mechanization. Roller mills became widely used in the milling industry, providing more precise control over the milling process. Roller mills are typically used for dry milling, where grains are crushed between a series of rollers to produce flour or other finely ground products. Today, both wet and dry milling methods are used in various industries. Wet milling is commonly used in the pharmaceutical, chemical, and food industries, where materials are ground in the presence of a liquid medium. Dry milling, on the other hand, is often employed in the production of flour, cereals, and other dry powders. It's important to note that the specific techniques and equipment used in wet and dry milling have continued to evolve over time, incorporating advancements in technology and engineering principles. However, the basic principles of crushing, grinding, and separating materials through milling have remained constant throughout history (Clermont & De Haas, 2010; Fuerstenau & Abouzeid, 2002; Ogonowski et al., 2018; Wang & Forssberg, 2001).

2. UNVEILING THE POTENTIAL OF WET MILLING AND DRY MILLING

The process of milling holds significant importance throughout many sectors, encompassing food processing, medicines, and materials manufacture. The process entails the lowering of particle size and the fragmentation of solid materials into tiny particles that are easier to handle. There are two basic methods employed in the milling process, namely wet milling and dry milling. Each of these processes possesses distinct benefits and finds application in certain contexts (Wołosiewicz-Głąb et al., 2017).

2.1 Wet Milling

Wet milling is a multifaceted approach that is frequently utilised in the food, pharmaceutical, and chemical sectors. The process entails the use of a liquid medium, commonly water or solvents, to facilitate the decrease in size and subsequent segregation of particles. During the wet milling process, materials undergo processing in a liquid medium. This method is commonly employed for commodities that require reduced particle sizes and dispersion or are vulnerable to elevated temperatures. The method often encompasses the inclusion of the subsequent steps (Figure.2).

2.1.1 Mixing

The initial phase of wet milling is the process of mixing, when the raw material is blended with a liquid, commonly water or a suitable solvent. The combination of substances gives rise to a suspension or slurry, which acts as the initial stage in the milling procedure. The primary aim of the mixing process is to achieve a uniform amalgamation of the solid substance and the liquid, hence guaranteeing optimal efficiency and achieving the necessary decrease in particle size during following milling stages. Mechanical agitators, such as paddles or impellers, are frequently employed in order to induce turbulence within the mixture, hence facilitating the dispersion and homogeneity of particles. High-shear mixers are capable of generating significant shear forces, which effectively disintegrate particle agglomerates and result in the formation of a more homogeneous solution. Ultrasonic waves have the ability to disperse and deagglomerate particles by means of cavitation, a phenomenon characterised by the fast production and subsequent collapse of bubbles that create localised high energy. In certain instances, the raw material and liquid are combined in a continuous manner as they are introduced into the milling machine, therefore facilitating prompt processing. The process of mixing plays a crucial role in establishing the fundamental basis for achieving effective wet milling. Ensuring a homogeneous slurry ensures that the milling forces are evenly distributed across all particles, leading to a constant reduction in particle size and maintaining product quality. Furthermore, the appropriate blending of materials plays a crucial role in managing the increase in temperature throughout the milling process, particularly for substances that are sensitive to heat. This aspect significantly enhances the overall efficiency and efficacy of the procedure.

2.1.2 Size Reduction

The process of size reduction is a fundamental operation in milling procedures, whereby the primary objective is to disintegrate bigger particles into smaller counterparts. Ensuring uniformity in particle size distribution is a crucial step for enhancing the overall quality and performance of the final product. The objective of size reduction is to achieve a uniform and limited range of particle sizes by minimising the variability in particle size. The presence of consistency contributes to the improvement of both product quality and functioning. The surface area of particles is enhanced by the reduction of particle size. Enhanced surface contact, such as dissolution, reaction rates, or absorption, is of utmost importance in situations where it is necessary. The achievement of a uniform particle size distribution has been found to result in enhanced product consistency and performance. In certain instances, it may also serve to hinder the segregation of particles inside the ultimate product. The process of grinding entails the application of mechanical forces to the material in order to facilitate the fragmentation of its particles into smaller sizes. This can be accomplished by several techniques, including impact, compression, and attrition. The design and application of grinding machines exhibit variations, encompassing ball mills, hammer mills, and jet mills. The final qualities of the milled material are significantly influenced by the process of size reduction. The optimisation of particle size distribution contributes to the enhancement of several aspects of the product, including its performance, solubility, reactivity, and visual characteristics. Furthermore, this technique guarantees that every individual particle experiences the milling pressures in a homogeneous manner, resulting in consistent outcomes across the whole treated material. The selection of a size reduction technique is contingent upon several criteria, including the inherent properties of the material, the intended range of particle sizes, and the particular demands of the industry or application.

2.1.3 Separation

The process of separation plays a crucial role in milling operations, particularly in the context of wet milling, when the slurry consists of a combination of solid particles and liquid. The primary aim of this study is to separate distinct components, particles, or fractions by utilising their size, density, or other inherent characteristics. Separation procedures are employed for the purpose of extracting distinct constituents from a slurry, including particles within a certain size range, valuable compounds, pollutants, or by-products. Particles can be fractionated based on several parameters, such as size, density, or other factors, depending on the specific separation process employed. The process of separation is helpful in enhancing the purity of the end product by the elimination of undesirable particles or constituents. The process of separation facilitates the retrieval of valuable substances that may possess distinct characteristics compared to the other components of the mixture. Centrifugation is a process that entails subjecting the slurry to rapid rotation within a centrifuge at elevated velocities. The centrifugal force segregates particles according to their density, causing denser particles to migrate towards the periphery while lighter particles tend to remain closer to the central region. Filtration is a process that entails the passage of a slurry through a filter media, which exhibits the ability to selectively permit the passage of smaller particles or liquid, while effectively keeping bigger particles. The process of sedimentation facilitates the settling of particles due to the gravitational force. The settling rate is determined by the size and density of particles. After the process of settling, the distinct layers can be separated using decantation or collection. The process of separation plays a crucial role in the purification of the processed slurry, allowing for the extraction of specific components or particle fractions that are needed. This process is crucial in ensuring that the end product conforms to predetermined quality standards, is devoid of any impurities, and possesses the desired features as intended. The use of efficient separation processes plays a significant role in enhancing the purity, uniformity, and yield of products, while simultaneously reducing waste formation. The selection of a separation process is contingent upon several criteria, including the characteristics of the particles, the distribution of their sizes, and the particular demands of the industry or application at hand (Buchczik et al., 2017; Rosentrater & Bucklin, 2022; Santra & Schoenlechner, 2017).

Figure 2. Wet milling

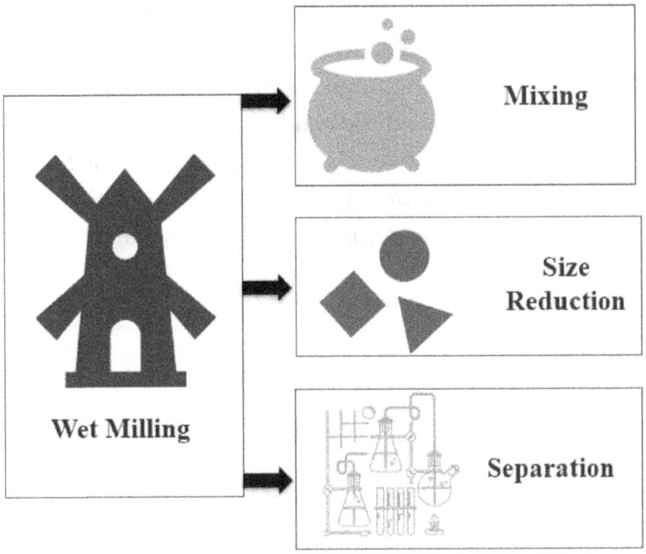

2.2 Dry Milling

As the name implies, dry milling entails processing materials in a dry state, often without the use of a liquid medium. When processing a material that is not heat-sensitive and when the finished product calls for larger particle sizes, this approach is frequently used. Usually, the procedure includes the following steps (Figure.3).

2.2.1 Size Reduction

The reduction in size is a fundamental operation inside dry milling procedures, with the objective of diminishing the particle size of the initial substance. Ensuring uniformity in particle size distribution is a crucial step in attaining consistency and improving the overall quality and usefulness of the end product. The objective of size reduction is to achieve a uniform distribution of particle sizes by the process of reducing bigger particles into smaller ones. Enhancing product quality and performance is achieved via the implementation of a restricted and consistent particle size distribution. The reduction in particle size leads to an increase in the surface area of the material, hence playing a significant role in several processes like dissolution, chemical reactions, and enhanced reactivity. The achievement of a uniform particle size distribution is advantageous as it promotes product consistency, mitigates segregation, and guarantees consistent performance throughout the material. The process of crushing entails the use of compressive pressures in order to disintegrate bigger particles into smaller bits. The earliest stage in the reduction of bulk materials generally involves this phase. The final qualities of the milled material are significantly influenced by the process of size reduction. The optimisation of particle size distribution is crucial in enhancing the performance, reactivity, and general appropriateness of the product for its intended use. The selection of a size reduction technique is contingent upon several criteria, including the inherent properties of the material, the intended range of particle sizes, and the specific demands imposed by the industry or application in question.

2.2.2 Mixing (Optional)

The technique of mixing or blending has significant importance in the dry milling procedure, particularly when the desired end result necessitates a precise amalgamation of constituents or characteristics. This procedure facilitates the formation of composite materials by the amalgamation of the desiccated milled substance with supplementary substances, fillers, or other constituents. The objective is to attain a state of uniform distribution and homogeneity among the different constituents present in the ultimate product. The primary aim of the process of mixing is to achieve a uniform distribution of all constituent elements inside the substance. As a consequence, there is a constant level of quality and performance observed in the product. The process of mixing facilitates the integration of various additives, such as colours, stabilisers, flavours, reinforcing agents, or functional elements, into the dry milled basic material. The process of blending provides the opportunity to tailor the characteristics of the ultimate product through the manipulation of component ratios and kinds. The incorporation of mixing or blending techniques in dry milling operations is necessary in order to get uniform and regulated characteristics in the ultimate result. This procedural measure guarantees the homogeneous distribution of additives or components, hence boosting the quality, stability, and functioning of the material. The process of blending is of considerable importance in the attainment of desired attributes, including but not limited to uniformity of colour, controlled release of active components, enhanced texture, and overall improvement of the product. The selection of a blending technique is contingent upon several criteria, including the inherent characteristics of the substances being mixed, the desired degree of uniformity, the magnitude of the manufacturing process, and the demands imposed by the industry or application in question.

2.2.3 Packaging

The act of packaging serves as the concluding phase inside the milling process, whereby the processed substance is encased within suitable receptacles to facilitate its storage, distribution, and further utilisation. The phase of packaging holds significant importance in terms of maintaining the quality of the product, safeguarding it from contamination, and delivering crucial information to customers. Packaging serves as a protective barrier that shields the processed material from various external elements, including moisture, air, light, and pollutants. This practise guarantees that the material maintains its inherent characteristics and level of excellence till it is received by the final consumer. The implementation of appropriate packing techniques serves to prolong the duration for which a product remains viable for consumption, as it effectively mitigates the potential negative impact of external factors that may contribute to its deterioration. The role of packaging is of utmost importance in guaranteeing the safety of the product, deterring tampering, and safeguarding customers from any risks. The selection of packing materials is influenced by several factors related to the processed material, including its moisture content, sensitivity to oxidation, and fragility. The determination of the appropriate packaging for a substance, whether it is in powder, granule, or other forms, is contingent upon preserving its integrity. The adherence to industry-specific requirements is crucial for packaging, particularly in areas such as food, medicines, and chemicals. The design of packaging should prioritise user-friendliness, ease of opening, and the option to reseal if required, in order to enhance the overall consumer experience. Packaging functions as the intermediary between the product and the consumer. In addition to maintaining the integrity of the product, it also acts as a means of conveying branding, usage instructions, and other pertinent information. The implementation of appropriate packing measures guarantees the delivery of a secure and superior product to the consumer, therefore satisfying their anticipated standards.

2.2.4 Sizing and Classification

The process of sizing and classification plays a pivotal role in the refining of milled materials, since it involves the segregation of particles into distinct size fractions. This procedure guarantees that the ultimate outcome possesses the intended particle size distribution, hence augmenting its quality, performance, and usefulness. The attainment of the specified particle size distribution and overall product quality is contingent upon the crucial aspects of sizing and classification. The appropriate sizing of particles is crucial in ensuring the desired functionality of the product, encompassing effective mixing, enhanced dissolving, and even aesthetic improvement. Particle size is a crucial metric that significantly impacts the performance, stability, and effectiveness of the final product in several sectors, particularly in medicines and materials research. Hence, precise determination of size and categorization play a pivotal role in adhering to regulatory and quality benchmarks. The selection of sizing and classification methods is contingent upon several criteria, including the required range of particle sizes, the properties of the particles, the volume of production, and the unique requirements of the industry (Chaira, 2014; Chaira, 2015; Rajendrachari et al., 2019; Shashanka & Chaira, 2015).

Figure 3. Dry milling

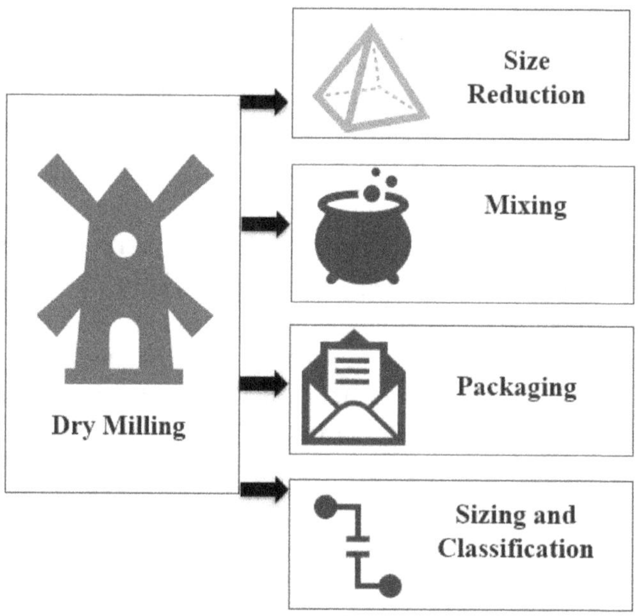

2.3 Comprehensive Look into The Process

Wet milling and dry milling are fundamental methodologies employed in the processing of materials across many industrial sectors. Both processes encompass the integration of mixing, size reduction, optional mixing or blending, sizing and classification (in the context of dry milling), and packing. The selection between the two options is contingent upon several criteria, including the inherent characteristics of the materials, the preferred distribution of particle sizes, and the specific demands of the industry. The comprehension of these procedures empowers firms to manufacture superior items that adhere to precise performance standards. Wet milling or dry milling should be chosen depending on the material's characteristics, the preferred particle size distribution, the likelihood of chemical reactions, and the intended application of the finished product. Dry milling is appropriate for materials that can withstand the mechanical forces involved in the process and where moisture is not a factor, but wet milling is frequently preferred for creating smaller particles and attaining greater dispersion in liquids. In conclusion, wet milling and dry milling are two different types of processing techniques used to change the size and characteristics of materials. Depending on the particular requirements of the application, each technique has its own benefits and considerations (Nayak et al., 2016; Rajendrachari, 2018).

2.4 Advantages of Wet Milling

For some materials and applications, wet milling is the preferable processing technique due to its many benefits. The following are some of the main benefits of wet milling:

When compared to dry milling, wet milling is known to produce finer and more uniform particle sizes. A liquid medium's presence aids in preventing agglomeration and encourages improved particle dispersion, resulting in a smaller particle size distribution. When wet milling, the use of a liquid media improves the dispersion of particles in the suspension. This is crucial for applications where uniformity and equal dispersion are essential, like the creation of medications, pigments, and coatings. Due to the utilisation of a liquid medium for processing, the probability of generating airborne particulate matter and contaminants is significantly reduced. The adherence to stringent hygiene and cleanliness standards is of utmost importance in industries such as food and pharmaceuticals. Wet milling is an appropriate processing method for materials that are susceptible to heat damage or high temperatures. The likelihood of thermal damage to the material is decreased by the liquid medium's ability to disperse heat produced during milling. When exposed to moisture or other reactive liquids, some materials experience chemical reactions. Wet milling can speed up and regulate these reactions, enabling the creation of unusual materials or compounds that dry milling would not be able to. Wet milling offers an efficient way to mix and combine various ingredients. This helps produce homogeneous compositions and consistent outcomes in goods like pastes, emulsions, and suspensions. Dry milling can be difficult to use when processing materials with adhesive qualities or high viscosities because agglomeration and clogging problems can occur. By keeping them floating inside the liquid medium, wet milling may manage such materials with effectiveness. The presence of a liquid medium can act as a lubricant, minimising the deterioration of milling machinery. This minimises maintenance expenses and increases the equipment's lifespan. Compared to some dry milling techniques, wet milling technologies can be scaled up from laboratory- to industrial-scale output with less disruption to the process. Liquid medium used in wet milling can offer a controlled processing environment, enabling improved control over variables like temperature, pH, and reaction kinetics.

Particle size control, dispersion, temperature sensitivity, and the capacity to handle materials with specific chemical reactivity or sticky qualities are all benefits of wet milling, to sum up. These benefits make wet milling an appropriate option for situations where obtaining fine particle sizes and homogeneity is essential and where the advantages of a liquid media outweigh the disadvantages (Bartos et al., 2016; Kelsey & Shafizadeh, 1980; Malamatari, Taylor, Malamataris, Douroumis, & Kachrimanis, 2018; Rausch et al., 2019; Yang et al., 2015).

Figure 4. Advantages of wet milling

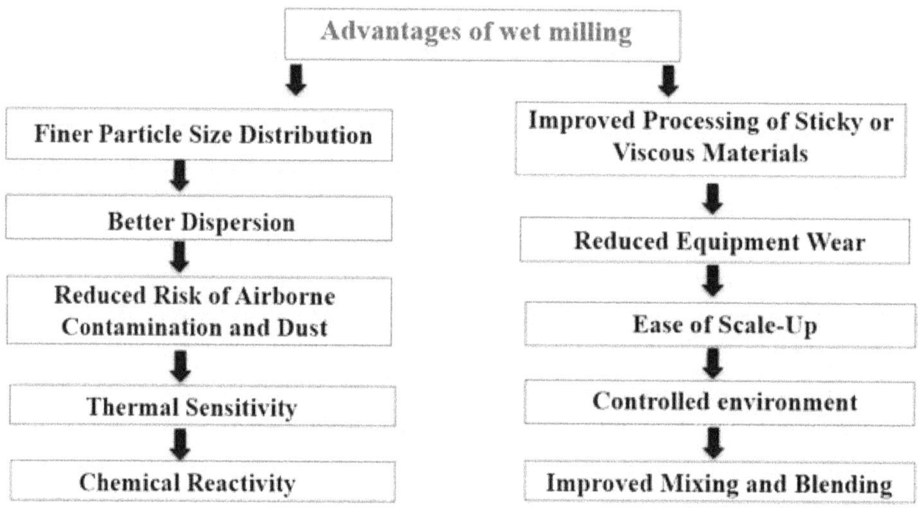

2.4 Advantages of Dry Milling

Dry milling is a better option for particular materials and applications due to its own set of benefits. The following are some of the main benefits of dry milling (Figure.5)

Since wet milling doesn't require the time-consuming drying step, dry milling is typically faster than wet milling. Particularly in high-volume production environments, this may result in greater productivity and efficiency. Compared to wet milling methods, which employ liquid media, dry milling typically uses less energy. Saving money and lessening the impact on the environment are possible outcomes. Dry

Figure 5. Advantages of dry milling

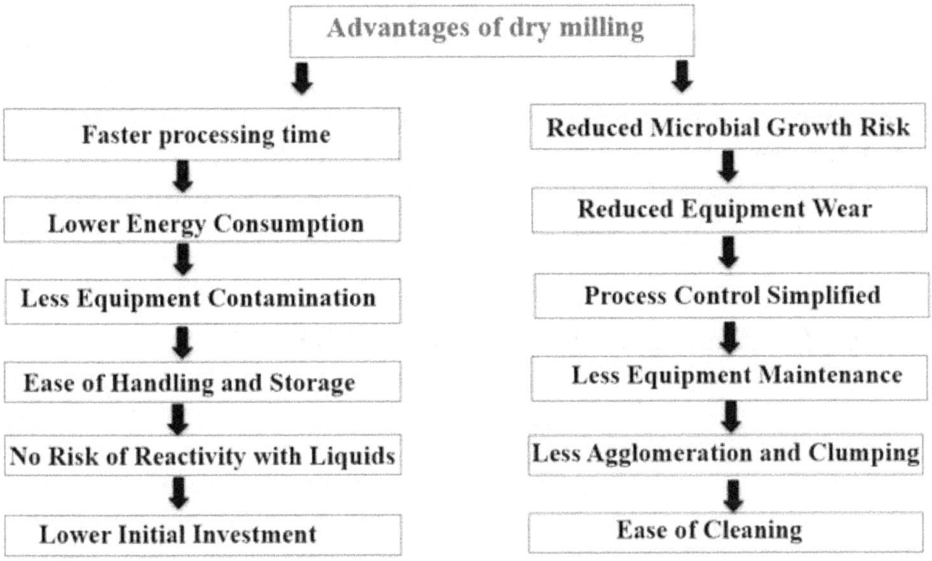

milling lowers the possibility of batch-to-batch contamination of various ingredients. There is less likelihood of residue buildup or material carryover between milling sessions when there is no liquid medium involved. Since dry milled materials don't contain additional moisture, they are frequently simpler to handle, package, and store. This can make logistics and transportation easier, especially for materials that are susceptible to deterioration in humid environments. Dry milling is appropriate for materials that might react with or degrade in the presence of liquids since there is no risk of reactivity with liquids. This makes it possible to treat materials that, in the past, could have undergone chemical changes upon coming into contact with water or other solvents. In some materials, moisture can encourage microbial growth and deterioration. Dry milling is a good option for products when microbial contamination is a concern because it helps to reduce this danger. The lack of a liquid medium may result in easier process control and monitoring. Processes are streamlined by removing factors like moisture content and liquid-to-solid ratios. Compared to wet milling equipment, which can be subjected to the corrosive effects of liquids over time, dry milling equipment typically experiences less wear and corrosion. Dry milling has the potential to be beneficial in reducing agglomerations and clumps in materials, improving product uniformity and quality. Compared to wet milling settings, which include pumps, filters, and tanks for liquid handling, dry milling configurations occasionally may require less complicated and expensive equipment. Cleaning wet milling equipment might be more difficult since liquid residues can be harder to entirely remove than dry milling equipment. In conclusion, dry milling has benefits such quicker processing, lower energy usage, less chance of cross-contamination, and the capacity to handle materials that are delicate to liquids. When working with materials that can be negatively impacted by moisture or when a streamlined process is sought, it is especially helpful. The exact material qualities, processing objectives, and end product will determine whether dry milling or wet milling is best (Adeyemi, 1983; Chen et al., 2011; Gupta et al., 2015; Rahman et al., 2001; Su et al., 2016; Uysal, 2016).

3. CHALLENGES AND LIMITATIONS OF WET MILLING

Wet milling has a number of benefits, but it also has some drawbacks and restrictions that must be taken into account while using it. Several of these difficulties include (Figure.6). The necessity to pump and circulate the liquid medium makes wet milling procedures frequently energy intensive. When compared to dry milling, this may lead to increased operating expenses. Wet milling arrangements can be more complicated and need specialised machinery to manage the liquid media, such as pumps, tanks, and filters. This intricacy can make operations more challenging overall and increase the need for maintenance. Before being used or going through more processing, wet milled products frequently need to go through a drying phase to get rid of extra moisture. This extra step can increase the cost and lengthen the production time overall. If correct hygiene and cleaning procedures are not followed, the presence of

Figure 6. Challenges and limitations of wet milling

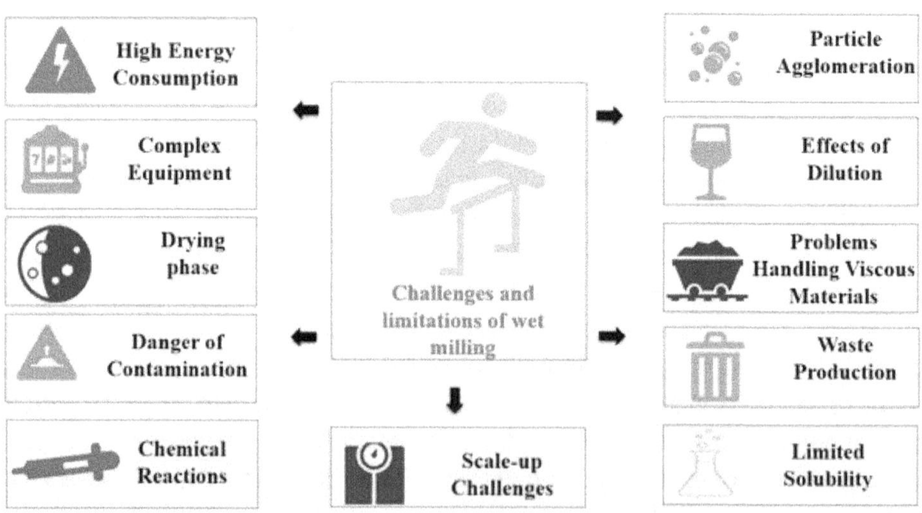

liquid might increase the danger of microbiological development and contamination. To ensure product safety, a clean environment must be maintained. Wet milling can cause undesirable chemical reactions in materials that are sensitive to moisture or reactive with the liquid media, even if it can be beneficial for some chemical processes. Wet milling can occasionally cause particle agglomeration or flocculation despite efforts to prevent it because of the attraction interactions that exist between particles in a liquid media. Wet milling can dilute the concentration of active substances or components, which may affect the performance of the finished product depending on the material and the liquid medium utilised. Due to their increased resistance to flow and mixing, processing very viscous materials in a liquid medium can be difficult. The wet milling process' separation step frequently produces waste in the form of liquid by products. These by products' proper disposal can raise issues from an environmental and legal standpoint. Some materials might not dissolve completely in the liquid medium of choice, which could make it challenging to get the required particle size distribution. Maintaining process efficiency, homogeneity, and quality can be difficult when scaling up wet milling techniques from the laboratory to the industrial level. Wet milling necessitates the use of water or other solvents, which can be expensive in areas with limited water supplies. This might affect how long the process lasts. In conclusion, wet milling has advantages like enhanced dispersion and finer particle size control, but it also has drawbacks like high energy consumption, complicated equipment, contamination hazards, and potential chemical reactions. To ensure the deployment is effective, these issues need to be carefully addressed and handled (Agimelen et al., 2018; Khot et al., 2020; Pavitra et al., 2022; Rivas-Cantu et al., 2013; Sarkar et al., 2012; Shashanka, 2017).

4. CHALLENGES AND LIMITATIONS OF DRY MILLING

While dry milling can be useful in some circumstances, it also has its own set of drawbacks. When determining if dry milling is the best processing option for a specific material and application, these criteria must be taken into account. The following are a few difficulties and restrictions with dry milling (Figure.7). Dry milling offers a wider range of particle size distribution in comparison to wet milling. Obtaining a consistent and narrow particle size distribution can pose challenges, especially for materials that have a tendency to agglomerate. The process of dry milling has the potential to generate particulate matter in the form of airborne dust, which poses a risk to the well-being of workers and has the potential to contaminate the surrounding environment. In order to mitigate these issues, it is imperative to implement suitable strategies for

Figure 7. Challenges and limitations of wet milling

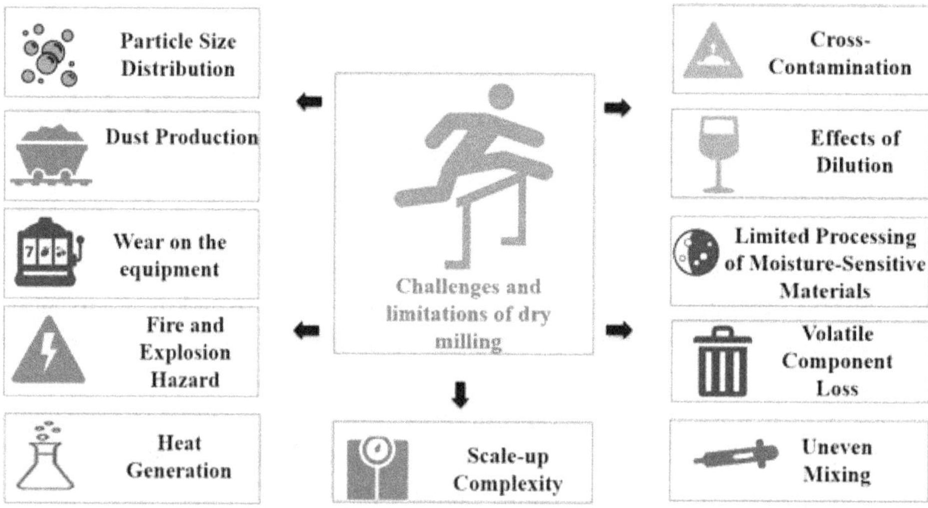

dust control. Dry milling procedures have been observed to exhibit a higher level of abrasiveness when compared to wet milling techniques. Dry milling is characterised by substantial impact and friction, leading to accelerated wear and tear of equipment components. Under certain conditions, it is possible for some materials utilised in the process of dry milling to undergo ignition or explosion when they are suspended in the air as minuscule dust particles. The prevention of fire and explosive hazards necessitates the implementation of sufficient safety protocols. The application of mechanical forces in the process of dry milling has the potential to generate thermal energy. The generation of excessive heat may lead to the deterioration of materials or the occurrence of undesirable chemical reactions, particularly for materials that are susceptible to thermal sensitivity. The dry milling apparatus has the potential to retain residual particles from previous batches, hence elevating the risk of cross-contamination among different substances. Dry milling is not advisable for materials that exhibit sensitivity to moisture, as prolonged exposure to air during the milling process may lead to unfavourable modifications in the material's characteristics. During the process of dry milling, evaporation leads to the loss of volatile components from some materials. The influence of this may affect the performance and composition of the final product. The presence of adhesive or cohesive materials can adhere to milling equipment, leading to operational disruptions characterised by an irregular distribution of particle sizes and blockages. Achieving complete homogeneity in the mixing of dry milled materials might provide challenges when dealing with powders that exhibit distinct properties or particle sizes. The utilisation of dry milling may not be suitable for processes that require a specific chemical reaction facilitated by a liquid medium. In contrast to wet milling, which employs a liquid medium, dry milling lacks the same level of control over environmental factors such as temperature and pH. The challenges of maintaining a consistent particle size distribution and preventing equipment issues might increase when transitioning from laboratory-scale to production-scale dry milling processes. In summary, it can be inferred that dry milling offers advantages such as expedited processing and reduced energy consumption. However, it is imperative to address certain challenges and limitations, namely particle size distribution, dust generation, and equipment deterioration, in a meticulous manner. When making the decision to utilise dry milling, it is imperative to take into account the distinct material characteristics, desired particle size distribution, and the ramifications of dust generation and equipment deterioration on both product quality and worker well-being (AboulFotouh et al., 2020; Bhirud & Gawande, 2017; Shashanka & Chaira, 2016; Shashanka & Chaira, 2017a; Vadnere & Kalpande, 2023).

5. COMPARATIVE ANALYSIS OF WET AND DRY MILLING TECHNIQUES

Various considerations, such as the parameters of the material, intended end product features, processing efficiency, and potential difficulties, must be taken into account when comparing wet versus dry milling procedures. To further understand the distinctions and factors to be aware of with each procedure, here is a comparison of wet and dry milling methods:

5.1 The Distribution of Particle Sizes

The utilisation of wet milling techniques often leads to the production of particles with finer and more uniform sizes, mostly due to enhanced dispersion within the liquid medium. Dry milling is known to have a tendency to generate particles that are coarser and form agglomerates, resulting in a greater dispersion of bigger particle sizes.

5.2 Energy Efficiency and Operational Costs

The energy consumption in wet milling is frequently higher due to the requirement for liquid handling, pumping, and additional drying activities. Dry milling is a process that does not necessitate the energy-intensive procedures of liquid handling and drying, hence frequently resulting in reduced energy consumption.

5.3 Reaction Time

The duration of wet milling may be extended due to the necessity of managing liquid substances and conducting supplementary drying procedures. Due to the absence of liquid management or drying procedures, dry milling often exhibits enhanced efficiency in terms of processing speed.

5.4 Sensitivity to Material

Wet milling is a suitable technique for processing materials that exhibit sensitivity to heat and moisture, as the utilisation of liquid medium serves to mitigate thermal degradation and facilitate heat dissipation. Dry milling is the method of choice for working with materials that are resistant to the effects of heat, moisture, and air reaction.

5.5 Dispersion and Agglomeration

The use of wet milling rather than dry milling helps provide higher dispersion in the liquid media and inhibits the agglomeration of particles. Dry milling can lead to agglomeration of particles, which makes it more difficult to obtain a dispersion that is uniform.

5.6 Equipment Complexity

The setups for wet milling can be more involved and require specialised apparatus for handling and separating liquids from the solids that are being milled. The configurations employed for dry milling operations are generally characterised by reduced complexity and the utilisation of less intricate machinery.

5.7 Cross-Examination

There is potentially a greater risk of cross-contamination because of the employment of complex machinery and a liquid media in the experiment. Dry milling is advantageous in terms of reducing the potential for cross-contamination due to the absence of liquid handling.

5.8 Environmental Effects

In regions characterised by restricted water resources, the wet milling methods can impose a significant demand on available resources. Dry milling processes are generally considered to have a reduced environmental impact because to their lower water and energy consumption.

5.9 Product Stability

Dry milling processes are generally considered to have a reduced environmental impact because to their lower water and energy consumption. The formation of dust during the dry milling process can potentially affect the stability and storage conditions of the milled goods.

5.10 Scaling Up

Scaling up wet milling procedures might provide challenges because to the complexities associated with liquid management and maintaining consistent dispersion. The ease of scaling up dry milling processes can be attributed to the utilisation of simpler equipment and a reduced number of variables. In summary, the determinants impacting the choice between wet and dry milling methodologies encompass the specific material, the desired particle size distribution, the efficacy of processing, and the potential ramifications related to elements such as energy consumption, agglomeration, and cross-contamination. Every approach possesses advantages and disadvantages, and the decision should be made through a meticulous evaluation of these factors in relation to the specific needs of the intended application (Ali et al., 2011; Cañellas, Al-Rifai, Padrela, Tajber, Khamiakova, Otava, & Geertman, 2023; Rajendrachari, 2022; Shashanka et al., 2016).

6. APPLICATION OF WET AND DRY MILLING PROCESSES

Wet milling and dry milling are distinct processing techniques employed in several industries, notably the food and pharmaceutical sectors, to effectively handle a diverse array of materials. Wet milling and dry milling find many applications in different industries.

6.1 Wet Milling

Wet milling is a processing technique that is heavily used in the production of maize ethanol. The process starts with soaking the maize kernels in water before removing the starch, germ, fibre, and protein components. The starch is then fermented, leading to the production of ethanol. Wet milling is a technique that can be used to reduce the particle size of many different substances, including pigments, dyes, and pharmaceutical components. Wet milling is a method for dissolving solid particles in a liquid medium, which ultimately produces smaller-sized particles. Wet milling is the preferred technique for distributing solid particles or oil droplets in a liquid medium. This approach is advantageous for the development of pharmaceuticals, artwork, inks, and personal care products. Wet milling is a method frequently used in the production of ceramic materials including porcelain tiles and pottery. It is beneficial to combine all of the raw ingredients when creating a slurry out of them, such as clay, feldspar, and water, which may then be moulded and dried before being fired.

6.1.1 Pharmaceuticals

Wet milling is used for drug formulation, decreasing particle size, and enhancing bioavailability, among other applications. Finer particle sizes are achieved, improving medication absorption and solubility. Controlling moisture content to stop deterioration or microbiological growth is a challenge.

6.1.2 Food and Drink

Wet milling is used to make pastes, emulsions, and suspensions. Guarantees consistency in texture, dispersion, and stability in goods like sauces, dressings, and beverages. Managing components with varying solubilities and avoiding spoiling are challenges.

6.1.3 Coatings and Paints

Wet milling is used for pigment dispersion and colour development in applications. Produces homogeneous pigment distribution and consistent colour in paints, inks, and coatings. Preventing pigment settling and achieving desired rheological characteristics.

6.2 Dry Milling

Dry milling is frequently used as one of the main processing techniques when producing flour from grains including wheat, maize, and rice. The grains are first washed, then tempered, and finally crushed into flour using dry milling techniques. By using dry milling, the endosperm, bran, and germ of the grain may be more easily separated. Dry milling is required for both the removal of the outer layers and the acquisition of the coffee beans that will be roasted during the processing of coffee beans. As part of the dry milling procedure, the coffee beans are sorted, polished, and hulled. Dry milling is widely used in the mining and mineral industry to grind a wide range of ores and minerals. The ore must first be broken down into smaller pieces before being processed in dry milling machinery, such as ball mills or hammer mills, until it is reduced to a fine powder in order to complete the processing. In the pharmaceutical industry, dry milling is a procedure used to create powdered versions of pharmaceuticals and active pharmaceutical ingredients (APIs). The process of dry milling aids in reducing the materials' particle size, which enhances their flow properties and speeds up their dissolution. It is important to be aware that depending on the industry and the chemical being processed, the specific applications of wet milling and dry milling may vary. The aforementioned examples offer a high-level review of the numerous applications for both techniques. The particular application and the demands of the finished product frequently have an impact on the decision between wet and dry milling procedures. Understanding these application-specific variables can help establish whether wet or dry milling is more appropriate because different industries and materials have different needs.

6.2.1 Minerals and Ceramics

Ceramics, minerals, and abrasive materials are all processed using dry milling. Prevents moisture-related problems and enables effective grinding of tough and abrasive materials.

Managing equipment wear and reducing dust production are challenges.

6.2.2 Products From Agriculture

Grain, seed, and agricultural items are processed by dry milling. Prevents moisture-related deterioration and maintains product quality and shelf life. Preventing dust during milling and achieving uniform particle sizes.

6.2.3 Metal Powders

Dry milling is used to create metal powders for applications such as additive manufacturing. Makes it possible to precisely control particle size and form. Dealing with particle aggregation and oxidation problems.

6.2.4 Excipient Used in Pharmaceuticals

The production of excipients for use in tablet formulations is a good application for dry milling. Offers regulated flow characteristics and particle sizes for reliable tablet compression. Maintaining quality standards and ensuring minimal particle attrition are challenges.

These illustrations show how the particular demands of each application affect the decision between wet and dry milling. The right milling procedure is determined in large part by variables such particle size management, material sensitivity, moisture content, product stability, and the requirement for uniform dispersion. Understanding the specific requirements of the application ultimately aids in guiding decision-making and guarantees the manufacture of high-quality products (Al-Hajji, 2019; Arndt et al., 2018; Mahale & Shashanka, 2022; Pradeep et al., 2022; Shashanka & Chaira, 2017b; Shashanka et al., 2020).

7. NEW DEVELOPMENTS AND PROBABLE FUTURE PATHS IN THE MILLING INDUSTRY

The demand for more sustainable and effective operations, shifting consumer preferences, and technological improvements all contribute to the ongoing evolution of the milling business. In the milling sector, the following new developments and future directions can be anticipated:

Automation, data analytics, and advanced sensor technologies are all being incorporated into milling operations. Systems for real-time monitoring and control increase productivity, save energy use, and improve product quality. By using Industry 4.0 ideas, mills are becoming smarter. Smart mills employ networked components and systems to streamline operations, anticipate maintenance requirements, and adapt to shifting production demands. Nano milling techniques are becoming more popular for creating nanoparticles with particular properties for use in electronics, medicines, coatings, and other applications. The production of feedstock materials for additive manufacturing, specifically 3D printing, has seen a growing significance attributed to milling processes. This is mostly due to the capability of milling to provide precise manipulation and regulation of particle size and composition. In order to mitigate the adverse impacts on the environment and attain sustainability goals, the adoption of sustainable milling practises is being welcomed. The aforementioned strategies encompass the utilisation of energy-efficient technologies, the reduction of water consumption, and the minimization of waste production. Milling processes are now utilised in the production of customised ingredients for the purpose of creating functional and personalised foods that cater to specific dietary needs and health concerns. Novel milling techniques, including as fortification and micronization for increased bioavailability, are being investigated to improve the nutritional profile of milled products. Milling is a widely employed technique for the extraction and isolation of useful chemical compounds from plant materials, hence facilitating the production of naturally occurring tastes, fragrances, and bioactive substances. Milling techniques are being used to create advanced materials for application in electronics, aerospace, and other industries, such as graphene, nanotubes, and high-performance composites. Customization and localised production are made possible by milling processes, allowing companies to produce goods that are tailored to certain markets and client tastes. To reduce food waste and promote sustainable practices, novel milling techniques are being investigated. Hybrid milling technologies are being developed to obtain particular product qualities. These hybrid methods incorporate wet and dry milling as well as additional processing techniques like ultrasonication and high-pressure homogenization. Energy-efficient milling technology advancements are made with the goal of reducing energy consumption while preserving or increasing milling efficiency. A growing number of businesses are recognising the value of being able to precisely mill particles to control their size, shape, and surface characteristics. The milling business will probably continue to develop in these ways and other areas as long as technology keeps advancing and industries continue to call for more specialised and effective goods. These trends' acceptance will be influenced by things like legal requirements, economic viability, and the particular requirements of each application (Adimule et al., 2023; Gong & Bassi, 2016; Rajendrachari, Adimule, Jayaprakash et al, 2023; Rajendrachari, Basavegowda, Vinaykumar et al, 2023).

8. AREAS OF FURTHER RESEARCH AND POTENTIAL APPLICATIONS OF WET AND DRY MILLING TECHNIQUES

The exploration of wet and dry milling processes is increasingly being conducted to explore novel applications across diverse sectors. The following discourse presents a range of prospective study domains and practical implementations for both methodologies:

8.1 Nano material production

Investigating the best ways to wet and dry mill materials to create nanoparticles with certain properties for use in electronics, catalysis, medication delivery, and other fields.

8.2 Energy Efficiency

Investigating strategies to make both wet and dry milling processes more energy-efficient, hence lowering overall energy use and having a smaller negative impact on the environment.

8.4 Process Optimization and Modelling

Developing cutting-edge modelling and simulation tools to forecast and optimise milling processes while taking into account elements like particle size distribution, energy usage, and equipment wear.

8.5 Hybrid Milling Approaches

Studying the possible advantages of mixing wet and dry milling processes or incorporating milling with additional processing techniques like ultrasonication or high-pressure homogenization.

8.6 Personalized Nutrition

Investigating how milling can be adapted to produce individualised and practical food ingredients that satisfy certain dietary demands and wellness objectives.

8.7 Waste Utilisation

Investigating methods to turn food waste or byproducts into useful components or materials using wet and dry milling.

8.8 Biodegradable Materials

Studying the use of milling to create biodegradable materials that can replace single-use plastics or other hazardous environmental pollutants.

9. POTENTIAL APPLICATIONS

Customising wet milling processes to provide patient-specific medication formulations with the highest possible bioavailability is the basis of precision medicine pharmaceuticals. Creating sophisticated materials with distinctive features through exact control over particle size and distribution, such as nanocomposites or nanostructured metals. Modification of food textures through milling, including the development of novel textures in processed foods and enhancement of product mouthfeel. Wet and dry milling techniques are used in the production of specialised cosmetic ingredients, such as pigments, exfoliants, and microencapsulated actives, in the cosmetics and personal care industry. Milled particles can be used as functional additives in a variety of goods, such as paints, coatings, adhesives, and polymers. Creating ecologically friendly processes by substituting wet and dry milling for inefficient processes or conventional chemical reactions. Using milled particles, advanced drug delivery systems can produce controlled-release drug delivery systems with exact release patterns. Examining the use of milled materials for energy storage applications, such as in supercapacitors and improved batteries. Creating milled biomaterials with enhanced mechanical and biocompatibility for use in biomedical implants. Using milled materials as controlled-release fertilisers or soil conditioners is referred to as precision agriculture. Using milling materials, creating specialised inks and coatings for electronics, wearable technology, and sensors. By using milling, bioactive components from plant materials can be extracted from medicines, nutraceuticals, and natural goods more effectively. The extensive potential of wet and dry milling techniques to influence numerous industries and contribute to technical improvements, superior products, and more sustainable processes is illustrated by these fields of study and applications (Kanaginahal et al., 2023; Malamatari, Taylor, Malamataris, Douroumis, & Kachrimanis, 2018; Shashanka & Chaira, n.d.-a; Shashanka & Chaira, n.d.-b).

10. CONCLUSION

Gaining a comprehensive comprehension of the distinctions between wet milling and dry milling is of paramount importance within the domain of industrial manufacturing. The aforementioned techniques, each possessing unique attributes and obstacles, serve as the fundamental principles for particle size reduction and material processing throughout diverse industrial sectors. The milling process encompasses the use of mechanical forces, including the application of grinding media and screens, to effectively disintegrate particles and augment the characteristics of the ultimate result. Wet milling is a process that entails the dispersion of materials within a liquid media, and the use of solid grinding components to accomplish the reduction of particle size. Wet milling, characterised by the incorporation of a liquid, has the distinct benefit of generating particles of reduced size and improving physical properties. In contrast, dry milling primarily depends on the interaction between particles to produce the decrease in size, resulting in a process that requires less manual labour and is frequently chosen as an initial step. Both procedures are designed to accommodate precise particle size specifications and may provide difficulties, especially when working with diverse materials and batch processing. Although wet milling can achieve the desired outcome of producing particles with a high degree of fineness, dry milling is often favoured due to its efficiency and simplicity. The proper selection of milling machinery is of utmost importance in order to preserve a competitive advantage, since nanoparticles continue to play an increasingly vital role in numerous product advancements. Although wet milling and dry milling have different characteristics, they both include some dangers, such as the progressive degradation of milling equipment caused by variables like carrier fluids in wet milling or abrasive particles in dry milling. Understanding the distinctions between wet and dry milling methods provides industries with the necessary information to enhance their production processes, enhance the quality of their products, and maintain flexibility in response to growing manufacturing requirements. The selection of wet milling versus dry milling as a method for achieving small particles or efficient processing is contingent upon the particular requirements and goals of each industrial application.

REFERENCES

AboulFotouh, K., Zhang, Y., Maniruzzaman, M., Williams, R. O. III, & Cui, Z. (2020, September 25). Amorphous solid dispersion dry powder for pulmonary drug delivery: Advantages and challenges. *International Journal of Pharmaceutics*, *587*, 119711. doi:10.1016/j.ijpharm.2020.119711 PMID:32739389

Adeyemi, I. A. (1983, July 1). Dry-milling of sorghum for ogi manufacture. *Journal of Cereal Science*, *1*(3), 221–227. doi:10.1016/S0733-5210(83)80024-1

Adimule, V., Keri, R., Rajendrachari, S., Kendrekar, P., & Kulkarni, C. V. (2023, February). Highly photoluminescence and wide band gap insulating metal hybrid nanoparticles array of samarium-doped SrO: CoO: synthesis, characterizations and sensor characteristics. *Journal of Materials Science Materials in Electronics*, *34*(5), 442. doi:10.1007/s10854-023-09899-2

Agimelen, O. S., Svoboda, V., Ahmed, B., Cardona, J., Dziewierz, J., Brown, C. J., McGlone, T., Cleary, A., Tachtatzis, C., Michie, C., Florence, A. J., Andonovic, I., Mulholland, A. J., & Sefcik, J. (2018, December 1). Multi-sensor inline measurements of crystal size and shape distributions during high shear wet milling of crystal slurries. *Advanced Powder Technology*, *29*(12), 2987–2995. doi:10.1016/j.apt.2018.09.003

Al-Hajji, L. (2019, May). A comparative study on the zinc metatitanate microstructure by ball milling and solvothermal approaches. *Journal of Structural Chemistry*, *60*(5), 830–837. doi:10.1134/S0022476619050172

Ali, H. S., York, P., Ali, A. M., & Blagden, N. (2011, January 20). Hydrocortisone nanosuspensions for ophthalmic delivery: A comparative study between microfluidic nanoprecipitation and wet milling. *Journal of Controlled Release*, *149*(2), 175–181. doi:10.1016/j.jconrel.2010.10.007 PMID:20946923

Arndt, O. R., Baggio, R., Adam, A. K., Harting, J., Franceschinis, E., & Kleinebudde, P. (2018, December 1). Impact of different dry and wet granulation techniques on granule and tablet properties: A comparative study. *Journal of Pharmaceutical Sciences*, *107*(12), 3143–3152. doi:10.1016/j.xphs.2018.09.006 PMID:30244008

Bartos, C., Szabó-Révész, P., Bartos, C., Katona, G., Jójárt-Laczkovich, O., & Ambrus, R. (2016, April 21). The effect of an optimized wet milling technology on the crystallinity, morphology and dissolution properties of micro-and nanonized meloxicam. *Molecules (Basel, Switzerland)*, *21*(4), 507. doi:10.3390/molecules21040507 PMID:27110752

Bhirud, N. L., & Gawande, R. R. (2017, December). Measurement and prediction of cutting temperatures during dry milling: Review and discussions. *Journal of the Brazilian Society of Mechanical Sciences and Engineering*, *39*(12), 5135–5158. doi:10.1007/s40430-017-0869-7

Buchczik, D., Wegehaupt, J., & Krauze, O. (2017). Indirect measurements of milling product quality in the classification system of electromagnetic mill. In *2017 22nd International Conference on Methods and Models in Automation and Robotics (MMAR)* (pp. 1039-1044). IEEE. 10.1109/MMAR.2017.8046973

Cañellas, F. M., Al-Rifai, N., Padrela, L., Tajber, L., Khamiakova, T., Otava, M., & Geertman, R. (2023, July 21). A comparative study of dry and wet top-down milling approaches for the preparation of microparticle suspensions. *Powder Technology*, *428*, 118829. doi:10.1016/j.powtec.2023.118829

Chaira, D. (2014, June 1). Phase transformation and microstructure study of nano-structured austenitic and ferritic stainless steel powders prepared by planetary milling. *Powder Technology*, *259*, 125–136. doi:10.1016/j.powtec.2014.03.061

Chaira, D. (2015, January 1). Development of nano-structured duplex and ferritic stainless steels by pulverisette planetary milling followed by pressure less sintering. *Materials Characterization*, *99*, 220–229. doi:10.1016/j.matchar.2014.11.030

Chen, M., Jiang, L., Guo, G., & An, Q. (2011, August). Experimental and FEM study of coated and uncoated tools used for dry milling of compacted graphite cast iron. *Transactions ofTianjin University.*, *17*(4), 235–241. doi:10.1007/s12209-011-1609-1

Clermont, B., & De Haas, B. (2010, March). Optimization of mill performance by using online ball and pulp measurements. *Journal of the Southern African Institute of Mining and Metallurgy*, *110*(3), 133–140.

Fuerstenau, D. W., & Abouzeid, A. Z. (2002, November 1). The energy efficiency of ball milling in comminution. *International Journal of Mineral Processing*, *67*(1-4), 161–185. doi:10.1016/S0301-7516(02)00039-X

Gong, M., & Bassi, A. (2016, December 1). Carotenoids from microalgae: A review of recent developments. *Biotechnology Advances*, *34*(8), 1396–1412. doi:10.1016/j.biotechadv.2016.10.005 PMID:27816618

Gupta, S., Shashanka, R., & Chaira, D. (2015). Synthesis of nano-structured duplex and ferritic stainless steel powders by planetary milling: An experimental and simulation study. In *IOP conference series: materials science and engineering* (Vol. 75, No. 1, p. 012033). IOP Publishing.

Kanaginahal, G. M., Mahale, R. S., Vasanth, S., Sharath, P. C., & Shashanka, R. (2023, June 30). Studies on Dielectric, Super-Capacitive and Photoluminescence Properties of Metal Organic Frameworks Subjected to Varying Temperature and Frequencies. *Nano Hybrids and Composites.*, *39*, 17–25. doi:10.4028/p-si6964

Kelsey, R. G., & Shafizadeh, F. (1980, May). Enhancement of cellulose accessibility and enzymatic hydrolysis by simultaneous wet milling. *Biotechnology and Bioengineering*, *22*(5), 1025–1036. doi:10.1002/bit.260220511

Khot, M., Raut, G., Ghosh, D., Alarcón-Vivero, M., Contreras, D., & Ravikumar, A. (2020, January 1). Lipid recovery from oleaginous yeasts: Perspectives and challenges for industrial applications. *Fuel*, *259*, 116292. doi:10.1016/j.fuel.2019.116292

Loh, Z. H., Samanta, A. K., & Heng, P. W. (2015, July 1). Overview of milling techniques for improving the solubility of poorly water-soluble drugs. *Asian Journal of Pharmaceutical Sciences*, *10*(4), 255–274. doi:10.1016/j.ajps.2014.12.006

Mahale, R. S., & Shashanka, R. (2022). Mechanical testing of spark plasma sintered materials: A review. In *AIP Conference Proceedings* (Vol. 2469, No. 1). AIP Publishing. 10.1063/5.0080183

Malamatari, M., Taylor, K. M., Malamataris, S., Douroumis, D., & Kachrimanis, K. (2018, March 1). Pharmaceutical nanocrystals: Production by wet milling and applications. *Drug Discovery Today*, *23*(3), 534–547. doi:10.1016/j.drudis.2018.01.016 PMID:29326082

Nayak, A. K., Shashanka, R., & Chaira, D. (2016). Effect of nanosize yittria and tungsten addition to duplex stainless steel during high energy planetary milling. In *IOP Conference Series: Materials Science and Engineering* (Vol. 115, No. 1, p. 012008). IOP Publishing. 10.1088/1757-899X/115/1/012008

Nykamp, G., Carstensen, U., & Müller, B. W. (2002, August 21). Jet milling—A new technique for microparticle preparation. *International Journal of Pharmaceutics*, *242*(1-2), 79–86. doi:10.1016/S0378-5173(02)00150-3 PMID:12176228

Ogonowski, S., Wołosiewicz-Głąb, M., Ogonowski, Z., Foszcz, D., & Pawełczyk, M. (2018, March 29). Comparison of wet and dry grinding in electromagnetic mill. *Minerals (Basel)*, *8*(4), 138. doi:10.3390/min8040138

Pavitra, V., Praveen, B. M., Nagaraju, G., & Shashanka, R. (2022, January 8). Energy storage, Photocatalytic and Electrochemical nitrite sensing of ultrasound-assisted stable Ta2O5 nanoparticles. *Topics in Catalysis*, 1–4. doi:10.1007/s11244-021-01553-7

Pradeep, N. B., Hegde, M. R., Rajendrachari, S., & Surendranathan, A. O. (2022, August 1). Investigation of microstructure and mechanical properties of microwave consolidated TiMgSr alloy prepared by high energy ball milling. *Powder Technology*, *408*, 117715. doi:10.1016/j.powtec.2022.117715

Protonotariou, S., Ritzoulis, C., & Mandala, I. (2021, April 1). Jet milling conditions impact on wheat flour particle size. *Journal of Food Engineering*, *294*, 110418. doi:10.1016/j.jfoodeng.2020.110418

Rahman, M., Senthil Kumar, A., & Not Available, N. A. (2001, August). Evaluation of minimal of lubricant in end milling. *International Journal of Advanced Manufacturing Technology*, *18*(4), 235–241. doi:10.1007/s001700170063

Rajendrachari, S. (2018, January 1). Effect of sintering temperature on the pitting corrosion of ball milled duplex stainless steel by using linear sweep voltammetry. *Analytical and Bioanalytical Electrochemistry.*, *10*(3), 349–361.

Rajendrachari, S. (2022, June 21). An overview of high-entropy alloys prepared by mechanical alloying followed by the characterization of their microstructure and various properties. *Alloys.*, *1*(2), 116–132. doi:10.3390/alloys1020008

Rajendrachari, S., Adimule, V. M., Jayaprakash, G. K., & Pandith, A. (2023, May 17). Electrochemical oxidation of methylene blue dye in wastewater using mechanically alloyed high entropy alloy modified carbon paste electrode using cyclic voltammetry. *Materials Research Express*, *10*(5), 054003. doi:10.1088/2053-1591/acd3d7

Rajendrachari, S., Basavegowda, N., Vinaykumar, R., Narsimhachary, D., Somu, P., & Lee, M. J. (2023, June 30). Electrocatalytic determination of methyl orange dye using mechanically alloyed novel metallic glass modified carbon paste electrode by cyclic voltammetry. *Inorganic Chemistry Communications*, *155*, 111010. doi:10.1016/j.inoche.2023.111010

Rajendrachari, S., Kamacı, Y., Taş, R., Ceylan, Y., Bülbül, A. S., Uzun, O., & Karaoğlanlı, A. C. (2019, December 1). Antimicrobial investigation of CuO and ZnO nanoparticles prepared by a rapid combustion method. *Physical Chemistry Research.*, *7*(4), 799–812.

Rausch, K.D., Hummel, D., Johnson, L.A., & May, J.B. (2019). Wet milling: the basis for corn biorefineries. In *Corn* (pp. 501-535). AACC International Press.

Rivas-Cantu, R. C., Jones, K. D., & Mills, P. L. (2013, April). A citrus waste-based biorefinery as a source of renewable energy: Technical advances and analysis of engineering challenges. *Waste Management & Research*, *31*(4), 413–420. doi:10.1177/0734242X13479432 PMID:23439875

Rosentrater, K.A., & Bucklin, R. (2022). Structural, physical, and engineering properties of cereal grains and grain products. In *Storage of Cereal Grains and Their Products* (pp. 135-178). Woodhead Publishing.

Santra, D.K., & Schoenlechner, R. (2017). Amaranth part 2—sustainability, processing, and applications of amaranth. In *Sustainable protein sources* (pp. 257-264). Academic Press.

Sarkar, N., Ghosh, S. K., Bannerjee, S., & Aikat, K. (2012, January 1). Bioethanol production from agricultural wastes: An overview. *Renewable Energy*, *37*(1), 19–27. doi:10.1016/j.renene.2011.06.045

Shashanka, R. (2017). Synthesis of nano-structured stainless steel powder by mechanical alloying-an overview. *International Journal of Scientific and Engineering Research*, *8*, 588–594.

Shashanka, R., & Chaira, D. (2015, July 1). Optimization of milling parameters for the synthesis of nano-structured duplex and ferritic stainless steel powders by high energy planetary milling. *Powder Technology*, *278*, 35–45. doi:10.1016/j.powtec.2015.03.007

Shashanka, R., & Chaira, D. (2016, January). Effects of nano-Y 2 O 3 and sintering parameters on the fabrication of PM duplex and ferritic stainless steels. *Acta Metallurgica Sinica*, *29*(1), 58–71. doi:10.1007/s40195-015-0362-1

Shashanka, R., & Chaira, D. (2017a, March 4). Effect of sintering temperature and atmosphere on nonlubricated sliding wear of nano-yttria-dispersed and yttria-free duplex and ferritic stainless steel fabricated by powder metallurgy. *Tribology Transactions*, *60*(2), 324–336. doi:10.1080/10402004.2016.1168897

Shashanka, R., & Chaira, D. (2017b). *Ball Milled Nano-Structured Stainless Steel Powders: Fabrication and Characterization*. Educreation Publishing.

Shashanka, R., & Chaira, D. (n.d.-a). Optimization of Milling Parameters. *Ball Milled Nano-Structured Stainless Steel Powders: Fabrication and Characterization, 32*.

Shashanka, R., & Chaira, D. (n.d.-b). Fabrication of NANO-Y2O3 Dispersed Stainless Steel By Conventional And Spark Plasma Sintering. *Ball Milled Nano-Structured Stainless Steel Powders: Fabrication and Characterization, 54*.

Shashanka, R., Chaira, D., & Chakravarty, D. (2016). Fabrication of nano-yttria dispersed duplex and ferritic stainless steels by planetary milling followed by spark plasma sintering and non-lubricated sliding wear behaviour study. *J Mater Sci Eng B.*, *6*, 111–125.

Shashanka, R., Uzun, O., & Chaira, D. (2020). Synthesis of nano-structured duplex and ferritic stainless steel powders by dry milling and its comparison with wet milling. *Archives of Metallurgy and Materials*, *65*(1).

Su, Y., Du, H., Huo, Y., Xu, Y., Wang, J., Wang, L., Zhao, S., & Xiong, S. (2016, June 1). Characterization of cationic starch flocculants synthesized by dry process with ball milling activating method. *International Journal of Biological Macromolecules*, *87*, 34–40. doi:10.1016/j.ijbiomac.2015.11.093 PMID:26905465

Uysal, A. (2016, June 13). Investigation of flank wear in MQL milling of ferritic stainless steel by using nano graphene reinforced vegetable cutting fluid. *Industrial Lubrication and Tribology*, *68*(4), 446–451. doi:10.1108/ILT-10-2015-0141

Vadnere, A. P., & Kalpande, S. D. (2023, May 9). A review on effects of process parameters in milling challenging materials under nanofluid-based MQL conditions. *Industrial Lubrication and Tribology*, *75*(4), 361–371. doi:10.1108/ILT-01-2023-0010

Wang, Y.M., & Forssberg, E. (2001). *New milling technology*. Technical report.

Wołosiewicz-Głąb, M., Foszcz, D., & Ogonowski, S. (2017, July 1). Design of the electromagnetic mill and the air stream ratio model. *IFAC-PapersOnLine*, *50*(1), 14964–14969. doi:10.1016/j.ifacol.2017.08.2554

Yang, Y., Song, L., Gao, T., & Nagy, Z. K. (2015, December 2). Integrated upstream and downstream application of wet milling with continuous mixed suspension mixed product removal crystallization. *Crystal Growth & Design*, *15*(12), 5879–5885. doi:10.1021/acs.cgd.5b01290

Chapter 2
Electrochemical Sensor Applications of Mechanically Alloyed Materials

A. Santhy
https://orcid.org/0000-0002-7748-7975
Department of Chemistry, Amrita Vishwa Vidyapeetham, Amritapuri, India

R. Rejithamol
Department of Chemistry, Amrita Vishwa Vidyapeetham, Amritapuri, India

R. Chandana
Department of Chemistry, Amrita Vishwa Vidyapeetham, Amritapuri, India

P. Sivanand
Department of Chemistry, Amrita Vishwa Vidyapeetham, Amritapuri, India

S. Beena
Department of Chemistry, Amrita Vishwa Vidyapeetham, Amritapuri, India

ABSTRACT

Development of electrochemical sensors and its research extends over a wide range of applications including clinical analysis, food safety, environmental monitoring, and many more. Electrochemical sensing devices are more convenient for the analytical determination of various electroactive analytes due to its simple fabrication procedure and the reliability in the obtained results. The sensitivity of the electrochemical sensors can be improved by using various chemically modified electrodes instead of traditional unmodified electrodes. Mechanically alloyed materials are one of the fascinating materials used for the modification of electrodes in the development of electrochemical sensors. Metal alloys and their nano porous structures provide good conductivity and high surface area making it more convenient for the sensing applications. The chapter summarizes the analytical characteristics of the mechanically synthesized metal alloys modified electrochemical sensors giving a special emphasis to various dyes, neurotransmitters, biological molecules, drugs, etc. as the target analytes.

DOI: 10.4018/978-1-6684-9385-4.ch002

INTRODUCTION

The term 'sensor' is indispensable as we come across various sensing devices in diverse areas such as clinical purposes, automobiles, weather forecasting, food sensing, environmental monitoring, pharmaceutical industries, forensics, etc. The sensors are analytical devices that can selectively detect a target analyte, both quantitatively and qualitatively, from a complex matrix by the different types of interaction with the target species and the sensing device (W. Wang, 2016). There are different types of sensors based on the method of operation and their working, and they can be divided mainly into chemical and physical sensors. Physical sensors respond to physical stimuli like temperature, pressure, etc., whereas chemical sensors need stimuli of chemical origin associated with a chemical or biochemical interaction with the target analyte and the sensor. The integral parts of a chemical sensor are the transducer and receptor. The biochemical or chemical interaction of target species occurs via the receptor part of the sensor, and the transducer takes up the chemical energy produced during the exchange. Further, the transducer converts this chemical information into a helpful signal by various operating principles. Based on the signal conversion by the transducers, the sensors are categorized into mass sensitive, optical, magnetic, electrochemical, etc. (Hulanicki et al., 1991).

Among the various types of chemical sensors, electrochemical sensors play a vital role in all aspects of our life, including healthcare monitoring, food safety, detection of various toxic pollutants from environmental samples, etc. (Krishnan et al., 2020)(Antherjanam & Saraswathyamma, 2021; Krishnan & Saraswathyamma, 2021)(Mahale et al., 2022). The electrochemical sensors are highly convenient due to their simple instrumentation, mainly consisting of a working electrode, a reference electrode, and a counter electrode dipped in a supporting electrolyte medium (Shashanka & Kumara Swamy, 2020a). The three-electrode setup is then connected to a potentiostat, and the instrumentation of an electrochemical sensing device is given in Figure 1.

The chemical modification of working electrodes can improve the electrochemical sensors' sensitivity and analytical performance (Nahirny et al., 2020). Nanomaterials, conducting polymers, metal-organic frameworks, MXenes, metal alloys, various hybrid materials, etc., were used for the chemical modification of electrodes in the fabrication of several electrochemical sensors(Qi et al., 2018)(Aravindan & M.V., 2017) (J. Zhang et al., 2018)(Yu & Breslin, 2020)(Lu, 2019)(Terbouche et al., 2022). (Shashanka & Kumara Swamy, 2020b) Metal alloys are one of the exciting materials for the surface modification of electrodes as these materials act as electrocatalysts for the electrochemical oxidation or reduction of the target analytes (Sakthivel et al., 2019). Mechanical alloying is the process of manufacturing alloyed materials, which was developed in the 1960s to satisfy industrial needs. It is a vastly growing area of

Figure 1. Instrumentation of an electrochemical sensor with a three-electrode system
Source: Santhy, Saraswathyamma, et al. (2020). (Reproduced with permission from Elsevier)

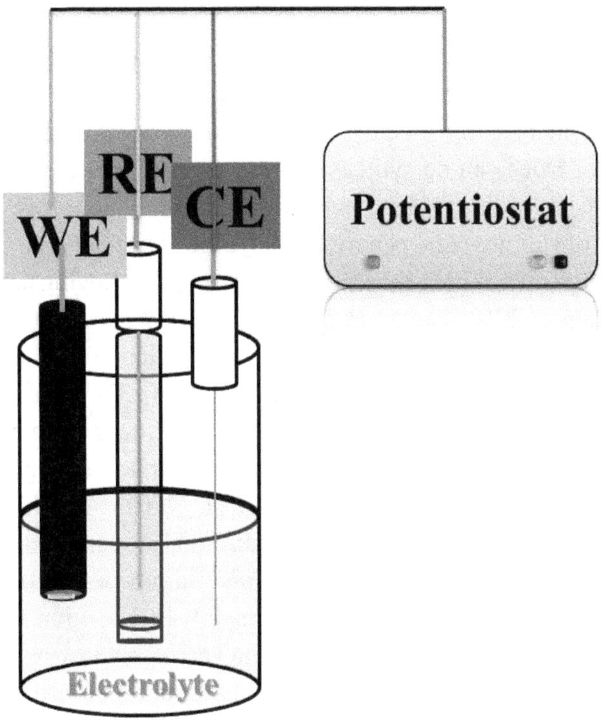

synthesizing high-strength alloys, in which the raw materials are subjected to high-energy collision using a ball mill to produce fine powders followed by heat treatment methods (Suryanarayana, 2019)(Lü & Lai, 1998). Highly energetic collisions by ball milling and the successive heat treatment method produce permanent deformation in the structure, which leads to the formation of High Entropy Alloys (HEA)(Geambazu et al., 2021)(Dastanpoor & Enayati, 2017). Mechanically alloyed materials find a wide range of electrochemical applications, including supercapacitors, hydrogen evolution reactions, corrosion resistance, electrochemical sensors, etc.(B. Li et al., 2016)(Panek et al., 2014)(Shang et al., 2017)(Winiarski et al., 2022)(Kumar et al., 2022)(Tian et al., 2019). The mechanically alloyed possess high strength and good oxidation resistance and was successfully employed to develop electrochemical sensors (Suryanarayana, 2019). The following sections will give the significance of mechanically alloyed materials in fabricating electrochemical sensors for various target analytes such as biomolecules, neurotransmitters, dyes, pharmaceuticals, etc. Also, this chapter will give the analytical performance of the different electrochemical sensors modified by mechanically developed alloys.

MECHANICAL ALLOYING PROCESS

Mechanical alloying is one of the most prominent processes used to synthesize various nanostructured alloys using the high energy ball milling technique (Suñol & Escoda, 2022). The properties of the alloys can be varied by using different compositions of metals. By mechanical alloying, nanostructured metallic alloys with unique properties can be synthesized for various applications ("Technology and Challenges in Additive Manufacturing of Duplex Stainless Steels," 2021). Multiple forms of stainless alloys have been synthesized, and their properties were studied in numerous literature reports on the mechanical alloying or powder milling process (Shashanka, 2017)(Nayak et al., 2016)(Shashanka et al., 2020). Shalabh Gupta and coworkers have reported synthesizing various forms of stainless steel by mechanical alloying process (S. Gupta et al., 2015). The mechanically alloyed materials can be used in both high temperatures and room temperatures, and this technique is widely used for synthesizing oxide dispersion-strengthened materials containing iron and nickel (Suryanarayana et al., 2001). The high-energy alloy formation by ball milling facilitates the formation of proper solid solutions due to the formation of defects and the movement of solid powder particles to the grain boundaries of the alloyed materials (R & Chaira, 2014). The optimization of milling parameters is an essential factor in synthesizing various nanostructured alloys with superior properties in terms of compressibility, sinterability, the microstructure of the sintered powder, etc. (P. Zhang et al., 2013)(Shashanka & Chaira, 2016). R. Shashanka and D. Chaira synthesized and studied the milling parameters of two types of nanostructured stainless steel, namely duplex and ferritic stainless steel, by powder metallurgy process (Shashanka & Chaira, 2015). The above work reported that the formation of nanostructured particles with good morphology and size had been highly governed by the parameters such as a ball-to-powder weight ratio, milling speed, etc. numerous research is going on in the synthesis and applications of mechanically alloyed materials, especially the high entropy alloys, oxide dispersed materials, super alloys, etc. (Rajendrachari, 2022)(Rajendrachari, 2022). In another report, the effect of yttria dispersion on mechanically alloyed stainless-steel materials was for the corrosion resistance properties (Shashanka et al., 2018). According to linear sweep voltammetric studies, the yttria dispersion highly increased the resistance to pitting corrosion. Mechanically alloyed materials have been applied for many electrochemical and biomedical applications, and it is a vast and intriguing area of research (Pradeep et al., 2022)(Hussain et al., 2021)(Plata-Torres et al., 2007)(Cheng et al., 2020). This chapter gives a detailed study of the electrochemical applications of mechanically alloyed materials emphasizing electrochemical sensing.

MECHANICALLY ALLOYED MATERIALS BASED ELECTROCHEMICAL SENSORS

Electrochemical sensing of biomolecules is an important area of research, as the deficiency or excess of many biomolecules in the human body is associated with clinical disorders. Also, several biomolecules act as biomarkers for many diseases. Sensitivity and selectivity are the essential parameters to be evaluated for the effective monitoring of the target analytes from complex matric such as urine, blood, etc. Electrochemical sensors offer more sensitivity, and these are one of the cost-effective devices for the electrochemical monitoring of various analytes (Rajendrachari, Basavegowda, et al., 2022)(Kshama et al., 2023). The modification of electrodes using mechanically developed metal alloys could significantly improve the sensors' sensitivity due to their electrocatalytic properties. The mechanically alloyed materials used for the electrochemical sensing applications from previous literature and the analytes determined using the developed sensors were tabulated in Table 1.

Folic acid is an essential water-soluble B vitamin and an antioxidant compound in our body, and its deficiency may cause severe neurological disorders (Hasan et al., 2022). Numerous electrochemical sensors were reported to determine folic acid in different real samples considering the significance of this biomolecule (Santhy, Beena, et al., 2020)(Alizadeh et al., 2022). An electrochemical sensor for the folic acid sensor was developed using a mechanically alloyed stainless-steel material by R. Shashanka and coworkers in 2015 (Shashanka et al., 2015a). Stainless steel powder was obtained using high-energy dual-drive planetary ball milling of elemental Fe, Cr, and Ni, adopting the procedure reported in previous literature (R & Chaira, 2014). Thus, The powdered alloy was in the nanostructured form and was used to modify the carbon paste electrode (CPE) for the sensor fabrication. The as-synthesized alloy was then characterized using X-Ray Diffraction (XRD), Scanning Electron Microscope (SEM), Transmission Electron Microscope (TEM), etc. The ball-milled alloy powder was then mixed with graphite powder and silicone and made a homogeneous dispersion with Yttrium (III) oxide (Y_2O_3), and the paste was then filled in the CPE to form the modified electrode YSTL/CPE. A CPE was modified with stainless steel powder without Y_2O_3 (STL/CPE), and another bare CPE with only the graphite powder and silicone oil was fabricated to compare the performances of the modified electrodes. The electrochemical oxidation of folic acid was studied on YSTL/CPE, STL/CPE, and bare CPE under optimized conditions in a phosphate buffer solution of pH 7.2. The anodic peak current for folic acid was maximum on YSTL/CPE, indicating the influence of modifying material in the electrooxidation of folic acid. Using Cyclic Voltammetry (CV), a linear range of 2 mM to 2.6 mM was obtained on YSTL/CPE. This electrode could be a successful device for monitoring folic acid in practical applications.

An electrochemical sensor was reported to determine Ascorbic Acid (AA), another water-soluble B vitamin based on a HEA synthesized by mechanical alloying (Rajendrachari, Adimule, et al., 2022). HEAs are made by mixing several alloying metals in a ball mill, resulting in alloys with complex compositions (Han et al., 2018). These mechanically alloyed HEAs possess unique electrocatalytic properties, making them a trending material in various electrochemical applications (K. Li & Chen, 2021). For the AA sensor development, Shashanka Rajendrachari et al. synthesized a HEA from the elements Fe, Cr, Ni, Ti, and Mn via ball milling process to form 23Fe-21Cr-18Ni-20Ti-18Mn. The HEA was analytically studied for morphological and structural determination using SEM, XRD, and Energy-dispersive X-ray spectroscopy (EDS). The graphical representation of the synthesis of HEA and the sensing of AA is given in Figure 2. The alloyed HEA material was then modified on a CPE to study the electrochemical characteristics of AA using CV. The concentration of HEA in the modification process was optimized for the better electrochemical response of AA in CV analysis.

Figure 2. Graphical representation of the synthesis of HEA and the sensing of AA
Source: Reproduced from Rajendrachari, Adimule, et al. (2022), Creative Commons Attribution (CC BY) license (https://creativecommons.org/licenses/by/4.0/).

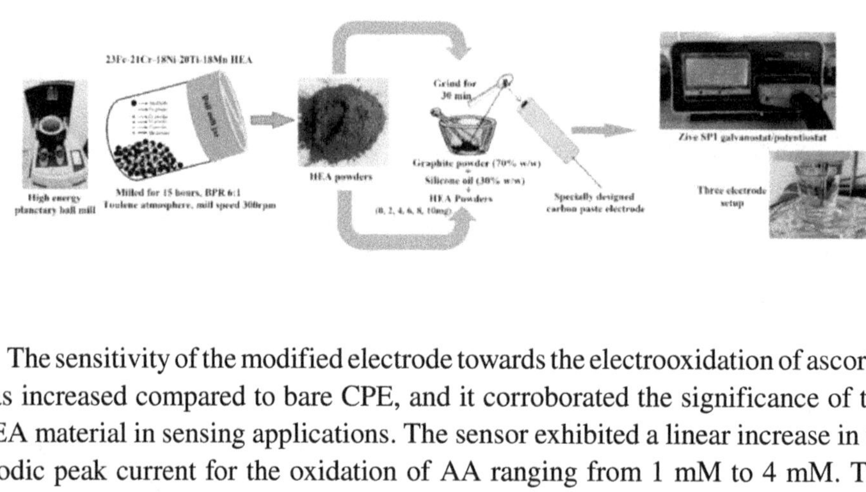

The sensitivity of the modified electrode towards the electrooxidation of ascorbic was increased compared to bare CPE, and it corroborated the significance of this HEA material in sensing applications. The sensor exhibited a linear increase in the anodic peak current for the oxidation of AA ranging from 1 mM to 4 mM. This developed sensor could be successfully used for ascorbic determination in various analytical applications.

In another work, Shashanka Rajendrachari and coworkers developed an electrochemical sensor for the determination of the biomolecules uric acid (UA), AA and Dopamine (DA) (Shashanka et al., 2015b). A duplex stainless material synthesized by mechanical alloying was employed for the sensing purpose. Stainless steel with composition Fe-18Cr-13Ni was modified on a CPE, and the modified material was characterized by various analytical methods such as XRD, EDS, TEM, etc. Fe-18Cr-13Ni /CPE was utilized for the electrochemical determination of UA, AA, and DA under optimized conditions. The sensor showed an increase in the anodic peak current with the increase in the concentration for all three analytes. Thus, this sensor could be an excellent analytical device for monitoring UA, AA, and DA in real samples.

The applicability of HEA for electrochemical sensing was reported in another work for the determination of urea (Ashwini et al., 2022). A composite of NiFeCrCoCu – HEA and graphene was prepared and characterized by TEM, XRD, Atomic force microscopy (AFM), Raman analysis, CV, etc. The HEA nanocomposite was modified on a glassy carbon electrode (GCE) by drop casting method. The electrochemical performance of the electrode towards the oxidation of urea was studied using CV analysis. Further, chronoamperometry was used for the analytical quantification of urea. A linear range of 1 mM – 50 mM was obtained for the electrochemical determination of urea using the HEA-nanocomposite-modified sensor.

Recently in 2023, Shashanka Rajendrachari et al. reported another novel HEA-based electrochemical sensor for the determination of methylene blue (MB), a toxic dye (Rajendrachari et al., 2023). A nanostructured HEA of composition 25Fe-19Cr-19Ni-18Ti-19Mn was synthesized by mechanical alloying and was modified on a CPE to determine MB electrochemically. MB is mostly discharged into waterbodies as discharge from various industries; hence, it is considered an environmental pollutant (Hassan et al., 2015). Due to the environmental impact of MB in water bodies, the analytical monitoring of MB is highly crucial. The schematic representation of the fabrication of the electrochemical sensor and its working using 25Fe-19Cr-19Ni-18Ti-19Mn modified CPE is given in Figure 3. The electrochemical oxidation of MB on 25Fe-19Cr-19Ni-18Ti-19Mn modified CPE and bare CPE were studied using CV. The CV results showed that the modification process significantly enhanced the anodic peak current of MB due to the high surface area of the HEA compared to the bare CPE surface. All the analytical parameters were optimized for the electrochemical determination of MB using the HEA-modified CPE, and a linear range of 1 mM- 5 mM was obtained in CV analysis.

An electrochemical sensor for the detection of methyldopa (MD) and Hydrochlorothiazide (HTZ) was developed using an iron-cobalt alloy as the sensing element (V. K. Gupta et al., 2015). MD and HTZ are used as anti-hypertensive drugs, and various electrochemical sensors have been reported for the electrochemical assay of these drug molecules (Halakoei et al., 2021)(Purushothama & Arthoba Nayaka, 2017). In the work reported by (V. K. Gupta et al., 2015), an alloy of Fe and Co was synthesized by mechanical alloying, and the Fe-Co alloy was modified on a CPE (Fe-Co/CPE). The alloy-modified CPE was then utilized for the electrochemical determination of MD with a linear range of 0.06- 600 μM. The electrode also showed

Figure 3. Schematic representation of the processes involved in the electrochemical sensing of MB
Source: Reproduced from Rajendrachari et al. (2023), Creative Commons Attribution (CC BY) license (https://creativecommons.org/licenses/by/4.0/).

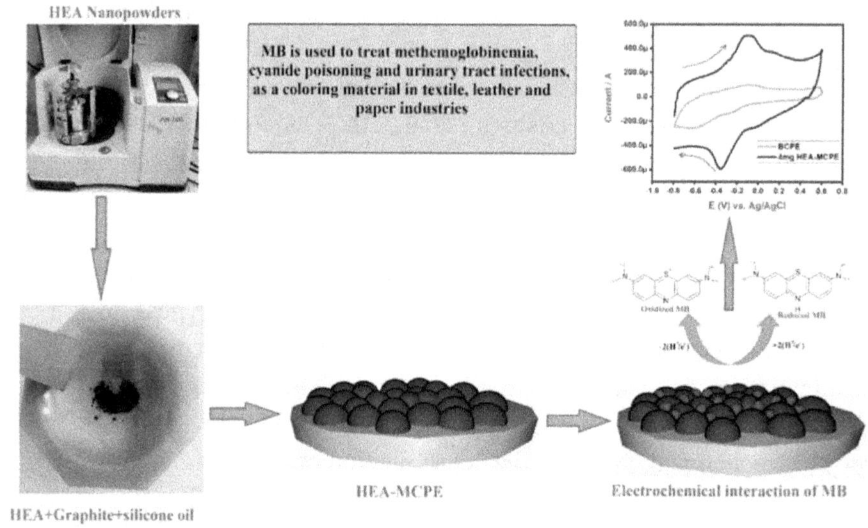

Table 1. Mechanically alloyed materials used in the electrochemical sensors from previous literature

Mechanically Alloyed Materials Used in Electrochemical Sensing	Analytes Used in Electrochemical Sensing	References
Y_2O_3 dopped stainless steel	Folic acid	(Hasan et al., 2022).
23Fe-21Cr-18Ni-20Ti-18Mn	Ascorbic acid	(Rajendrachari, Adimule, et al., 2022)
Fe-18Cr-13Ni	Uric acid	(Shashanka et al., 2015b)
NiFeCrCoCu	Urea	(Ashwini et al., 2022)
25Fe-19Cr-19Ni-18Ti-19Mn	Methylene blue	(Rajendrachari et al., 2023)
Iron cobalt alloy	Methyldopa and hydrochlorothiazide	(V. K. Gupta et al., 2015)
Ni50–Fe50	Tiopronin	Baghayeri et al., 2014
γ-$CoTe_2$	Ferulic acid	Malagutti et al., 2022
$NiTe_2$	Morin	de Fatima Ulbrich, Winiarski, Jost, & de Campos, 2020
CrO/CrN/C	Uric acid and dopamine	Gong et al., 2022
PtxAu100-x alloy	Dopamine and ascorbic acid	Oko et al., 2015
PtPdRhRuIrNi	Serotonin	M. Li et al., 2021
Ni3-xTe2	Dopamine and adrenaline	de Fatima Ulbrich, Winiarski, Jost, & Maduro de Campos, 2020
Fe-based alloy obtained from the steel industry	Dopamine, serotonin, and tryptophan	Winiarski et al., 2022

simultaneous resolution of the oxidation peaks for the MD and HTZ using square wave voltammetry (SWV). Thus, the Fe-Co/CPE could be used to successfully detect MD and HTZ in pharmaceutical analysis.

Tiopronin (TP) is a drug used for the treatment of liver-related disorders, arthritis, etc. (P. Zhang et al., 2019), and a voltammetric sensor for the determination of TP was reported by Mehdi Baghayeri and coworkers (Baghayeri et al., 2014). The sensor was fabricated based on a mechanically alloyed material of the composition, Ni50–Fe50, as the modifying material on a CPE. The Ni50–Fe50 modified CPE (Ni50–Fe50/CPE) showed a dynamic range of 0.01–50.0 μM for the electrooxidation of TP in Phosphate Buffer Solution (PBS) pH7. The oxidation over the potential of TP on the Ni50–Fe50/CPE was reduced to 500 mV than on the bare CPE, and hence it corroborated the electrocatalytic effect of the Ni50–Fe50 alloy in the electrooxidation of TP. The practical application of the sensor was made in pharmaceutical and urine samples with good selectivity.

A cobalt and tellurium nanocrystal alloy of composition γ-$CoTe_2$, prepared by mechanical synthesis, was applied for the electrochemical determination of ferulic acid, a compound with antioxidant properties (Malagutti et al., 2022)(Liu et al., 2014). γ-$CoTe_2$ was modified on a GCE to study the electrochemical behavior of ferulic acid using CV. The synthesized alloys were then characterized by XRD, SEM, TEM, etc. Ferulic acid showed well-defined redox peaks on γ-$CoTe_2$ modified GCE in Briton Robinson (BR) buffer of pH 3 in CV experiments. The rate of electron transfer was highly enhanced on the modified GCE compared to bare GCE, and hence the γ-$CoTe_2$ showed some electrocatalytic properties for the electrochemical oxidation of ferulic acid. The electrochemical response for converting quinone to hydroquinone is studied for the quantitative analysis of ferulic acid using the modified sensor. The γ-$CoTe_2$ /GCE showed a linear range of 0.03 μM – 28 μM for the determination of ferulic acid in SWV analysis. The stability studies of the sensor were also conducted, and the real sample analysis was done in cosmetic samples with good recovery.

Morin is another therapeutic drug and antioxidant agent used to prevent coronary artery diseases, protect human red blood cells, etc. (Yola et al., 2014) (Sukanya et al., 2018). An electrochemical method for the detection of morin was developed by Kelli de Fatima Ulbrich and their coworkers (de Fatima Ulbrich, Winiarski, Jost, & de Campos, 2020). A mechanical alloying process synthesized an alloy of nanocrystalline nickel ditelluride and was used to develop a morin sensor. A CPE was modified with the $NiTe_2$ nanocrystalline material to study the electrochemical properties of the analyte morin using the voltammetric method. The CV analysis of morin in BR buffer of pH 2 showed an increase in the oxidation peak current and lower oxidation overpotential for the $NiTe_2$/CPE compared to the unmodified electrode. The higher surface area of the nanocrystalline particles may facilitate the electrooxidation of morin with greater sensitivity, and hence $NiTe_2$ can be considered an electrocatalyst for the determination of morin. The analytical parameters of the sensor were optimized, and the quantification of morin was done by Differential Pulse Voltammetry (DPV) analysis. The sensor showed linearity over a range of 0.014–32.0 μM concentrations for morin determination. The $NiTe_2$/CPE sensor was then successfully employed to determine morin in red wine samples with good recovery.

DA and UA are the two most common physiologically active chemicals in the human body (J. Wang et al., 2017). DA, an essential neurotransmitter that regulates blood pressure and motor function, must be monitored in real-time, necessitating more sensitive and focused sensors across a dynamic concentration range (Tawade et al., 2020). The outcome of purine metabolism in the human body is UA, which possesses significant neuroprotective and antioxidant properties (Krishnan et al., 2020). Human mental and physical well-being are directly correlated with the release of DA and UA in blood serum. Since DA and UA typically coexist in biological fluids, their simultaneous detection in human blood serum is crucial for early diagnosis and therapy monitoring of linked disorders.

A mechanical milling method has been employed to create a low-cost, high-entropy porous CrO/CrN/C biosensor for the simultaneous detection of UA and DA (Gong et al., 2022). With the assistance of a high-energy ball milling process, the HEA was mixed with jasmine flowers and DA to form the biorecognition element (HEA-JF-DA) on a GCE and was used for biosensing UA and DA. The simultaneous determination of DA and UA in the 0.05 - 4 μM range was performed using the HEA-JF-DA modified GCE using DPV analysis. Human serum samples were used to assess the analytical performance of the Cr- HEA-JF-DA modified GCE, and the electrode showed good anti-interference properties and high stability.

Tavares et al. developed PtxAu100-x alloy-based NPs, and the electrochemical properties of the sensor were examined by CV for the detection of DA in the presence of AA (Oko et al., 2015). Pulsed laser ablation synthesized nanoparticles in an alkaline electrolyte without surfactants or reagents. The alloy of Au and Pt significantly improved AA's oxidation current, and the alloyed NPs' peak current is between two and four times higher than that of Au NPs alone. DPV demonstrated that the mechanically alloyed NPs have high sensitivities and low detection limits and can detect DA in solutions containing AA at concentrations 1000 times greater. Additionally, because the sensitivity and detection limit remained unaltered, the presence of AA had no impact on the nanoparticles' performance.

HEA (PtPdRhRuIrNi) nanoparticle-based printed sensor chips have been employed for the electrochemical sensing of serotonin or 5-Hydroxytryptophan (5-HT) molecules by Xinxin et al. (M. Li et al., 2021). 5-HT is a crucial neurotransmitter that plays a role in many biological and psychopathological processes. It is also thought to be a biomarker for diagnosing several disorders. The sensing material has excellent sensitivity on the active catalyst with a lower detection limit (LOD) of 10 nM. The selectivity of the sensor from other interfering species facilitates the electrooxidation of 5-HT and the successful detection of 5-HT in serum samples. This sensing material has tremendous potential in the area of biosensing research.

In conclusion, nanomaterials made of HEA can be effectively employed as catalysts in electrochemical sensing applications. The other remarkable highlight of this work is the selective quantification of 5-HT from the two coexisting molecules DA and epinephrine, which consistently interfere with the signal of 5-HT. The sensor also successfully detected serotonin in human serum samples that had been spiked, demonstrating its enormous potential for clinical diagnosis and neuroscience research.

The simultaneous detection of DA and adrenaline (AD) using the supporting electrolyte of PBS (pH 7.0) was developed by Carlos et al. (de Fatima Ulbrich, Winiarski, Jost, & Maduro de Campos, 2020). Starting with the Ni75Te25 powder mix, the composite, including Ni nanocrystals and Ni3-xTe2, was made using mechanochemical synthesis employing ball milling techniques in inert conditions. Different spectroscopic techniques have been applied to characterize the modifier. As determined by Rietveld analysis, the average crystallite size of Ni3-xTe2 and Ni nanocrystals, was confirmed by transmission electron microscopy pictures and chosen area electron diffraction data to be 10 nm and 4 nm, respectively. The composite exhibits ferromagnetic and superparamagnetic behavior controlled by various Ni counterpart nanocrystalline crystal sizes. With two distinct peaks, the CPE/Ni3-xTe2 electrode demonstrated the electrocatalytic capabilities of DA and AD. The linear range for both DA and AD determination was 4 μM -31 μM with detection limits for DA and AD, respectively, 0.15 μM and 0.35 μM. The preparation and operation method completely employed green chemistry approaches, and the modifier represents a potential electrochemical platform for DA and AD measurement in various matrices.

J.P. Winiarski et al. studied the electrochemical properties of DA, 5-HT, and tryptophan (TRP) on a Fe-based alloy (Winiarski et al., 2022). A sustainable electrode modifier material from the steel waste industry was employed in a single-step modification approach. The mill scale (MS) was made up of the different oxide ores of iron, such as magnetite (Fe_3O_4), hematite (Fe_2O_3), and wurtzite (FeO). This alloy was produced by the hot rolling method, and a high-energy shaker mill was used to ball mill it later. Powder XRD was used for the microstructural analysis of the developed alloy material. Controlling the milling duration can reduce the crystalline size of solids, which impacts their unique chemical reactivity and material characteristics.

The magnetic moment measurements of Fe_2O_3 crystals revealed the superparamagnetic properties in these materials, which are primarily used for many applications, especially in electrochemical studies. Ball milling was applied for the development of these materials for sensor studies. A one-step electrode modification method was employed to fabricate the MS carbon paste electrode. Overlapping peaks for DA and 5-HT were seen when using the unmodified electrode. But after modifying the carbonaceous substrate with metal oxides, the magnetic characteristics and selectivity for detecting several analytes were attained. Peak current intensity and analyte concentrations were shown to be linearly correlated under ideal square wave voltammetry conditions. For DA, 5-HT, and TRP, linear responses were found in the concentration ranges of 0.49-21, 0.49-13.6, and 2.9 μM -25 μM, with detection limits of 0.21, 0.13, and 2.18 μM, respectively. The use of an environmentally friendly electrode modification material with enhanced selectivity and sensibility is presented for clinical and biological applications.

CONCLUSION AND FUTURE OUTLOOK

This chapter provides information on an alloy's mechanical and physical properties, as strength, flexibility, and wear resistance influence the selective and sensitive quantification of neurotransmitters, dyes, vitamins, etc. The chapter highlights the mechanically assisted synthesis of alloys that formed between transition metals and p-block elements, specifically the carbon and oxygen family, for the electrochemical sensing application of various molecules. The introduction of electronegative elements into the crystal lattice of transition metals generates structural defects in them, thereby increasing the diffusivity of the analyte molecules to the modified materials on the electrode surface.

The disadvantages of the mechanical alloying process are the chemical purity of the starting powders, the milling equipment, the process control agents, as well as other methods of contamination. During mechanochemical processing, chemical reactions and phase changes may generate very high mechanical energy supplied for the process. Minimizing these mechanical errors is one of the challenges in this area of research in the current scenario. The electrochemical sensors based on mechanically alloyed materials are highly superior in analytical performance compared to traditional electrodes. The development of fabrication electrochemical sensing devices employing mechanically synthesized alloys is an emerging research area. Literature reports showed that only a few works had been published for fabricating the voltammetric sensing device incorporating mechanically alloyed materials. This chapter will enthuse the researchers to emphasize the development of different alloyed materials for the fabrication of highly advanced electrochemical sensors for various applications beneficial to human beings and society.

REFERENCES

Alizadeh, M., Garkani Nejad, F., Dourandish, Z., Tajik, S., Karimi, F., Mohammadzadeh Jahani, P., Aghaei Afshar, A., Zaimbashi, R., Sheikhshoaie, I., & Beitollahi, H. (2022). A brief review on the recent achievements in electrochemical detection of folic acid. In Journal of Food Measurement and Characterization (Vol. 16, Issue 5). doi:10.1007/s11694-022-01421-2

Antherjanam, S., & Saraswathyamma, B. (2021, September). Simultaneous electrochemical determination of hydrazine and hydroxylamine on a thiadiazole derivative modified pencil graphite electrode. *Materials Chemistry and Physics*, *275*, 125223. doi:10.1016/j.matchemphys.2021.125223

Aravindan, N., & M.V., S. (2017). Differential pulse voltammetry as an alternate technique for over oxidation of polymers: Application of electrochemically synthesized over oxidized poly (Alizarin Red S) modified disposable pencil graphite electrodes for simultaneous detection of hydroquinone and catechol. *Journal of Electroanalytical Chemistry (Lausanne, Switzerland)*, *789*. Advance online publication. doi:10.1016/j.jelechem.2017.02.037

Ashwini, R., Kumar, M. K. P., Rekha, M. Y., Santosh, M. S., & Srivastava, C. (2022). Optimization of NiFeCrCoCu high entropy alloy nanoparticle – graphene (HEA-G) composite for the enhanced electrochemical sensitivity towards urea oxidation. *Journal of Alloys and Compounds*, *903*, 163846. Advance online publication. doi:10.1016/j.jallcom.2022.163846

Baghayeri, M., Maleki, B., & Zarghani, R. (2014). Voltammetric behavior of tiopronin on carbon paste electrode modified with nanocrystalline Fe50Ni50 alloys. *Materials Science and Engineering C*, *44*, 175–182. Advance online publication. doi:10.1016/j.msec.2014.08.023 PMID:25280694

Cheng, J., Liu, P., Peng, T., Liu, Q., Chen, W., Liu, B., Yuan, Y., Zhang, W., Song, F., Gu, J., & Zhang, D. (2020). Mechanically alloyed NiTiO3/transition metal heterostructures: Introducing oxygen vacancies for exceptionally enhanced hydrogen evolution reaction activity. *Journal of Materials Chemistry. A, Materials for Energy and Sustainability*, *8*(30), 14908–14914. Advance online publication. doi:10.1039/D0TA05030J

Dastanpoor, E., & Enayati, M. H. (2017). Role of impurities in mechanical alloying amorphization of Cu-Zr and Cu-Zr-Al systems. *Indian Journal of Engineering and Materials Sciences*, *24*(1).

de Fatima Ulbrich, K., Winiarski, J. P., Jost, C. L., & de Campos, C. E. M. (2020). Green and facile solvent-free synthesis of NiTe2 nanocrystalline material applied to voltammetric determination of antioxidant morin. *Materials Today. Communications, 25*, 101251. Advance online publication. doi:10.1016/j.mtcomm.2020.101251

de Fatima Ulbrich, K., Winiarski, J. P., Jost, C. L., & Maduro de Campos, C. E. (2020). Mechanochemical synthesis of a Ni3-xTe2 nanocrystalline composite and its application for simultaneous electrochemical detection of dopamine and adrenaline. *Composites. Part B, Engineering, 183*, 107649. Advance online publication. doi:10.1016/j.compositesb.2019.107649

Geambazu, L. E., Cotruţ, C. M., Miculescu, F., & Csaki, I. (2021). Mechanically alloyed CoCrFeNiMo0.85 high-entropy alloy for corrosion resistance coatings. *Materials (Basel), 14*(14), 3802. Advance online publication. doi:10.3390/ma14143802 PMID:34300721

Gong, W., Li, J., Chu, Z., Yang, D., Subhan, S., Li, J., Huang, M., Zhang, H., & Zhao, Z. (2022). A low-cost high-entropy porous CrO/CrN/C biosensor for highly sensitive simultaneous detection of dopamine and uric acid. *Microchemical Journal, 175*, 107188. Advance online publication. doi:10.1016/j.microc.2022.107188

Gupta, S., Shashanka, R., & Chaira, D. (2015). Synthesis of nanostructured duplex and ferritic stainless steel powders by planetary milling: An experimental and simulation study. *IOP Conference Series. Materials Science and Engineering, 75*(1), 012033. Advance online publication. doi:10.1088/1757-899X/75/1/012033

Gupta, V. K., Khosravi, S., Karimi-Maleh, H., Alizadeh, M., & Sharafi, S. (2015). A voltammetric sensor for determination of methyldopa in the presence of hydrochlorothiazide using fe: Co Nanoalloy modified carbon paste electrode. *International Journal of Electrochemical Science, 10*(4).

Halakoei, H., Ghalkhani, M., Sobhani-Nasab, A., & Rahimi-Nasrabadi, M. (2021). An efficient electrochemical sensor based on CeVO4-CuWO4nanocomposite for methyldopa. *Materials Research Express, 8*(8), 085001. Advance online publication. doi:10.1088/2053-1591/ac16f1

Han, Z. D., Luan, H. W., Liu, X., Chen, N., Li, X. Y., Shao, Y., & Yao, K. F. (2018). Microstructures and mechanical properties of TixNbMoTaW refractory high-entropy alloys. *Materials Science and Engineering A, 712*, 380–385. Advance online publication. doi:10.1016/j.msea.2017.12.004

Hasan, I. M. A., Abd-Elsabur, K. M., Assaf, F. H., & Abd-Elsabour, M. (2022). Folic Acid Determination in Food Samples Using Green Synthesized Copper Oxide Nanoparticles and Electro-Poly (Methyl Orange) Sensor. *Electrocatalysis (New York), 13*(6), 759–772. Advance online publication. doi:10.1007/s12678-022-00756-0

Hassan, S. S., Nafady, A., Sirajuddin, Solangi, A. R., Kalhoro, M. S., Abro, M. I., & Sherazi, S. T. H. (2015). Ultra-trace level electrochemical sensor for methylene blue dye based on nafion stabilized ibuprofen derived gold nanoparticles. *Sensors and Actuators. B, Chemical, 208*, 320–326. Advance online publication. doi:10.1016/j.snb.2014.11.021

Hulanicki, A., Glab, S., & Ingman, F. (1991). Chemical sensors definitions and classification. *Pure and Applied Chemistry, 63*(9), 1247–1250. Advance online publication. doi:10.1351/pac199163091247

Hussain, I., Lamiel, C., Ahmad, M., Chen, Y., Shuang, S., Javed, M. S., Yang, Y., & Zhang, K. (2021). High entropy alloys as electrode material for supercapacitors: A review. In Journal of Energy Storage (Vol. 44). doi:10.1016/j.est.2021.103405

Krishnan, R. G., Rejithamol, R., & Saraswathyamma, B. (2020). Non-enzymatic electrochemical sensor for the simultaneous determination of adenosine, adenine and uric acid in whole blood and urine. *Microchemical Journal, 155*, 104745. Advance online publication. doi:10.1016/j.microc.2020.104745

Krishnan, R. G., & Saraswathyamma, B. (2021). Disposable electrochemical sensor for coumarin induced milk toxicity in raw milk samples. *Measurement, 170*, 108709. Advance online publication. doi:10.1016/j.measurement.2020.108709

Kshama, S. S., Swamynathan, K., & Shashanka, R. (2023). Electrochemical Devices for Soil Analysis. In *ACS Symposium Series* (Vol. 1437). 10.1021/bk-2023-1437.ch006

Kumar, A., Singh, A., & Suhane, A. (2022). Mechanically alloyed high entropy alloys: Existing challenges and opportunities. *Journal of Materials Research and Technology, 17*, 2431–2456. Advance online publication. doi:10.1016/j.jmrt.2022.01.141

Li, B., Hu, Y., Li, J., Liu, M., Kong, L., Hu, Y., & Kang, L. (2016). Mechanical alloying synthesis of Co9S8 particles as materials for supercapacitors. *Metals, 6*(6), 142. Advance online publication. doi:10.3390/met6060142

Li, K., & Chen, W. (2021). Recent progress in high-entropy alloys for catalysts: synthesis, applications, and prospects. In Materials Today Energy (Vol. 20). doi:10.1016/j.mtener.2021.100638

Li, M., Wang, X., Cheng, Y., Su, L., Xu, P., & Li, X. (2021). High-Entropy Alloy Nanoparticles As Catalyst for Nanomolar-Level Detection of Neurotransmitter Serotonin in Serum. *Proceedings of the IEEE International Conference on Micro Electro Mechanical Systems (MEMS), 2021-January*. 10.1109/MEMS51782.2021.9375269

Liu, L., Gou, Y., Gao, X., Zhang, P., Chen, W., Feng, S., Hu, F., & Li, Y. (2014). Electrochemically reduced graphene oxide-based electrochemical sensor for the sensitive determination of ferulic acid in A. sinensis and biological samples. *Materials Science and Engineering C*, *42*, 227–233. Advance online publication. doi:10.1016/j.msec.2014.05.045 PMID:25063114

Lu, L. (2019). Nanoporous noble metal-based alloys: a review on synthesis and applications to electrocatalysis and electrochemical sensing. In Microchimica Acta (Vol. 186, Issue 9). doi:10.1007/s00604-019-3772-3

Lü, L., & Lai, M. O. (1998). Introduction to Mechanical Alloying. In Mechanical Alloying. doi:10.1007/978-1-4615-5509-4_1

Mahale, R. S., Shashanka, R., Vasanth, S., & Vinaykumar, R. (2022). Voltammetric determination of various food azo dyes using different modified carbon paste electrodes. In Biointerface Research in Applied Chemistry (Vol. 12, Issue 4). doi:10.33263/BRIAC124.45574566

Malagutti, M. A., de Fátima Ulbrich, K., Winiarski, J. P., Paes, V. Z. C., Geshev, J., Jost, C. L., & Maduro de Campos, C. E. (2022). Mechanochemical synthesis of γ-CoTe2 nanocrystals and their application for determination of ferulic acid. *Materials Today. Communications*, *31*, 103481. Advance online publication. doi:10.1016/j.mtcomm.2022.103481

Nahirny, E. P., Bergamini, M. F., & Marcolino-Junior, L. H. (2020). Improvement in the performance of an electrochemical sensor for ethanol determination by chemical treatment of graphite. *Journal of Electroanalytical Chemistry (Lausanne, Switzerland)*, *877*, 114659. Advance online publication. doi:10.1016/j.jelechem.2020.114659

Nayak, A. K., Shashanka, R., & Chaira, D. (2016). Effect of Nanosize Yittria and Tungsten Addition to Duplex Stainless Steel during High Energy Planetary Milling. *IOP Conference Series. Materials Science and Engineering*, *115*(1), 012008. Advance online publication. doi:10.1088/1757-899X/115/1/012008

Oko, D. N., Garbarino, S., Zhang, J., Xu, Z., Chaker, M., Ma, D., Guay, D., & Tavares, A. C. (2015). Dopamine and ascorbic acid electrooxidation on Au, AuPt and Pt nanoparticles prepared by pulse laser ablation in water. *Electrochimica Acta*, *159*, 174–183. Advance online publication. doi:10.1016/j.electacta.2015.01.192

Panek, J., Kubisztal, J., & Bierska-Piech, B. (2014). Ni50Mo40Ti10 alloy prepared by mechanical alloying as electroactive material for hydrogen evolution reaction. *Surface and Interface Analysis*, *46*(10–11), 716–720. Advance online publication. doi:10.1002/sia.5501

Plata-Torres, M., Torres-Huerta, A. M., Domínguez-Crespo, M. A., Arce-Estrada, E. M., & Ramírez-Rodríguez, C. (2007). Electrochemical performance of crystalline Ni-Co-Mo-Fe electrodes obtained by mechanical alloying on the oxygen evolution reaction. *International Journal of Hydrogen Energy*, *32*(17), 4142–4152. Advance online publication. doi:10.1016/j.ijhydene.2007.05.023

Pradeep, N. B., Hegde, M. M. R., Rajendrachari, S., & Surendranathan, A. O. (2022). Investigation of microstructure and mechanical properties of microwave consolidated TiMgSr alloy prepared by high energy ball milling. *Powder Technology*, *408*, 117715. Advance online publication. doi:10.1016/j.powtec.2022.117715

Purushothama, H. T., & Arthoba Nayaka, Y. (2017). Electrochemical study of hydrochlorothiazide on electrochemically pre-treated pencil graphite electrode as a sensor. *Sensing and Bio-Sensing Research*, *16*, 12–18. Advance online publication. doi:10.1016/j.sbsr.2017.09.004

Qi, P., Wang, J., Wang, X., Wang, X., Wang, Z., Xu, H., Di, S., Wang, Q., & Wang, X. (2018). Sensitive determination of fenitrothion in water samples based on an electrochemical sensor layered reduced graphene oxide, molybdenum sulfide (MoS2)-Au and zirconia films. *Electrochimica Acta*, *292*, 667–675. Advance online publication. doi:10.1016/j.electacta.2018.09.187

R, S., & Chaira, D. (2014). Phase transformation and microstructure study of nanostructured austenitic and ferritic stainless steel powders prepared by planetary milling. *Powder Technology, 259*. doi:10.1016/j.powtec.2014.03.061

Rajendrachari, S. (2022). An Overview of High-Entropy Alloys Prepared by Mechanical Alloying Followed by the Characterization of Their Microstructure and Various Properties. *Alloys*, *1*(2), 116–134. Advance online publication. doi:10.3390/alloys1020008

Rajendrachari, S., Adimule, V., Gulen, M., Khosravi, F., & Somashekharappa, K. K. (2022). Synthesis and Characterization of High Entropy Alloy 23Fe-21Cr-18Ni-20Ti-18Mn for Electrochemical Sensor Applications. *Materials (Basel)*, *15*(21), 7591. Advance online publication. doi:10.3390/ma15217591 PMID:36363181

Rajendrachari, S., Adimule, V. M., Jayaprakash, G. K., & Pandith, A. (2023). Electrochemical oxidation of methylene blue dye in wastewater using mechanically alloyed high entropy alloy modified carbon paste electrode using cyclic voltammetry. *Materials Research Express*, *10*(5), 054003. Advance online publication. doi:10.1088/2053-1591/acd3d7

Rajendrachari, S., Basavegowda, N., Adimule, V. M., Avar, B., Somu, P., Saravana, S. K., & Baek, K. H. (2022). Assessing the Food Quality Using Carbon Nanomaterial Based Electrodes by Voltammetric Techniques. In Biosensors (Vol. 12, Issue 12). doi:10.3390/bios12121173

Sakthivel, R., Kubendhiran, S., Chen, S. M., & Kumar, J. V. (2019). Rational design and facile synthesis of binary metal sulfides VS2-SnS2 hybrid with functionalized multiwalled carbon nanotube for the selective detection of neurotransmitter dopamine. *Analytica Chimica Acta*, *1071*, 98–108. Advance online publication. doi:10.1016/j.aca.2019.04.058 PMID:31128761

Santhy, A., Beena, S., Veena, & Shahina, S. (2020). Electrochemical quantification of vitamin B9 on poly tyrosine modified pencil graphite electrode. *IOP Conference Series. Materials Science and Engineering*, *872*(1), 012128. Advance online publication. doi:10.1088/1757-899X/872/1/012128

Santhy, A., Saraswathyamma, B., Parvathy Krishnan, A., & Luscious, L. (2020). Nanomaterials incorporated electrochemical sensors for the monitoring of pyridoxine: A mini review. *Materials Today: Proceedings*, *46*, 2998–3004. Advance online publication. doi:10.1016/j.matpr.2020.12.703

Shang, C., Axinte, E., Sun, J., Li, X., Li, P., Du, J., Qiao, P., & Wang, Y. (2017). CoCrFeNi(W1 − xMox) high-entropy alloy coatings with excellent mechanical properties and corrosion resistance prepared by mechanical alloying and hot pressing sintering. *Materials & Design*, *117*, 193–202. Advance online publication. doi:10.1016/j.matdes.2016.12.076

Shashanka, R. (2017). Synthesis of nanostructured stainless steel powder by mechanical alloying-an overview. *International Journal of Scientific and Engineering Research*, *8*(6).

Shashanka, R., & Chaira, D. (2015). Optimization of milling parameters for the synthesis of nanostructured duplex and ferritic stainless steel powders by high energy planetary milling. *Powder Technology*, *278*, 35–45. Advance online publication. doi:10.1016/j.powtec.2015.03.007

Shashanka, R., & Chaira, D. (2016). Effects of nano-Y2O3 and sintering parameters on the fabrication of PM duplex and ferritic stainless steels. *Acta Metallurgica Sinica, 29*(1), 58–71. Advance online publication. doi:10.1007/s40195-015-0362-1

Shashanka, R., Chaira, D., & Kumara Swamy, B. E. (2015a). Electrocatalytic response of duplex and yittria dispersed duplex stainless steel modified carbon paste electrode in detecting folic acid using cyclic voltammetry. *International Journal of Electrochemical Science, 10*(7), 5586–5598. doi:10.1016/S1452-3981(23)17279-8

Shashanka, R., Chaira, D., & Kumara Swamy, B. E. (2015b). Electrochemical investigation of duplex stainless steel at carbon paste electrode and its application to the detection of dopamine, ascorbic and uric acid. *International Journal of Scientific and Engineering Research, 6*(9).

Shashanka, R., Chaira, D., & Kumara Swamy, B. E. (2018). Effect of Y2O3 nanoparticles on corrosion study of spark plasma sintered duplex and ferritic stainless steel samples by linear sweep voltammetric method. *Archives of Metallurgy and Materials, 63*(2). Advance online publication. doi:10.24425/122401

Shashanka, R., & Kumara Swamy, B. E. (2020a). Biosynthesis of silver nanoparticles using leaves of Acacia melanoxylon and their application as dopamine and hydrogen peroxide sensors. *Physical Chemistry Research, 8*(1). Advance online publication. doi:10.22036/pcr.2019.205211.1688

Shashanka, R., & Kumara Swamy, B. E. (2020b). Simultaneous electro-generation and electro-deposition of copper oxide nanoparticles on glassy carbon electrode and its sensor application. *SN Applied Sciences, 2*(5), 956. Advance online publication. doi:10.1007/s42452-020-2785-1

Shashanka, R., Uzun, O., & Chaira, D. (2020). Synthesis of nanostructured duplex and ferritic stainless steel powders by dry milling and its comparison with wet milling. *Archives of Metallurgy and Materials, 65*(1). Advance online publication. doi:10.24425/amm.2019.131091

Sukanya, R., Sakthivel, M., Chen, S. M., & Chen, T. W. (2018). A new type of terbium diselenide nano octagon integrated oxidized carbon nanofiber: An efficient electrode material for electrochemical detection of morin in the food sample. *Sensors and Actuators. B, Chemical, 269*, 354–367. Advance online publication. doi:10.1016/j.snb.2018.05.013

Suñol, J. J., & Escoda, L. (2022). Novel Materials Synthesis by Mechanical Alloying/ Milling. In Materials (Vol. 15, Issue 19). doi:10.3390/ma15196973

Suryanarayana, C. (2019). Mechanical Alloying: A Novel Technique to Synthesize Advanced Materials. *Research, 2019*, 2019/4219812. Advance online publication. doi:10.34133/2019/4219812 PMID:31549063

Suryanarayana, C., Ivanov, E., & Boldyrev, V. V. (2001). The science and technology of mechanical alloying. *Materials Science and Engineering A, 304–306*(1–2), 151–158. Advance online publication. doi:10.1016/S0921-5093(00)01465-9

Tawade, A. K., Kamble, B. B., Sharma, K. K. K., & Tayade, S. N. (2020). Simultaneous electrochemical investigations of dopamine and uric acid by in situ amino functionalized reduced grahene oxide. *SN Applied Sciences, 2*(6), 1082. Advance online publication. doi:10.1007/s42452-020-2806-0

Technology and Challenges in Additive Manufacturing of Duplex Stainless Steels. (2021). *Biointerface Research in Applied Chemistry, 12*(1), 1110–1119. Advance online publication. doi:10.33263/BRIAC121.11101119

Terbouche, A., Boulahia, S., Mecerli, S., Ait-Ramdane-Terbouche, C., Belkhalfa, H., Guerniche, D., Sehailia, M., Bachari, K., Mezaoui, D., & Hauchard, D. (2022). A novel hybrid carbon materials-modified electrochemical sensor used for detection of gallic acid. *Measurement, 187*, 110369. Advance online publication. doi:10.1016/j.measurement.2021.110369

Tian, Y., Lu, C., Shen, Y., & Feng, X. (2019). Microstructure and corrosion property of CrMnFeCoNi high entropy alloy coating on Q235 substrate via mechanical alloying method. *Surfaces and Interfaces, 15*, 135–140. Advance online publication. doi:10.1016/j.surfin.2019.02.004

Wang, J., Yang, B., Zhong, J., Yan, B., Zhang, K., Zhai, C., Shiraishi, Y., Du, Y., & Yang, P. (2017). Dopamine and uric acid electrochemical sensor based on a glassy carbon electrode modified with cubic Pd and reduced graphene oxide nanocomposite. *Journal of Colloid and Interface Science, 497*, 172–180. Advance online publication. doi:10.1016/j.jcis.2017.03.011 PMID:28284071

Wang, W. (2016). Introductory Chapter: What is Chemical Sensor? In Progresses in Chemical Sensor. doi:10.5772/64626

Winiarski, J. P., Tavares, B. F., de Fátima Ulbrich, K., de Campos, C. E. M., Souza, A. A. U., Souza, S. M. A. G. U., & Jost, C. L. (2022). Development of a multianalyte electrochemical sensor for depression biomarkers based on a waste of the steel industry for a sustainable and one-step electrode modification. *Microchemical Journal, 175*, 107141. Advance online publication. doi:10.1016/j.microc.2021.107141

Yola, M. L., Gupta, V. K., Eren, T., Şen, A. E., & Atar, N. (2014). A novel electro analytical nanosensor based on graphene oxide/silver nanoparticles for simultaneous determination of quercetin and morin. *Electrochimica Acta*, *120*, 204–211. Advance online publication. doi:10.1016/j.electacta.2013.12.086

Yu, T., & Breslin, C. B. (2020). Review—Two-Dimensional Titanium Carbide MXenes and Their Emerging Applications as Electrochemical Sensors. *Journal of the Electrochemical Society*, *167*(3), 037514. Advance online publication. doi:10.1149/2.0142003JES

Zhang, J., Xu, X., & Chen, L. (2018). An ultrasensitive electrochemical bisphenol A sensor based on hierarchical Ce-metal-organic framework modified with cetyltrimethylammonium bromide. *Sensors and Actuators. B, Chemical*, *261*, 425–433. Advance online publication. doi:10.1016/j.snb.2018.01.170

Zhang, P., Jia, C., Zhao, Y., Luo, H., Tan, X., Ma, X., & Wang, Y. (2019). Detection of tiopronin in body fluids and pharmaceutical products using red-emissive DNA-stabilized silver nanoclusters as a fluorescent probe. *Mikrochimica Acta*, *186*(9), 609. Advance online publication. doi:10.1007/s00604-019-3730-0 PMID:31392427

Zhang, P., Jia, D., Yang, Z., Duan, X., & Zhou, Y. (2013). Influence of ball milling parameters on the structure of the mechanically alloyed SiBCN powder. *Ceramics International*, *39*(2), 1963–1969. Advance online publication. doi:10.1016/j.ceramint.2012.08.047

Chapter 3
Fabrication of Ceramic Material

Tabrez Qureshi
https://orcid.org/0009-0005-8194-3709
National Institute of Technology, Srinagar, India

Mohammad Mohsin Khan
https://orcid.org/0000-0001-9465-5144
National Institute of Technology, Srinagar, India

Harveer Singh Pali
National Institute of Technology, Srinagar, India

ABSTRACT

This chapter introduces the emerging field of high entropy ceramics, which are characterized by their complex and disordered structures resulting from the amalgamation of multiple elements in nearly equal amounts. Unlike traditional ceramics composed of one or a few elements, high entropy ceramics exhibit exceptional properties such as high strength, hardness, and thermal stability. The chapter delves into high-entropy ceramics' synthesis, properties, and potential applications, highlighting their versatility in various fields. Additionally, the chapter outlines the fabrication techniques involved in producing high-entropy ceramics, including material selection, powder preparation, powder compaction, and sintering.

1. INTRODUCTION

Ceramics are an essential class of materials used for thousands of years due to their unique properties. They are inorganic non-metallic materials widely used in applications such as construction (Maged et al., 2023), electronics (Cao et al., 2023), biomedical (Lakshmi Priya et al., 2017), and energy due to their excellent thermal

DOI: 10.4018/978-1-6684-9385-4.ch003

(Smith & Naït-Ali, 2021), mechanical (Ke et al., 2023), and electrical properties (Guo et al., 2023). Traditional ceramics are typically made of a single or a few elements, such as alumina or zirconia, and exhibit a relatively ordered structure (Wright & Luo, 2020). In recent years, a new class of ceramics called high entropy ceramics (HECs) has emerged. HECs are characterized by a high degree of structural complexity and disorder due to the mixing of multiple elements in almost equal proportions (Gild et al., 2016). Mixing multiple elements in high entropy ceramics results in a significant entropy increase, leading to their distinctive properties, including elevated hardness (Qin et al., 2021), enhanced strength (Iwan et al., 2022), and robust thermal stability (Li et al., 2019), setting them apart from conventional ceramics.

The concept of high-entropy materials (HEMs) finds its roots in high-entropy alloys (HEAs), which are multi-principal element alloys (MPEAs) characterized by an equal or near-equal atomic fraction of constituent elements (Cantor et al., 2004; Rajendrachari, 2022b). The distinction between HEAs and traditional alloys extends beyond properties and their design strategy. In traditional alloy design, a single base element, for instance, Fe, Mg or Ti, serves as the principal element, with minor alloy elements added to enhance properties, following the "base element concept" (Zhang et al., 2014). In contrast, the HEA design strategy involves mixing five or more multi-principal elements into a single lattice, creating a solid solution (Kale et al., 2022). In phase diagrams, traditional alloy compositions are generally found at the corners, limiting compositional design space. At the same time, HEAs occupy the central region, significantly expanding the design space through the combination of multi-principal elements and microstructures (Cantor, 2014).

Inspired by the breakthroughs in the field of metals, the high-entropy concept has been extended to composites (HECos), polymers (HEPs) and ceramics (HECs). Like HEAs, all high-entropy materials are characterized by long-range structural order but compositional disorder. However, high-entropy ceramics (HECs) offer more opportunities for tuning properties and addressing material application challenges. Ceramics boast diverse crystal structures beyond the simple FCC, BCC, and HCP structures found in metals (Sharma, Yadav, Biswas et al, 2018). Moreover, the variety in band structure and chemical bonding in ceramics opens new avenues for material design and property tuning through band structure and phonon engineering. HECs, with their numerous possibilities in composition design and property tuning paradigms, have demonstrated improved stability, enhanced mechanical properties, thermal conductivity (Braun et al., 2018) and potential in various functional and structural applications, ranging from ultrahigh-temperature thermal protection and insulation in hypersonic vehicles to thermal/environmental barrier protection of engine materials, high-speed and dry cutting, Li-ion batteries, thermoelectrics, and electromagnetic wave absorption & interference shielding (Chen, Xiang, Dai, Liu, Lei et al, 2019).

Despite rapid development, the field of HECs is still in its early stages, both in terms of academic research and practical applications. Additionally, the novel design concept and the underlying mechanisms responsible for these unusual phenomena present theoretical and experimental challenges.

The preparation of ceramic materials by ball milling followed by the sintering method has been widely accepted because it is simple, cost-effective, and can yield highly homogeneous and fine powders (Harrington et al., 2019). The synthesis and characterization of high entropy ceramics are still in their early stages, and much research is currently underway to understand their properties and potential applications. This chapter aims to provide an overview of the various synthesis techniques, properties, and applications of high entropy ceramics, focusing on their unique properties and potential applications in various fields.

2. DEFINITIONS

Traditional ceramics, also known as "whiteware," refer to ceramic products made from naturally occurring raw materials such as clay, feldspar, quartz, and kaolin. These raw materials are mixed together and formed into a shape, such as a pot or a vase, by processes such as wheel throwing, pressing, or casting (Yahya et al., 2022). Traditional ceramics are fired at high temperatures in a kiln to produce a dense, hard, and durable product. These products are typically glazed to enhance their appearance and functionality (Luo et al., 2023). Traditional ceramics have been produced for thousands of years and are still used today in many applications, such as tableware, tiles, and sanitary ware.

High Entropy Ceramics are a new class of advanced ceramics that consist of at least four or more elements in equimolar or near-equimolar ratios, which gives rise to a high degree of structural complexity and disorder (Oses et al., 2020). These materials are characterized by their unique properties, such as high strength, hardness, toughness, and excellent thermal and chemical stability, making them ideal for high-temperature applications (Qiao et al., 2022).

Additionally, high-entropy ceramics draws inspiration from high-entropy alloys (HEAs). HEAs are defined in two distinct ways based on their composition and configuration entropy. **In terms of composition**, HEAs are characterized as alloys comprising over five primary elements mixed in approximately equal proportions (Rajendrachari, 2022a).

In terms of entropy, they are identified as alloys with configuration entropies exceeding 1.5 times the gas constant (represented as 1.5R) when forming random solid solutions.

3. FABRICATION TECHNIQUES OVERVIEW

The fabrication of ceramic materials is categorized into two aspects.

3.1. Synthesis

Synthesis refers to the process of creating or producing a high-entropy ceramic material from its constituent elements or compounds. It involves mixing the starting materials in the desired proportions, followed by a high-temperature reaction or sintering process. Synthesis aims to obtain a high entropy ceramic material with desired properties and composition (Gild, 2019). To date, three approaches have been employed for HEC synthesis: solid-state reaction, wet chemical processes, and epitaxial growth. This section primarily delves into solid-state reactions, specifically exploring the formation kinetics of various HECs. The remaining two methods have primarily been utilized for oxides and will be briefly addressed later in the same section.

Solid-state synthesis is the primary method for preparing HECs. Typically, powders are generated through mechanochemistry or by mixing precursor powders during ball milling (Kale et al., 2022). In cases involving low melting point precursors like chalcogenides, mechanochemical synthesis occurs during ball milling, along with the disappeared XRD peaks of precursors. Conversely, high melting point precursors like oxides and carbides only undergo mixing during ball milling, and HECs synthesis primarily occurs during sintering, where component inter-diffusion occurs.

Under uniform processing conditions, ZrC emerged as a second phase in HEC (Hf– Zr –Ta–Ti)C, while (Hf–Ta–Zr–Nb)C remained a single phase (Castle et al., 2018). This was attributed to ZrC diffusing more slowly in HfC–TaC–ZrC–TiC mixed phases than in HfC–TaC–ZrC–NbC. The order of inter-diffusing rates suggested was TaC < ZrC < HfC < NbC < TiC, based on various factors. TiC–TaC or NbC–TaC occurred initial inter-diffusion by affecting ZrC's kinetics. In a similar study, Hong et al. (Hong et al., 2019) examined (Mg,Co,Ni,Cu,Zn)O high entropy oxides. XRD analysis of HECs heat-treated from 650 to 825°C revealed inter-diffusion starting at 650°C, with MgO and CoO peaks merging and others weakening. NiO, ZnO, and CuO peaks disappeared with increasing temperature, indicating faster inter-diffusion of CoO and MgO than ZnO and CuO, as shown in Figure 1. The incorporation rate may be linked to metal vacancy formation energies, while cation radii showed no correlation with inter-diffusion.

Wet chemical techniques employed in HEC synthesis include nebulized or flame spray pyrolysis and reverse co-precipitation techniques. These approaches have been utilized to produce oxide powders at the nanoscale, including (Mg,Co,Cu,Zn,Ni)O (Cheng et al., 2020; Sarkar et al., 2017), $(PtNiMgCuCoZn)O_x$ (Chen et al., 2018),

Figure 1. (a) The XRD spectra for High entropy Ceramics samples subjected to different heat treatment temperatures ranging from 650°C to 825°C. (b) An enlarged view of the spectra for the (200) peak. Reprint with permission from (Hong et al., 2019) © John Wiley and Sons

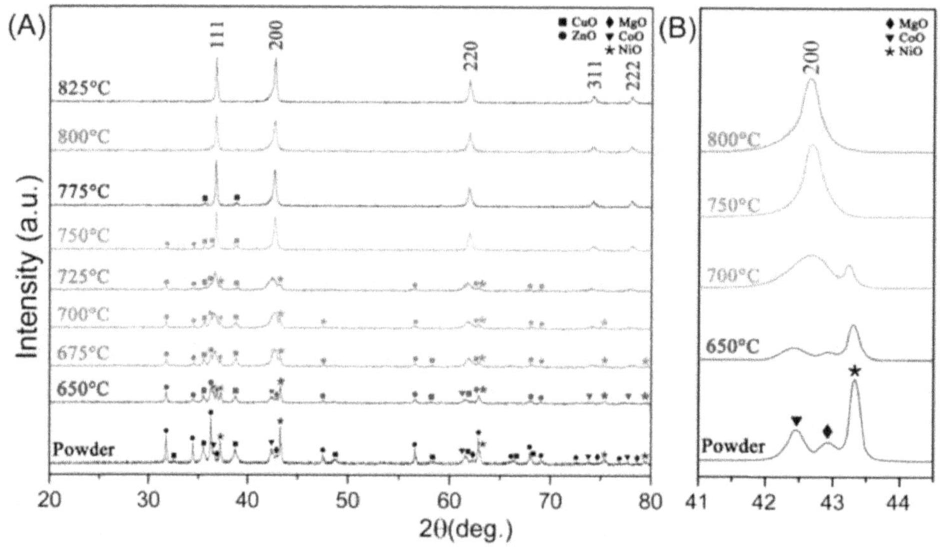

($Li_x(Co_{0.2}Cu_{0.2}Mg_{0.2}Zn_{0.2}Ni_{0.2})OF_x$) (Wang et al., 2019), and HE-oxides. High entropy metal oxide nanotube arrays were fabricated via anodic oxidation of TaTiNbHfZr HEA precursors.

Epitaxial growth is also valuable for crafting structures that cannot be achieved through solid-state reactions or wet chemical techniques, such as superlattices and heterostructures (Meisenheimer et al., 2017). The epitaxial growth of thin films represents another approach for synthesizing HECs. In the existing literature, exclusively pulsed laser deposition (PLD) has been utilized to fabricate HEC oxides. This m2ethod offers a distinct advantage in cultivating single crystals (Sharma, Musico, Gao et al, 2018), consequently reducing grain boundary segregation and facilitating precise management of elemental ratios.

3.2. Densification

Densification refers to the process of reducing the porosity and increasing the density of a high-entropy ceramic material (Guo et al., 2023). This process typically involves heating the material at high temperatures in a controlled atmosphere to promote the rearrangement of atoms and the formation of strong atomic bonds (Gild et al.,

2020). The goal of densification is to improve the mechanical, thermal, and electrical properties of the material (Zhang, Guo, Jiang et al, 2019).

The Spark Plasma Sintering (SPS) method is widely favoured for achieving the densification of High-Entropy Ceramics (HECs) due to its exceptional performance, which is facilitated by its ability to provide higher heating rates and pressures than other sintering methods (Dwivedi et al., 2023). In early studies, substantial progress was made, attaining approximately 92% relative density by utilizing commercial diboride powders under SPS conditions at 2,000°C and 30 MPa (Gild et al., 2016). However, oxide impurities hindered densification. SPS was also applied to consolidate powders synthesized through SHS, reaching ~92.5% relative density at 1,950°C and 30 MPa, though B was present as a secondary phase (Tallarita et al., 2020; Tallarita et al., 2019).

High-entropy ceramics (HECs), known for achieving remarkable relative densities exceeding ~99% at elevated temperatures ranging from 2,200°C to 2,300°C (Castle et al., 2018; Harrington et al., 2019; Qin et al., 2022; Sarker et al., 2018) are featured in Figure 2. The SEM micrographs reveal that the $(Ti_{0.2}Zr_{0.2}Hf_{0.2}Ta_{0.2}Er_{0.2})B_2$

Figure 2. SEM micrographs and EDS elemental mapping of $((Ti_{0.2}Zr_{0.2}Hf_{0.2}Ta_{0.2}Er_{0.2})B_2)$ specimen at various magnifications, fabricated via reactive SPS in (a)(c) and borocarbothermal reduction in (b)(d) respectively (Qin et al., 2022) (Open access, reprinting permitted without permission)

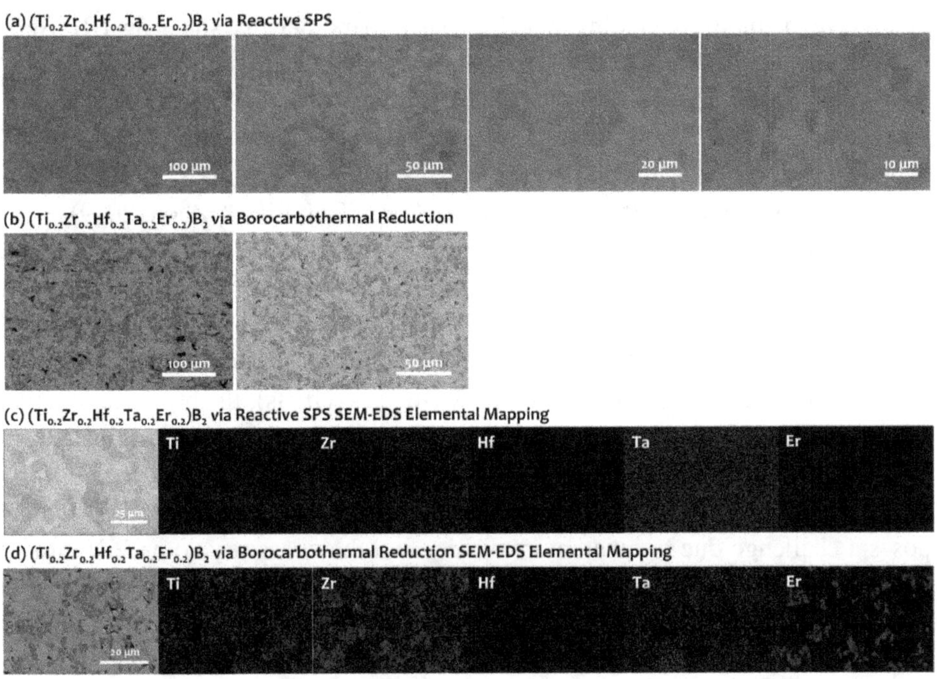

specimen produced through reactive SPS exhibits exceptional density (99.1%) with a homogeneous solid solution and minimal porosity. The EDS analysis confirms a composition close to the nominal equimolar composition. Tungsten contamination, a concern with HEBM, was minimal due to shorter milling duration and softer precursor powders. In contrast, the borocarbothermal reduction method yielded a specimen with multiple phases, porosity, and Er-rich secondary phases.

Whereas Reactive flash spark plasma sintering (RSPS) can also yield dense HECs, one study revealed residual graphite and oxides, indicating incomplete carbon transfer reactions (Gild et al., 2019). Conversely, pressureless sintering produced porous HECs in a reported study (Chen, Xiang, Dai, Liu, Lei et al, 2019).

Moreover, Incorporating in situ reactions and densification processes enhances densification compared to solid-state formation of individual carbide powders. While relative densities of ~98% were achieved by reacting metals with carbon, challenges arose due to uneven distribution of Zr and Nb resulting from oxidation and self-ignition of metal powders (Wei et al., 2019). Alternatively, dense high-energy ceramics (HECs) can be produced from powders synthesized through combustion-triggered reactions (CTR) (Feng et al., 2019; Wei et al., 2019). Combining CTR and solid-state formation in a single step proved challenging for producing single-phase HECs, potentially due to substantial CO gas formation. More successful results were obtained using two-step processes with separate CTR and solid-state densification steps.

Furthermore, High-Entropy Boride (HEB) ceramics have been densified using alternative methods, including reactive flash spark plasma sintering (RF-SPS) (Cologna et al., 2010), and additive manufacturing (AM) (Failla et al., 2020) and RF-SPS produced dense $(Hf,Nb,Ta,Ti,Zr)B_2$ ceramics with fine grain sizes. However, incorporating CrB2 via AM was challenging due to its lower melting point and high vapour pressure. Pressureless sintering resulted in less dense $(Hf,Nb,Ta,Ti,Zr)B_2$ ceramics, approximately 75% dense (Chen, Xiang, Dai, Liu, & Zhou, 2019).

Valuable findings from studies on densification can be briefly summarized. Achieving full densification in High-Entropy Ceramics (HECs) can be realized at relatively lower temperatures, as low as 1,850°C, provided that submicron particles with oxygen content below 1 wt% are employed. Nevertheless, similar to the challenges faced in the densification of High-Entropy Borides (HEBs), the presence of oxide impurities poses an obstacle to HEC consolidation. These impurities originate from inherent surface oxides and are further introduced during milling. Creating fully dense High-Entropy Ceramics (HECs) using commercial carbide starting powders poses a challenge due to inherent impurities and the necessary comminution and mixing steps. A more promising strategy involves integrated processes that generate submicron particles and achieve direct densification, thereby reducing the requirement for additional processing steps to attain complete density in HECs. However, similar to High-Entropy Borides (HEBs), there is a current lack of fundamental studies investigating sintering mechanisms and kinetics in the context of HECs.

4. SEQUENTIAL STEPS OF FABRICATING CERAMIC MATERIAL

In general, the fabrication of a ceramic material involves several steps.

4.1. Material Selection

High entropy ceramics comprise multiple elements with equal or nearly equal atomic percentages (Xie, 2022). Therefore, carefully selecting the right materials (transition metal elements) for the fabrication of High-Entropy Ceramics (HECs) involves careful consideration of various parameters to ensure the desired properties and performance. Here are the key parameters to follow when selecting materials for the fabrication of HECs:

4.1.1. Atomic Size Effects

The increase in entropy in high-energy ceramics (HECs) primarily results from an elevation in the degree of atomic disorder. Consequently, the atomic size effect (represented as δ) serves as the primary criterion to explain the development of high entropy solid solutions within these multi-component systems (Euchner & Mayrhofer, 2015). Atomic sizes play a crucial role in analyzing crystal structures, particularly when employing close-packing models such as hexagonal close-packed (HCP) or cubic closest packing (CCP) arrangements. In this context, the size of the constituent atoms influences the dimensions of the interstitial spaces within these packing structures. The size of these interstitial spaces ultimately dictates whether atomic elements can be arranged in HCP or CCP configurations. In a multi-component system, the average atomic size difference (δ) is determined by various factors and is given by an appropriate mathematical expression (Dubrovinskaia et al., 2007).

$$\delta = \sqrt{\sum_{i=1}^{i=N} X_i \left(1 - \frac{r_i}{\sum_{i=1}^{i=N} X_i r_i}\right)} \quad (1)$$

The parameter δ, defined as you've mentioned with X_i representing molar content and ri representing atomic radii for the ith component, is a key factor in characterizing multi-component systems. According to the Hume-Rothery rule, which considers the similarity of atomic radii in High-Entropy Ultra-High-Temperature Ceramics (HE-UHTCs), the atomic size misfit parameter δ should be ≤ 6.6%. However, it's important to note that this criterion is not an absolute requirement. For instance, in the case of HEUHTCs studied by Gild et al. (2020), computed δ values ranged from 3.5% to 8.0%. Notably, even the sample with an δ value of around 8%, which exceeds the suggested limit of δ ≤ 6.6%, still forms a single-phase solid solution with high entropy. This underscores that while atomic size misfit can be a desirable criterion, it is unnecessary for high-entropy solid solution formation.

4.1.2. Lattice Constants

In solids, ideal atomic placement is disrupted by atomic size, electronegativity, and neighboring atoms, causing lattice distortion—a fundamental effect in High-Entropy Materials (HEMs) (Yeh, 2006). In borides with an AlB2 structure, research indicates that the atomic radii difference of metals is inadequate for characterizing lattice distortion. Instead, accurately representing lattice misfit (δ) in multi-component Ultra-High-Temperature Ceramics (UHTCs) involves considering the lattice constants of individual constituents, surpassing reliance on atomic radii:

$$\delta_a = \sqrt{\sum_{i=1}^{i=N} X_i \left(1 - \frac{a_i}{\sum_{i=1}^{i=N} X_i a_i}\right)} \tag{2}$$

$$\delta_c = \sqrt{\sum_{i=1}^{i=N} X_i \left(1 - \frac{c_i}{\sum_{i=1}^{i=N} X_i c_i}\right)} \tag{3}$$

The sample exhibited a computed δ value of approximately 8% (via Eq (1)), resulting in an δC value of about 5.2% (via Eq (3)). This fulfills the high-entropy formation criteria (δ ≤ 6.6%), indicating successful formation (Gild et al., 2016).

4.1.3. Phase Stability

The quest for novel materials remains a cornerstone of materials science, and the exploration of uncharted territory for stable compounds with unexpected properties is a central pursuit within the realm of High-Entropy Ceramics (HECs). Although the design of HECs is still in its infancy, many existing manuscripts on HECs take a straightforward approach: selecting a group of elements from a specific region of the periodic table. Numerous HECs have been synthesized and tested experimentally without ab initio computations to ascertain whether the chosen constituents would yield a single-phase material. Yet, this process can be expedited through theoretical predictions regarding the stability of potential candidates. Thus far, simulation techniques have primarily been harnessed to assess thermal and phase stability. Notably, Density Functional Theory (DFT) calculations and the CALPHAD method have become popular for determining mixing enthalpy and configurational entropy. These descriptors aid in pinpointing the optimal synthesis temperature for HECs based on negative Gibbs free energy of mixing, advancing our understanding of these intriguing materials (Ye et al., 2019).

The Calculation of Phase Diagrams (CALPHAD) method is crucial for investigating high-entropy ceramics' stability. It describes a system's Gibbs free energy using extensive experimental and computational thermodynamic relations databases. These databases enable numerical techniques like interpolation and extrapolation to calculate phase diagrams and predict new compositions (Zhong et al., 2019). However, high-entropy ceramics pose challenges due to the numerous phases requiring accurate data. Missing compounds in databases reduce prediction reliability, necessitating alternative stability forecasting methods (Gorsse & Senkov, 2018). Density Functional Theory (DFT) is commonly used for exploring thermodynamic descriptors in High-Entropy Borides (HEBs). Qiao et al. employed first-principle calculations using the Cambridge Sequential Total Energy Package (CASTEP) based on DFT, coupled with the plane-wave pseudopotential method and Virtual Crystal Approximation (VCA) method to predict microstructures and mechanical properties of HEBs. Their results affirmed the suitability of the VCA system in HEB studies. Researchers employ DFT and methods like the Virtual Crystal Approximation (VCA) to predict HEB microstructures and mechanical properties, demonstrating the VCA's suitability (Qiao et al., 2022).

Numerous descriptors and simulation techniques exist to pinpoint thermodynamically stable High-Entropy Ceramics (HECs). One such model, Entropy Forming Ability (EFA), introduced by Sarker et al. (Sarker et al., 2018), summarizes mixing enthalpy using descriptors. EFA facilitates the discovery of new HEC systems by addressing synthesizability through high-throughput first-principle calculations. Where EFA can be expressed as:

$$EFA = \sqrt{\frac{\sum_{i=1}^{n} g_i \left(H_i - H_{mix}\right)^2}{\left(\sum_{i=1}^{n} g_i\right) - 1}} \tag{4}$$

The standard deviation of energy distribution spectra for calculated energies impacts the synthesizability of atom configurations within the smallest unit cell. It's determined by n (the total geometrical configurations sampled), g_i (degeneracies), H_i (enthalpy of formation at zero temperature of a supercell), and H_{mix} (average enthalpies of all possible supercell configurations). And can be expressed as:

$$H_{mix} = \frac{\sum_{i=1}^{n} g_i H_i}{\sum_{i=1}^{n} g_i} \tag{5}$$

A larger Entropy Forming Ability (EFA) value corresponds to higher entropy in a five-component metal carbide solid solution selected from eight refractory metals. Another study introduced a descriptor to predict 56 HE-Carbide combinations, with only nine successfully synthesized. This approach streamlines research, saving time and resources compared to trial and error methods, which are critical for discovering and designing HECs with desired characteristics (Harrington et al., 2019).

4.1.4. Melting Point of Diborides

High-entropy ceramics (HECs) are complex mixtures comprising numerous constituents with distinct properties that significantly influence the final ceramic's characteristics. Most synthesis methods subject these precursors to extremely high temperatures, approaching their melting points. Unfortunately, these elevated temperatures often lead to grain coarsening and declining mechanical properties (Nisar, Khan, Bajpai et al, 2019). Residual carbon or impurities like oxide phases further affect the final ceramics (Akrami et al., 2021). HECs can be manufactured using self-synthesized powders via various processing routes, including elemental reactions, carbothermal reduction, and borocarbothermal reduction, to combat these issues (Zhang, Jiang, Sun et al, 2019). This approach lowers sintering temperatures and refines grain sizes, resulting in ceramics with enhanced elevated-temperature flexural strength and superior oxidation resistance (Nisar et al., 2020).

Understanding the melting points of individual constituents within HECs is crucial for achieving the desired properties in the final sintered samples. Table 1 compares the melting points of metals and their corresponding diborides, along with the ratio of their melting points, aiding in optimizing HEC fabrication for specific applications.

The ratios highlight the decreasing thermal stability of diborides compared to pure metals from Group IV to Group VI and within each group from lower to

Table 1. Melting points of metals and their diborides

T. Metal Ti Zr Hf V Nb Ta Cr Mo W
MP°C
Metal 1700 1850 2250 1735 2500 2990 1850 2620 3410
MP°C
TMB_2 2920/3225 3050/ 3546 /3240 2400/2980 3050 3200/ 3140 1900 /2200 2100/ 2180 2200/ 2900
Ratio in $^{\circ}K$
$\frac{MP(TMB_2)}{MP(M_e)}$ 1.62 1.57 1.40 1.33 1.20 1.06 1.02 0.83 0.67
(Bhatt et al., 2016; Golla et al., 2014; Hafnium Boride Powder, n.d.; Laszkiewicz-Łukasik et al., 2016; Okada et al., 1995; Post et al., n.d.; Sairam et al., 2014; Shanghai Greenearth Chemicals Co, n.d.; Stanfield et al., 2021)

higher atomic numbers. Conversely, pure metal melting temperatures follow an inverse pattern.

5. POWDER PREPARATION VIA HIGH ENERGY BALL MILLING

The individual elements are weighed, mixed, and milled to form a homogeneous powder (Cheng et al., 2023). High Energy Ball Milling (HEBM) is commonly used for this step to ensure a uniform mixture of the powders. In this method, the ball mill is operated at high speed and high energy is applied to the raw materials through the grinding media, reducing particle size and increasing specific surface area (Feng et al., 2017; Shashanka et al., 2020). Figure 3 illustrates a schematic diagram of the HEBM process. Additionally, it is widely reported in numerous research articles that the High-Energy Ball Milling (HEBM) process can introduce surface oxides onto particle surfaces. It may result in the incorporation of impurities from the wear of the milling media (Feng et al., 2017).

Here's a brief description of the significance of HEBM.

1. **Powder Blending**: Planetary ball milling allows the mixing of diverse precursor powders containing different elements or compounds, ensuring even distribution and homogeneity (Dippo et al., 2020; Fourmont et al., 2020; Šolcová et al., 2021).
2. **Size Reduction**: During milling, the collision and grinding of particles lead to a reduction in their size. This finer powder is more reactive during subsequent processing (Feng et al., 2019; Wang et al., 2020).

Figure 3. Schematic diagram of the HEBM process, the inset provides a visual representation of crack formation, refinement of raw powders and agglomeration during high-energy ball milling

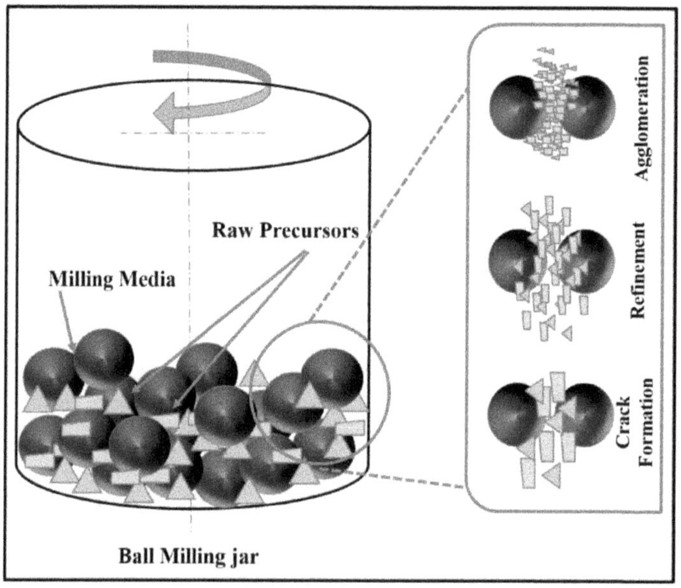

3. **Enhanced Reactivity**: Smaller particle sizes and increased surface area enhance the reactivity of the powders, promoting the formation of HEC phases at lower sintering temperatures (K et al., 2021; Yuan et al., 2022).
4. **Customized Composition**: Researchers can precisely control the composition by adjusting the proportions of constituent powders (Teng et al., 2021). This enables tailored material properties to meet specific application needs.
5. **Uniform Microstructure**: Achieving a uniform microstructure is crucial in HECs to ensure consistent mechanical, thermal, and electrical properties (Li et al., 2022).

Several types of ball milling methods can be used for the preparation of ceramic materials, including.

5.1. Planetary Ball Milling

Planetary ball milling is a pivotal technique in the realm of High-Entropy Ceramics (HECs). It involves the mechanical alloying of multiple constituent powders in a high-energy planetary ball mill (Shashanka & Chaira, 2015). Figure 2 shows the schematic view of the motion of the balls and powder mixture. In this method, the grinding media is placed in a planetary ball mill that rotates on its axis, causing the grinding media to collide with the raw materials (Gaur et al., 2023). Figure 4(a) illustrates the graphical representation of planetary ball milling. Although planetary ball milling introduces surface oxides on particles and milling media wear-derived impurities, affecting material purity. Kovalev et al. (2021) employed a two-step synthesis method, initially utilizing high-energy ball milling (HEBM) but restricted to 15 minutes to prevent α-Fe impurity introduction due to abrasive wear of milling balls and bowls, to obtain equiatomic (TiZrHfNbTa)C high-entropy carbide in powder form. Whereas Chicardi et al. (2019) reported a room-temperature synthesis of equiatomic (TiZrHfVNb)C high-entropy carbide via a mechanically-induced carbon diffusion route. This method involves simultaneous alloying of transition metals and carbon diffusion during milling, offering a simple and reproducible alternative to SPS consolidation at high temperatures.

On the other hand, Dusza et al. (2018) achieved a density of 99% and an average grain size of 12 μm in the high-entropy carbide (Hf-Ta-Zr-Nb)C through ball milling and spark plasma sintering (SPS), resulting in an Fm-3m crystal structure. Castle et al. (2018) explored (Hf-Ta-Zr-Ti)C and (Hf-Ta-Zr-Nb)C high-entropy monocarbides by planetary ball milling and SPS. They highlighted that the lattice parameter mismatch among component monocracies significantly affects single-phase solid solution formation.

5.2. Attrition Ball Milling

Attrition ball milling is a significant technique in the context of high-entropy ceramics. It involves the mechanical grinding of materials to create fine powders and enhance their homogeneity. Figure 4(b) illustrates the graphical representation of attrition ball milling. In the synthesis of high entropy ceramics, attrition ball milling is particularly valuable for achieving a uniform distribution of multiple elements, ensuring the desired phase formation, and optimizing material properties. Promoting efficient mixing and particle refinement contributes to the successful development of high entropy ceramics with tailored compositions and improved performance characteristics, making it a crucial step in their fabrication. In this method, the grinding media is made of small beads that are constantly agitated by the rotating ball mill, resulting in particle size reduction through attrition (Go et al., 2023). Song et al. (2021) investigate how the milling duration impacts the oxide formation behaviour. They observed that as the milling time increased, the composites exhibited a reduction in pore size and a higher number of particle boundaries.

Figure 4. Graphical illustration of (a) planetary ball milling, (b) attrition ball milling, (c) vibratory ball milling

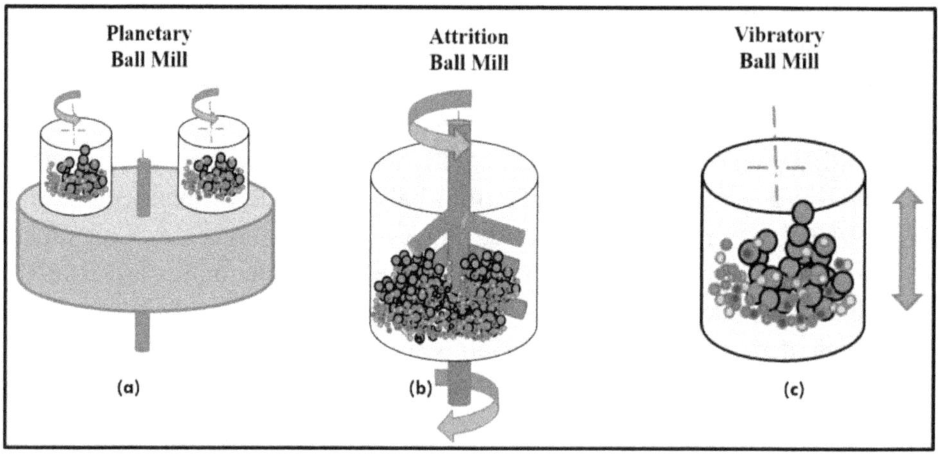

5.3. Vibratory Ball Milling

In vibratory ball milling, a high-frequency vibration is applied to the ball mill, causing the grinding media to collide with the raw materials and effectively reduce their particle size (Daskalakis et al., 2023). Figure 4(c) illustrates the graphical representation of Vibratory ball milling. It offers several notable advantages for processing HECs, including high energy efficiency, uniform mixture formation, reduced contamination risk, and rapid milling. However, it has limitations related to scalability for large-scale production and the significant overheating issue. In the study conducted by Kavak et al. (2022), the fabrication process of High-Entropy Carbides (HECs) with the composition (HfMoTiWZr)C involved the utilization of monocarbides powders (HfC, TiC, WC, ZrC) and dimolybdenum carbide (Mo2C) powders. These powders were initially blended and subsequently subjected to milling using a vibratory high-energy ball mill (HEBM) for a duration of 3 hours. The milling process maintained a consistent clamp speed of 1060 cycles per minute. To mitigate the overheating concern during milling, the mill was periodically halted after each operation hour, allowing for a 10-minute cooling period before resuming milling activities.

Each of these methods has its advantages and disadvantages, and the choice of method depends on the specific requirements of the prepared ceramic material.

6. FINAL SOLIDIFICATION

The mixed powder is then pressed into a desired shape and size using a hydraulic press (Keerthivasan & Anandakrishnan, 2022). This step is important to ensure that the final product has the desired density and porosity.

6.1. Sintering

The compacted powder is sintered at high temperatures to fuse the individual particles, forming a dense and homogeneous material. The sintering process involves heating the compacted powder in a furnace under controlled temperature, time, and atmosphere (Feng et al., 2023). High entropy ceramics can be synthesized using several techniques, each with advantages and disadvantages. The choice of synthesis technique depends on the application and desired properties of the final high entropy ceramic material (Storr et al., 2022). Some of the commonly used synthesis techniques for high-entropy ceramics include:

6.1.1. Solid-State Sintering

Solid-state sintering is the most commonly used technique for fabricating high-entropy ceramics. It involves mixing the elemental powders in equimolar or near-equimolar ratios and then compacting them into a desired shape. Figure 5 illustrates the process of interdiffusion of Raw precursors during solid-state sintering. The compacted material is sintered at high temperatures, typically between 1200°C and 1700°C, to produce the final ceramic material (Tallarita et al., 2020). This

Figure 5. Formulation of solid-state sintering

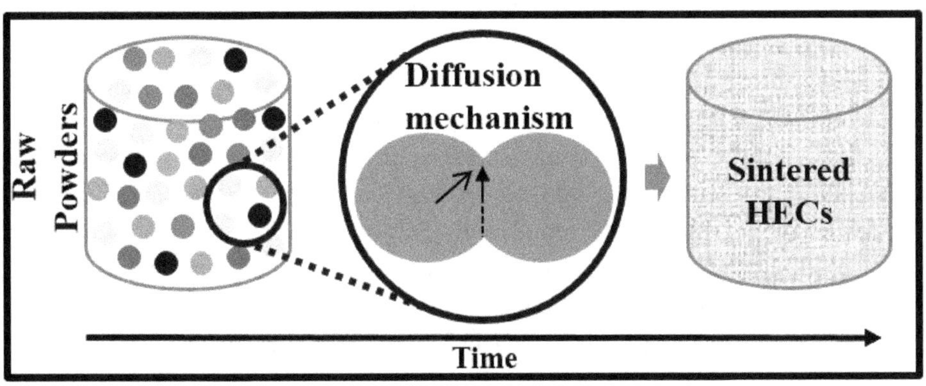

technique is relatively simple, and the resulting high-entropy ceramics exhibit excellent mechanical properties.

6.1.2. Spark Plasma Sintering

Spark plasma sintering (SPS) is a high-energy, low-temperature technique used to fabricate high-entropy ceramics (Saengvong et al., 2023). The process involves the application of a high-voltage, low-current electrical pulse to the ceramic powder, which causes localized heating and sintering, as shown in Figure 8. In SPS, the powder is placed in graphite dies, enclosed with punches, and the assembly is sealed within the SPS chamber with the chosen sintering atmosphere (e.g., vacuum or Argon) applied. This technique can produce high entropy ceramics with a fine microstructure and excellent mechanical properties (Zhang et al., 2021).

Spark Plasma Sintering (SPS) is a novel technique that significantly reduces the time required for the sintering process compared to conventional methods, which can take many hours or even days. SPS achieves this rapid sintering by enabling high heating rates through internal sample heating instead of the external heating used in conventional sintering (Oğuzyer & Rajendrachari, 2023). Additionally, SPS minimizes sintering time with short holding periods at the sintering temperature, typically ranging from 5 to 10 minutes, while conventional sintering may extend to several hours (Sharma et al., 2019). In conventional furnaces, heating rates typically reach 5 to 8°C per minute, with a maximum of 10°C per minute. Consequently, achieving a temperature of 1200°C can take 2 to 4 hours or more (Pradeep et al., 2022). In contrast, SPS easily achieves heating rates exceeding 300°C per minute, allowing a temperature of 1200°C to be reached in just 4 minutes (Exner & Arzt, 1996).

The simultaneous application of temperature and pressure in Spark Plasma Sintering (SPS) results in significant densification, allowing for the creation of dense compacts at sintering temperatures 200 to 250°C lower than those required in conventional sintering. Unlike conventional sintering, SPS prevents coarsening and grain growth, achieving high relative densities rapidly, even when using nano-sized powders, without substantial grain growth (Mahale et al., 2020). As a result, SPS facilitates the preparation of nano-structured ceramics and composites with enhanced densification and reduced defects, leading to outstanding mechanical properties such as increased strength and hardness.

Overall, synthesizing high entropy ceramics requires careful consideration of the desired properties and the choice of fabrication technique. The resulting high entropy ceramics exhibit exceptional properties and performance, making them attractive for various applications.

6.1.3. Reactive Sintering

Reactive sintering involves the reaction of the elemental powders during sintering to produce the final ceramic material. The process involves mixing the elemental powders in the desired stoichiometric ratio, followed by compaction and sintering at high temperatures in a reactive atmosphere (Huo et al., 2022), as shown in Figure 6. This technique can produce high entropy ceramics with high chemical homogeneity and a fine microstructure (Liu et al., 2020).

Figure 6. Graphical illustration of spark plasma sintering technique

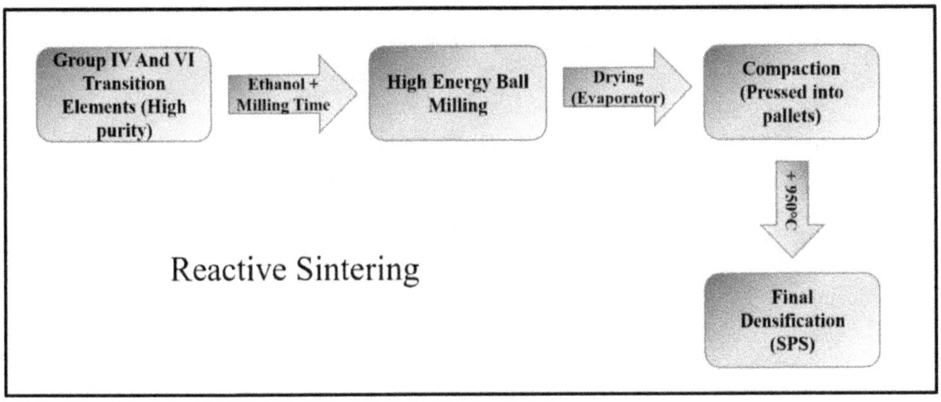

Reactive Spark Plasma Sintering (Reactive SPS) differs significantly from Spark Plasma Sintering (SPS) in terms of the materials they handle, their objectives, the conditions under which they operate, and their intended applications. SPS is a versatile method used to consolidate various materials, such as ceramics, metals, and composites, with the primary goal of achieving densification and enhancing mechanical properties without involving chemical reactions (Mahale et al., 2022). In contrast, Reactive SPS is specially designed for materials requiring chemical reactions during sintering, typically applied in synthesizing advanced ceramics and intermetallic compounds to generate new compounds or phases with specific functionalities. This process demands precise control of temperature, pressure, and reactive atmospheres (Gild et al., 2019). While SPS finds widespread applications across various industries, Reactive SPS is specifically focused on materials designed for improved functional properties, particularly in the realm of energy-related applications.

6.2. Chemical Vapor Deposition

Chemical vapor deposition (CVD) is a vapor-phase technique used to deposit high entropy ceramic coatings on a substrate (Ito & Matsumoto, 2023). The process involves the thermal decomposition of metalorganic precursors, which then react to form the desired ceramic material on the substrate surface (Han, 2022), as shown in Figure 7. This technique can produce high-quality coatings with a high degree of control over the composition and microstructure.

Figure 7. Graphical illustration of reactive sintering method

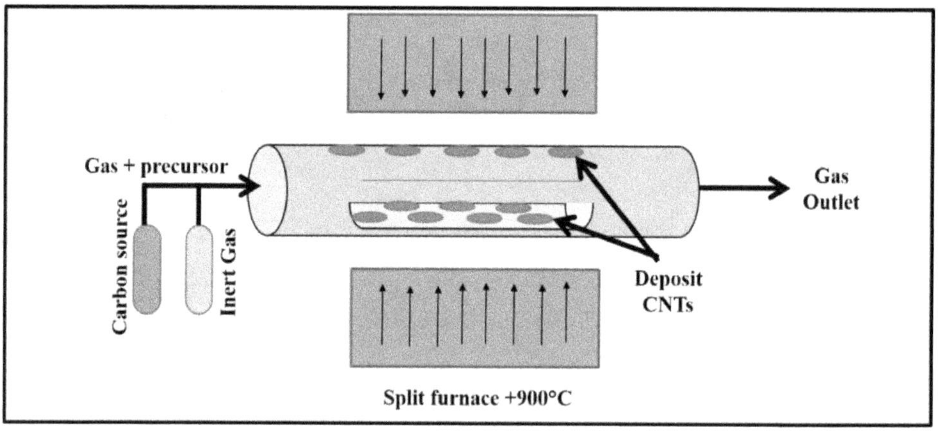

6.3. Hot Pressing

Sometimes, hot pressing further densifies the high entropy ceramics (Xiang et al., 2022). This process involves applying pressure and heat simultaneously to the sintered material, resulting in a product with improved mechanical properties (Lee et al., 2013). Figure 9 illustrates the graphical representation of the hot pressing method.

6.4. Post-Processing

The final step involves finishing the product, such as polishing or grinding, to achieve the desired surface finish and dimensional accuracy (Tan et al., 2021).

Overall, the fabrication process of high entropy ceramics is complex and requires careful control of each step to achieve the desired microstructure and properties. However, the resulting material has unique properties (Yeh et al., 2016), such as

Figure 8. Graphical illustration of chemical vapor deposition method

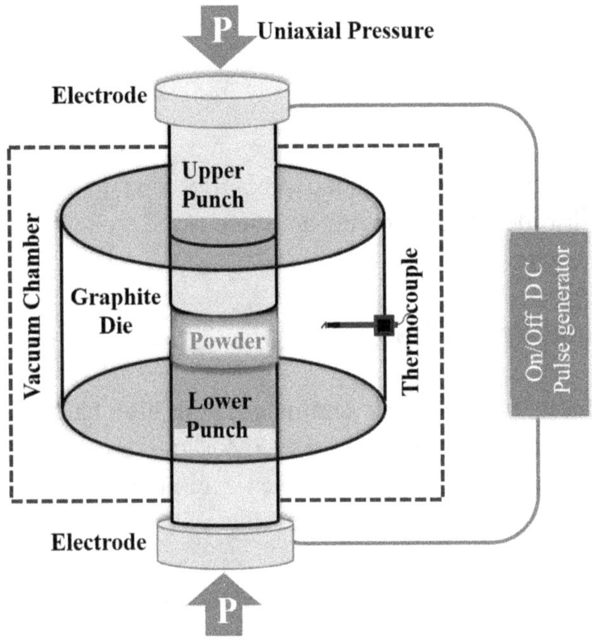

Figure 9. Graphical representation of the hot pressing method

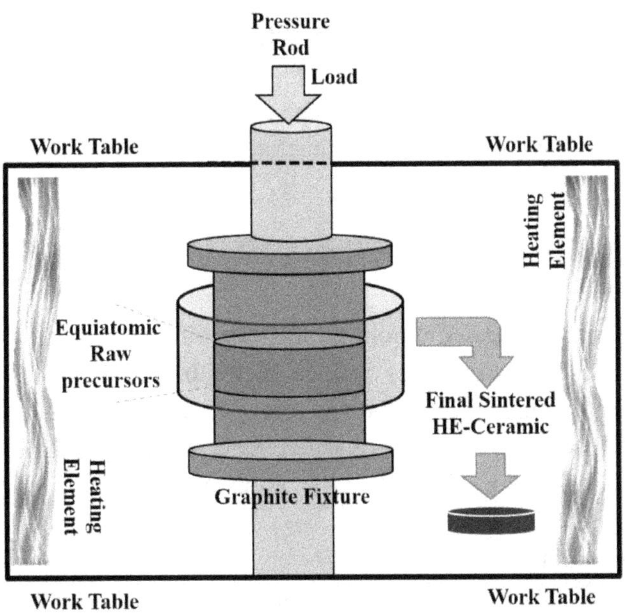

high strength, hardness, and thermal stability, that make it suitable for various applications in various industries.

7. PROPERTIES

High entropy ceramics exhibit unique properties due to their high structural complexity and disorder. The following are some of the properties and applications of high-entropy ceramics:

7.1. Mechanical Properties

1. High Hardness: High entropy ceramics exhibit high hardness, making them suitable for wear-resistant applications (Moskovskikh et al., 2020).
2. High Strength: High entropy ceramics exhibit high strength and fracture toughness, making them suitable for structural and engineering applications (Liu et al., 2021).
3. High Thermal Stability: High entropy ceramics exhibit high thermal stability, making them suitable for high-temperature applications (Li et al., 2019).

7.2. Physical Properties

1. Good Electrical Insulation: High entropy ceramics exhibit good electrical insulation properties, making them suitable for electronic applications (Meng et al., 2021).
2. Low Thermal Expansion: High entropy ceramics exhibit low thermal expansion, making them suitable for high-temperature applications (Qiu et al., 2022).

8. APPLICATIONS

High-entropy ceramics (HECs) hold promise for various applications due to their unique properties. Potential uses include high-temperature structural materials, thermal barrier coatings, superhard materials, and even as candidates for next-generation nuclear fuels, offering enhanced performance and durability in extreme environments. The graphical illustration of various applications is shown in Figure 10.

1. Wear-Resistant Coatings: High entropy ceramics can be used as coatings to improve the wear resistance of materials such as cutting tools, engine components, and biomedical implants (Zhao et al., 2023).

Figure 10. Graphical illustration of prospective applications

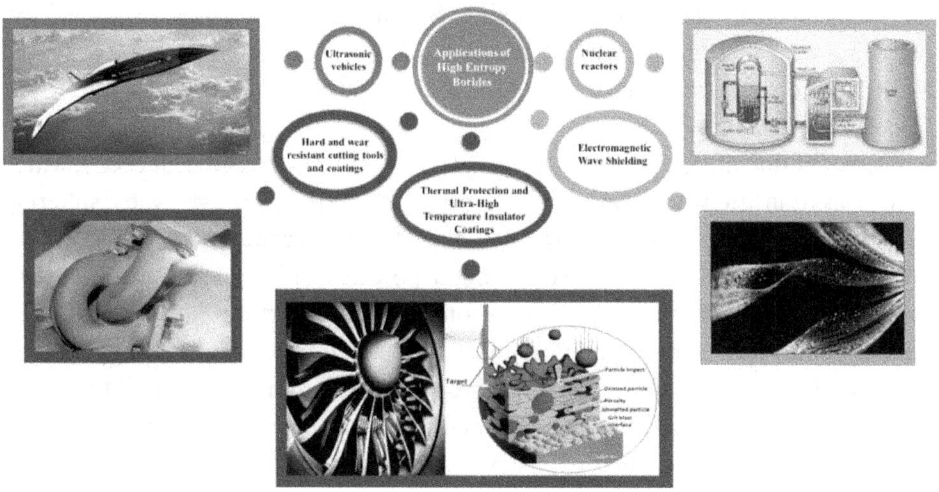

2. Structural Materials: High entropy ceramics can be used as structural materials in aerospace and engineering applications due to their high strength and fracture toughness (Zhang, Jiang, Sun et al, 2019).
3. Electronic Applications: High entropy ceramics can be used in electronic applications, such as insulators in microelectronic devices, hypersonic vehicles and substrates for thin film deposition (Kenfaui et al., 2023).
4. High-Temperature Applications: High entropy ceramics can be used in high-temperature applications, such as furnace components and thermal barrier coatings (Demirskyi, 2020).

In conclusion, high entropy ceramics exhibit unique properties that make them suitable for various applications. They have the potential to replace traditional materials in many high-performance applications due to their excellent mechanical, thermal, and electrical properties. The development of high-entropy ceramics is an area of ongoing research and is expected to lead to exciting new applications.

9. CONCLUSION

In conclusion, high entropy ceramics are a new class of ceramic materials attracting much attention due to their unique properties and potential applications (Nisar, Khan, & Balani, 2019). They have complex structures characterized by a high degree of structural disorder and mixing of multiple elements, which give them unique properties (Vega et al., 2023). High entropy ceramics offer superior mechanical, thermal, and electrical properties, making them attractive for various applications, including wear-resistant coatings (Zhao et al., 2023), structural materials (Kumar et al., 2023), electronic applications (Lin et al., 2023), and high-temperature applications (Feng et al., 2023).

Densification of High-entropy ultrahigh temperature ceramics (UHTCs), which encompass transition metal carbides and borides, typically requires elevated sintering temperatures (Rajendrachari, 2018). A beneficial approach to aid in this process involves reducing the particle size of the initial powders with higher surface energies, as it generates a more substantial driving force for densification (Nayak et al., 2016). Various techniques have been employed to produce ultrafine high-entropy carbide and boride powders to attain this objective. Among these methods, ball milling is one of the most widely adopted techniques in academic literature. High-Energy Ball Milling (HEBM) is renowned for its capacity to increase surface area and ensure homogeneous material mixing. However, it comes with the drawback of introducing surface oxides onto particle surfaces and the potential inclusion of impurities due to wear and tear of the milling media (Feng et al., 2017).

While high entropy ceramics offer significant advantages over traditional ceramics, some challenges still need to be addressed. These include synthesizing high entropy ceramics with precise control over the composition, developing new characterization techniques to understand better their complex structure, and exploring new high entropy ceramic compositions (Miracle et al., 2014). Addressing these challenges will pave the way for exciting advancements in this field.

REFERENCES

Akrami, S., Edalati, P., Fuji, M., & Edalati, K. (2021). High-entropy ceramics: Review of principles, production and applications. *Materials Science and Engineering R Reports*, *146*, 100644. doi:10.1016/j.mser.2021.100644

Bhatt, B., Murthy, T. S. R. C., Limaye, P. K., Nagaraj, A., Singh, K., Sonber, J. K., Sairam, K., Sashanka, A., Rao, G. V. S. N., & Rao, T. S. (2016). Tribological studies of monolithic chromium diboride against cemented tungsten carbide (WC–Co) under dry condition. *Ceramics International*, *42*(14), 15536–15546. doi:10.1016/j.ceramint.2016.06.208

Braun, J. L., Rost, C. M., Lim, M., Giri, A., Olson, D. H., Kotsonis, G. N., Stan, G., Brenner, D. W., Maria, J.-P., & Hopkins, P. E. (2018). Charge-induced disorder controls the thermal conductivity of entropy-stabilized oxides. *Advanced Materials*, *30*(51), 1805004. doi:10.1002/adma.201805004 PMID:30368943

Cantor, B. (2014). Multicomponent and high entropy alloys. *Entropy (Basel, Switzerland), 16*(9), 4749–4768. doi:10.3390/e16094749

Cantor, B., Chang, I. T. H., Knight, P., & Vincent, A. J. B. (2004). Microstructural development in equiatomic multicomponent alloys. *Materials Science and Engineering A, 375–377*, 213–218. doi:10.1016/j.msea.2003.10.257

Cao, W., Chen, P., Lin, R., Li, F., Ge, B., Song, D., Cheng, Z., & Wang, C. (2023). Boosting energy-storage performance in lead-free ceramics via polyphase engineering in the superparaelectric state. *Composites. Part B, Engineering, 255*, 110630. doi:10.1016/j.compositesb.2023.110630

Castle, E., Csanádi, T., Grasso, S., Dusza, J., & Reece, M. (2018). Processing and Properties of High-Entropy Ultra-High Temperature Carbides. *Scientific Reports, 8*(1), 8609. doi:10.1038/s41598-018-26827-1 PMID:29872126

Chen, H., Fu, J., Zhang, P., Peng, H., Abney, C. W., Jie, K., Liu, X., Chi, M., & Dai, S. (2018). Entropy-stabilized metal oxide solid solutions as CO oxidation catalysts with high-temperature stability. *Journal of Materials Chemistry. A, Materials for Energy and Sustainability, 6*(24), 11129–11133. doi:10.1039/C8TA01772G

Chen, H., Xiang, H., Dai, F.-Z., Liu, J., Lei, Y., Zhang, J., & Zhou, Y. (2019). High porosity and low thermal conductivity high entropy (Zr0.2Hf0.2Ti0.2Nb0.2Ta0.2)C. *Journal of Materials Science and Technology, 35*(8), 1700–1705. doi:10.1016/j.jmst.2019.04.006

Chen, H., Xiang, H., Dai, F. Z., Liu, J., & Zhou, Y. (2019). Porous high entropy (Zr0.2Hf0.2Ti0.2Nb0.2Ta0.2)B2: A novel strategy towards making ultrahigh temperature ceramics thermal insulating. *Journal of Materials Science and Technology, 35*(10), 2404–2408. doi:10.1016/j.jmst.2019.05.059

Cheng, B., Lou, H., Sarkar, A., Zeng, Z., Zhang, F., Chen, X., Tan, L., Glazyrin, K., liermann, H.-P., Yan, J., Wang, L., Djenadic, R., Hahn, H., & Zeng, Q. (2020). Lattice distortion and stability of (Co0.2Cu0.2Mg0.2Ni0.2Zn0.2)O high-entropy oxide under high pressure. *Materials Today. Advances, 8*, 100102. doi:10.1016/j.mtadv.2020.100102

Cheng, C., Zhang, F., Cheng, F., Hou, J., Ren, B., Miao, Y., Xin, L., & Wang, X. (2023). The effect of plasma-assisted ball milling on preparation and sintering behavior of (Zr0.1429Hf0.1429Ce0.1429Y0.2857La0.2857)O2-δ high entropy fluorite oxide. *Ceramics International, 49*(8), 13092–13101. doi:10.1016/j.ceramint.2022.12.185

Chicardi, E., García-Garrido, C., & Gotor, F. J. (2019). Low temperature synthesis of an equiatomic (TiZrHfVNb)C5 high entropy carbide by a mechanically-induced carbon diffusion route. *Ceramics International, 45*(17), 21858–21863. doi:10.1016/j.ceramint.2019.07.195

Cologna, M., Rashkova, B., & Raj, R. (2010). Flash sintering of nanograin zirconia in <5 s at 850°C. *Journal of the American Ceramic Society*, *93*(11), 3556–3559. doi:10.1111/j.1551-2916.2010.04089.x

Daskalakis, E., Scott, A., & Jha, A. (2023). Enhancing the ignitability of the Al–TiO2–B2O3 powder mixture through intensive vibratory ball milling (C:P12). *Ceramics International*, *49*(14), 24142–24155. Advance online publication. doi:10.1016/j.ceramint.2022.12.242

Demirskyi. (2020). Synthesis of medium-entropy (Zr1/3Hf1/3Ta1/3)B2 using the spark plasma consolidation of diboride powders. *Journal of the Ceramic Society of Japan*, *128*, 977–980. doi:10.2109/jcersj2.20151

Dippo, O. F., Mesgarzadeh, N., Harrington, T. J., Schrader, G. D., & Vecchio, K. S. (2020). Bulk high-entropy nitrides and carbonitrides. *Scientific Reports*, *10*(1), 1–11. doi:10.1038/s41598-020-78175-8 PMID:33277546

Dubrovinskaia, N., Dubrovinsky, L. & Solozhenko, V. L. (2007). Comment on 'synthesis of ultra-incompressible superhard rhenium diboride at ambient pressure'. *Science, 318.*

Dusza, J., Švec, P., Girman, V., Sedlák, R., Castle, E. G., Csanádi, T., Kovalčíková, A., & Reece, M. J. (2018). Microstructure of (Hf-Ta-Zr-Nb)C high-entropy carbide at micro and nano/atomic level. *Journal of the European Ceramic Society*, *38*(12), 4303–4307. doi:10.1016/j.jeurceramsoc.2018.05.006

Dwivedi, A., Khan, M. M., & Pali, H. S. (2023). A comprehensive review of thermal enhancement techniques in microchannel heat exchangers and heat sinks. *Journal of Thermal Analysis and Calorimetry*, *148*(23), 1–43. doi:10.1007/s10973-023-12451-3

Euchner, H., & Mayrhofer, P. H. (2015). Designing thin film materials-Ternary borides from first principles. *Thin Solid Films*, *583*, 46–49. doi:10.1016/j.tsf.2015.03.035 PMID:26082562

Exner, H. E. & Arzt, E. (1996). Sintering Processes. *Phys. Metall.,* 2627–2662. doi:10.1016/B978-044489875-3/50036-3

Failla, S., Galizia, P., Fu, S., Grasso, S., & Sciti, D. (2020). Formation of high entropy metal diborides using arc-melting and combinatorial approach to study quinary and quaternary solid solutions. *Journal of the European Ceramic Society*, *40*(3), 588–593. doi:10.1016/j.jeurceramsoc.2019.10.051

Feng, L., Fahrenholtz, W. G., & Hilmas, G. E. (2019). Low-temperature sintering of single-phase, high-entropy carbide ceramics. *Journal of the American Ceramic Society*, *102*(12), 7217–7224. doi:10.1111/jace.16672

Feng, L., Fahrenholtz, W. G., Hilmas, G. E., & Curtarolo, S. (2023). Boro/carbothermal reduction co-synthesis of dual-phase high-entropy boride-carbide ceramics. *Journal of the European Ceramic Society, 43*(6), 2708–2712. doi:10.1016/j.jeurceramsoc.2022.12.056

Feng, L., Lee, S. H., & Kim, H. N. (2017). Effects of high-energy ball milling and reactive spark plasma sintering on the densification of HfC-SiC composites. *Journal of the European Ceramic Society, 37*(5), 1891–1898. doi:10.1016/j.jeurceramsoc.2016.12.036

Fourmont, A., Le Gallet, S., Politano, O., Desgranges, C., & Baras, F. (2020). Effects of planetary ball milling on AlCoCrFeNi high entropy alloys prepared by Spark Plasma Sintering: Experiments and molecular dynamics study. *Journal of Alloys and Compounds, 820*, 153448. doi:10.1016/j.jallcom.2019.153448

Gaur, A., Chauhan, V. S., & Vaish, R. (2023). Planetary ball milling induced piezocatalysis for dye degradation using BaTiO3 ceramics. *Environmental Science. Advances, 2*(3), 462–472. doi:10.1039/D2VA00210H

Gild, J. (2019). A high-entropy silicide: (Mo0.2Nb0.2Ta0.2Ti0.2W0.2)Si2. *Journal of Materials, 5*, 337–343.

Gild, J., Kaufmann, K., Vecchio, K., & Luo, J. (2019). Reactive flash spark plasma sintering of high-entropy ultrahigh temperature ceramics. *Scripta Materialia, 170*, 106–110. doi:10.1016/j.scriptamat.2019.05.039

Gild, J., Zhang, Y., Harrington, T., Jiang, S., Hu, T., Quinn, M. C., Mellor, W. M., Zhou, N., Vecchio, K., & Luo, J. (2016). High-Entropy Metal Diborides: A New Class of High-Entropy Materials and a New Type of Ultrahigh Temperature Ceramics. *Scientific Reports, 6*(1), 37946. doi:10.1038/srep37946 PMID:27897255

Go, S. H., Kim, K. S., Kim, J. S. & Cheon, C. (2023). Effects of attrition milling on the microstructure and piezoelectric properties of BiFeO3–BaTiO3 ceramics. *J. Korean Ceram. Soc.* doi:10.1007/s43207-023-00291-8

Golla, B. R., Bhandari, T., Mukhopadhyay, A., & Basu, B. (2014). Titanium Diboride. *Ultra-High Temp. Ceram. Mater. Extrem. Environ. Appl., 9781118700*, 316–360.

Gorsse, S., & Senkov, O. N. (2018). About the reliability of CALPHAD predictions in multicomponent systems. *Entropy (Basel, Switzerland), 20*(12), 899. doi:10.3390/e20120899 PMID:33266623

Guo, C., Chen, W., & Deng, T. (2023). Fundamental study on the preparation of insulating ceramics via the phase reconstruction of phosphate tailings. *Ceramics International, 49*(6), 9419–9431. doi:10.1016/j.ceramint.2022.11.107

Hafnium Boride Powder (HfB2 Powder), CAS 12007-23-7. (n.d.). Heeger Materials. https://heegermaterials.com/boride-powders/1244-hafnium-boride-hfb2-powder-cas-12007-23-7.html

Han, J. (2022). Microstructural evolution of chemically vapor-deposited tantalum carbide at elevated temperature. *Journal of Ceramic Processing Research*, *23*, 751–757.

Harrington, T. J., Gild, J., Sarker, P., Toher, C., Rost, C. M., Dippo, O. F., McElfresh, C., Kaufmann, K., Marin, E., Borowski, L., Hopkins, P. E., Luo, J., Curtarolo, S., Brenner, D. W., & Vecchio, K. S. (2019). Phase stability and mechanical properties of novel high entropy transition metal carbides. *Acta Materialia*, *166*, 271–280. doi:10.1016/j.actamat.2018.12.054

Hong, W., Chen, F., Shen, Q., Han, Y.-H., Fahrenholtz, W. G., & Zhang, L. (2019). Microstructural evolution and mechanical properties of (Mg,Co,Ni,Cu,Zn)O high-entropy ceramics. *Journal of the American Ceramic Society*, *102*(4), 2228–2237. doi:10.1111/jace.16075

Huo, S., Chen, L., Liu, X., Kong, Q., Wang, Y., Gu, H., & Zhou, Y. (2022). Reactive sintering of dual-phase high-entropy ceramics with superior mechanical properties. *Journal of Materials Science and Technology*, *129*, 223–227. doi:10.1016/j.jmst.2022.03.035

Ito, A., & Matsumoto, S. (2023). Chemically vapor deposited oxide-based thick film scintillators. *Japanese Journal of Applied Physics*, *62*(1), 010612. doi:10.35848/1347-4065/aca249

Iwan, S., Lin, C.-M., Perreault, C., Chakrabarty, K., Chen, C.-C., Vohra, Y., Hrubiak, R., Shen, G., & Velisavljevic, N. (2022). High-Entropy Borides under Extreme Environment of Pressures and Temperatures. *Materials (Basel)*, *15*(9), 3239. doi:10.3390/ma15093239 PMID:35591574

K, V., Gorle, R., S, A., & Bakshi, S. R. (2021). Novel single phase (Ti0.2W0.2Ta0.2Mo0.2V0.2)C0.8 high entropy carbide using ball milling followed by reactive spark plasma sintering. *Journal of the European Ceramic Society*, *41*(13), 6756–6762. doi:10.1016/j.jeurceramsoc.2021.06.009

Kale, S., Adimule, V., Shashanka, R., Tanaka, M., & Wei, P. (2022). Synthesis and Modifications of Materials and its Properties. *Trans Tech Publ. Ltd. Switz.*, *134*, 1–134.

Kavak, S., Bayrak, K. G., Bellek, M., Mertdinç, S., Muhaffel, F., Gökçe, H., Ayas, E., Derin, B., Öveçoğlu, M. L., & Ağaoğulları, D. (2022). Synthesis and characterization of (HfMoTiWZr)C high entropy carbide ceramics. *Ceramics International*, *48*(6), 7695–7705. doi:10.1016/j.ceramint.2021.11.317

Ke, B., Ji, W., Zou, J., Wang, W., & Fu, Z. (2023). Densification mechanism, microstructure and mechanical properties of ZrC ceramics prepared by high-pressure spark plasma sintering. *Journal of the European Ceramic Society*, *43*(8), 3053–3061. doi:10.1016/j.jeurceramsoc.2023.02.038

Keerthivasan, N. & Anandakrishnan, V. (2022). Effect of recycled TiC addition in the compaction and sintering behavior of Al-Mg-Cu-Zn + r-TiC powder metallurgy composites. *Part. Sci. Technol.* . doi:10.1080/02726351.2022.2104187

Kenfaui, D., Guillemet-Fritsch, S., Valdez-Nava, Z., Laudebat, L., Tenailleau, C., Dufour, P., & Locatelli, M.-L. (2023). Tuning electrical conductivity in AlN-based ceramics by incorporating graphene. *Journal of the European Ceramic Society*, *43*(5), 1887–1896. doi:10.1016/j.jeurceramsoc.2022.12.039

Kovalev, D. Y., Kochetov, N. A., & Chuev, I. I. (2021). Fabrication of high-entropy carbide (TiZrHfTaNb)C by high-energy ball milling. *Ceramics International*, *47*(23), 32626–32633. doi:10.1016/j.ceramint.2021.08.158

Kumar, A., Kim, J.-W., Yadav, N., Vij, A., Kumar, S., Sharma, M. K., Kumari, K., Huh, S.-H., & Koo, B. H. (2023). Structural insights into the rare-earth modified La1.4-xSmxSr1.6Mn2O7 layered perovskites and probing their complex magnetic and magnetocaloric properties. *Materials Chemistry and Physics*, *302*, 127695. doi:10.1016/j.matchemphys.2023.127695

Lakshmi Priya, M., Rana, D. & Ramalingam, M. (2017). Ceramic nanofiber composites. *Nanofiber Compos. Biomed. Appl.* 33–54. doi:10.1016/B978-0-08-100173-8.00030-2

Laszkiewicz-Łukasik, J., Jaworska, L., Putyra, P., Klimczyk, P., & Garzeł, G. (2016). The influence of SPS heating rates on the synthesis reaction of tantalum diboride. *Boletín de la Sociedad Española de Cerámica y Vidrio*, *55*(4), 159–168. doi:10.1016/j.bsecv.2016.03.001

Lee, W. E., Gilbert, M., Murphy, S. T., & Grimes, R. W. (2013). Opportunities for advanced ceramics and composites in the nuclear sector. *Journal of the American Ceramic Society*, *96*(7), 2005–2030. doi:10.1111/jace.12406

Li, F., Zhou, L., Liu, J. X., Liang, Y., & Zhang, G. J. (2019). High-entropy pyrochlores with low thermal conductivity for thermal barrier coating materials. *J. Adv. Ceram.*, *8*(4), 576–582. doi:10.1007/s40145-019-0342-4

Li, Z., Chen, L., Chang, F., Hong, C., Zhao, X., Fang, Y., & Dai, P. (2022). Synthesis, microstructure and properties of Ti(C,N)-(HfZrTaNbTi)C5-HEA high-entropy cermets by high-energy ball milling and spark plasma sintering. *Ceramics International*, *48*(20), 30826–30837. doi:10.1016/j.ceramint.2022.07.036

Lin, F. L., Liu, B., Hu, C. C., & Song, K. X. (2023). Novel high-entropy microwave dielectric ceramics Sr(La0.2Nd0.2Sm0.2Eu0.2Gd0.2)AlO4 with excellent temperature stability and mechanical properties. *Journal of the European Ceramic Society*, *43*(6), 2506–2512. doi:10.1016/j.jeurceramsoc.2023.01.028

Liu, B., Zhao, J., Liu, Y., Xi, J., Li, Q., Xiang, H., & Zhou, Y. (2021). Application of high-throughput first-principles calculations in ceramic innovation. *Journal of Materials Science and Technology*, *88*, 143–157. doi:10.1016/j.jmst.2021.01.071

Liu, J., Shao, G., Liu, D., Chen, K., Wang, K., Ma, B., Ren, K., & Wang, Y. (2020). Design and synthesis of chemically complex ceramics from the perspective of entropy. *Materials Today. Advances*, *8*, 100114. doi:10.1016/j.mtadv.2020.100114

Luo, G., Zhang, G., Zhang, Y., Li, A., Sun, Y., Tu, R., & Shen, Q. (2023). Wide temperature range of stable dielectric properties in relaxor BaTiO3-based ceramics by co-doping synergistic engineering. *Materials Chemistry and Physics*, *127629*. Advance online publication. doi:10.1016/j.matchemphys.2023.127629

Maged, A., Abu El-Magd, S. A., Radwan, A. E., Kharbish, S., & Zamzam, S. (2023). Evaluation insight into Abu Zenima clay deposits as a prospective raw material source for ceramics industry: Remote Sensing and Characterization. *Scientific Reports*, *13*(1), 58. doi:10.1038/s41598-022-26484-5 PMID:36593265

Mahale, R. S., Shamanth, V., Hemanth, K., Sharath, P. C., & Shashanka, R. (2022). Mechanical testing of spark plasma sintered materials: A review. *AIP Conference Proceedings*, *2469*, 020026. doi:10.1063/5.0080183

Mahale, R. S., Shamanth, V., Sharath, P. C., Shashanka, R., & Hemanth, K. (2020). A review on spark plasma sintering of duplex stainless steels. *Materials Today: Proceedings*, *45*, 138–144. doi:10.1016/j.matpr.2020.10.357

Meisenheimer, P. B., Kratofil, T. J., & Heron, J. T. (2017). Giant enhancement of exchange coupling in entropy-stabilized oxide heterostructures. *Scientific Reports*, *7*(1), 13344. doi:10.1038/s41598-017-13810-5 PMID:29042610

Meng, X., Xu, J., Zhu, J., Zhang, P., Rydosz, A., Reece, M. J., & Gao, F. (2021). Pyrochlore-fluorite dual-phase high-entropy ceramic foams with extremely low thermal conductivity from particle-stabilized suspension. *Scripta Materialia*, *194*, 113714. doi:10.1016/j.scriptamat.2020.113714

Miracle, D. B., Miller, J., Senkov, O., Woodward, C., Uchic, M., & Tiley, J. (2014). Exploration and development of high entropy alloys for structural applications. *Entropy (Basel, Switzerland)*, *16*(1), 494–525. doi:10.3390/e16010494

Moskovskikh, D., Vorotilo, S., Buinevich, V., Sedegov, A., Kuskov, K., Khort, A., Shuck, C., Zhukovskyi, M., & Mukasyan, A. (2020). Extremely hard and tough high entropy nitride ceramics. *Scientific Reports*, *10*(1), 19874. doi:10.1038/s41598-020-76945-y PMID:33199795

Nayak, A. K., Shashanka, R., & Chaira, D. (2016). Effect of Nanosize Yittria and Tungsten Addition to Duplex Stainless Steel during High Energy Planetary Milling. *IOP Conf. Ser. Mater. Sci. Eng.*, *115*, 012008. 10.1088/1757-899X/115/1/012008

Nisar, A., Bajpai, S., Khan, M. M., & Balani, K. (2020). Wear damage tolerance and high temperature oxidation behavior of HfB2:ZrB2–SiC composites. *Ceramics International*, *46*(13), 21689–21698. doi:10.1016/j.ceramint.2020.05.276

Nisar, A., Khan, M. M., Bajpai, S., & Balani, K. (2019). Processing, microstructure and mechanical properties of HfB 2 -ZrB 2 -SiC composites: Effect of B 4 C and carbon nanotube reinforcements. *International Journal of Refractory & Hard Metals*, *81*, 111–118. doi:10.1016/j.ijrmhm.2019.02.014

Nisar, A., Khan, M. M., & Balani, K. (2019). Enhanced thermo-mechanical damage tolerance of functionally graded ZrB2-20SiC ceramic reinforced with carbon nanotubes. *Ceramics International*, *45*(5), 6198–6208. doi:10.1016/j.ceramint.2018.12.097

Oğuzyer, A., & Rajendrachari, S. (2023). An Overview of Nanocomposites with Recent Advancements. *J. Metastable Nanocrystalline Mater.*, *37*, 45–56. doi:10.4028/p-VW2fCP

Okada, S., Kudou, K., & Lundström, T. (1995). Preparations and some properties of w2b, δ-wb and wb2crystals from high-temperature metal solutions. *Japanese Journal of Applied Physics*, *34*(1R), 226–231. doi:10.1143/JJAP.34.226

Oses, C., Toher, C., & Curtarolo, S. (2020). High-entropy ceramics. *Nature Reviews. Materials*, *5*(4), 295–309. doi:10.1038/s41578-019-0170-8

Post, B., Glaserf, F. W. & Moskowitz, D. (n.d.). Transition Metal Diborides. *Acta Metall.*

Pradeep, N. B., Hegde, M. M. R., Rajendrachari, S., & Surendranathan, A. O. (2022). Investigation of microstructure and mechanical properties of microwave consolidated TiMgSr alloy prepared by high energy ball milling. *Powder Technology*, *408*, 117715. doi:10.1016/j.powtec.2022.117715

Qiao, L., Liu, Y., Gao, Y., Bi, J., Li, Y., Liu, C., Gao, J., Wang, W., & Qian, Z. (2022). First-principles prediction, fabrication and characterization of (Hf0.2Nb0.2Ta0.2Ti0.2Zr0.2)B2 high-entropy borides. *Ceramics International*, *48*(12), 17234–17245. doi:10.1016/j.ceramint.2022.02.281

Qin, M., Shivakumar, S., Lei, T., Gild, J., Hessong, E. C., Wang, H., Vecchio, K. S., Rupert, T. J., & Luo, J. (2022). Processing-dependent stabilization of a dissimilar rare-earth boride in high-entropy (Ti0.2Zr0.2Hf0.2Ta0.2Er0.2)B2 with enhanced hardness and grain boundary segregation. *Journal of the European Ceramic Society*, *42*(12), 5164–5171. doi:10.1016/j.jeurceramsoc.2022.05.034

Qin, M., Yan, Q., Liu, Y., Wang, H., Wang, C., Lei, T., Vecchio, K. S., Xin, H. L., Rupert, T. J., & Luo, J. (2021). Bulk high-entropy hexaborides. *Journal of the European Ceramic Society*, *41*(12), 5775–5781. doi:10.1016/j.jeurceramsoc.2021.05.027

Qiu, S., Xiang, H., Dai, F.-Z., Wang, H., Huang, M., Wan, C., Meng, Q., Li, J., Wang, X., & Zhou, Y. (2022). Medium-entropy (Me,Ti)0.1(Zr,Hf,Ce)0.9O2 (Me = Y and Ta): Promising thermal barrier materials for high-temperature thermal radiation shielding and CMAS blocking. *Journal of Materials Science and Technology*, *123*, 144–153. doi:10.1016/j.jmst.2022.01.019

Rajendrachari, S. (2018). Effect of sintering temperature on the pitting corrosion of ball milled duplex stainless steel by using linear sweep voltammetry. *Anal. Bioanal. Electrochem.*, *10*, 349–361.

Rajendrachari, S. (2022a). An Overview of High-Entropy Alloys Prepared by Mechanical Alloying Followed by the Characterization of Their Microstructure and Various Properties. *Alloys*, *1*(2), 116–134. doi:10.3390/alloys1020008

Rajendrachari, S. (2022b). New Advances in Powder Technology. *New Adv. Powder Technol.* . doi:10.5772/intechopen.102106

Saengvong, P., Boonlakhorn, J., Jumpatam, J., Chanlek, N., Prasongkit, J., Putasaeng, B., Moontragoon, P., Srepusharawoot, P., & Thongbai, P. (2023). Effects of sintering condition on giant dielectric and nonlinear current-voltage properties of Na1/2Y1/2Cu3Ti3.975Ta0.025O12 ceramics. *Heliyon*, *9*(1), e12946. doi:10.1016/j.heliyon.2023.e12946 PMID:36704279

Sairam, K., Sonber, J. K., Murthy, T. S. R. C., Subramanian, C., Fotedar, R. K., & Hubli, R. C. (2014). Reaction spark plasma sintering of niobium diboride. *International Journal of Refractory & Hard Metals*, *43*, 259–262. doi:10.1016/j.ijrmhm.2013.12.011

Sarkar, A., Djenadic, R., Usharani, N. J., Sanghvi, K. P., Chakravadhanula, V. S. K., Gandhi, A. S., Hahn, H., & Bhattacharya, S. S. (2017). Nanocrystalline multicomponent entropy stabilised transition metal oxides. *Journal of the European Ceramic Society*, *37*(2), 747–754. doi:10.1016/j.jeurceramsoc.2016.09.018

Sarker, P., Harrington, T., Toher, C., Oses, C., Samiee, M., Maria, J.-P., Brenner, D. W., Vecchio, K. S., & Curtarolo, S. (2018). High-entropy high-hardness metal carbides discovered by entropy descriptors. *Nature Communications*, *9*(1), 4980. doi:10.1038/s41467-018-07160-7 PMID:30478375

Shanghai Greenearth Chemicals Co. (n.d.). *Excellent properties of Vanadium diboride*. https://www.greenearthchem.com/excellent-properties-of-vanadium-diboride.html

Sharma, A. S., Yadav, S., Biswas, K., & Basu, B. (2018). High-entropy alloys and metallic nanocomposites: Processing challenges, microstructure development and property enhancement. *Materials Science and Engineering R Reports*, *131*, 1–42. doi:10.1016/j.mser.2018.04.003

Sharma, N., Alam, S. N. & Ray, B. C. (2019). Fundamentals of Spark Plasma Sintering (SPS): An Ideal Processing Technique for Fabrication of Metal Matrix Nanocomposites. *Spark Plasma Sinter. Mater.*, 21–59. . doi:10.1007/978-3-030-05327-7_2

Sharma, Y., Musico, B. L., Gao, X., Hua, C., May, A. F., Herklotz, A., Rastogi, A., Mandrus, D., Yan, J., Lee, H. N., Chisholm, M. F., Keppens, V., & Ward, T. Z. (2018). Single-crystal high entropy perovskite oxide epitaxial films. *Physical Review Materials*, *2*(6), 060404. doi:10.1103/PhysRevMaterials.2.060404

Shashanka, R., & Chaira, D. (2015). Development of nano-structured duplex and ferritic stainless steels by pulverisette planetary milling followed by pressureless sintering. *Materials Characterization*, *99*, 220–229. doi:10.1016/j.matchar.2014.11.030

Shashanka, R., Uzun, O., & Chaira, D. (2020). Synthesis of nano-structured duplex and ferritic stainless steel powders by dry milling and its comparison with wet milling. *Archives of Metallurgy and Materials*, *65*, 5–14.

Smith, D. S., & Naït-Ali, B. (2021). Thermal properties of ceramic materials. *Encycl. Mater. Tech. Ceram. Glas.*, *1–3*, 855–866. doi:10.1016/B978-0-12-818542-1.00010-2

Šolcová, P., Nižňanský, M., Schulz, J., Brázda, P., Ecorchard, P., Vilémová, M., & Tyrpekl, V. (2021). Preparation of High-Entropy (Ti, Zr, Hf, Ta, Nb) Carbide Powder via Solution Chemistry. *Inorganic Chemistry*, *60*(11), 7617–7621. doi:10.1021/acs.inorgchem.1c00776 PMID:34028253

Song, Y., Kim, D., Nam, S., Lee, K. A., & Choi, H. (2021). Effect of milling duration on oxide-formation behavior of oxide-dispersion-strengthened high-entropy alloys. *Archives of Metallurgy and Materials*, *66*, 735–740.

Stanfield, A. D., Manara, D., Robba, D., Hilmas, G. E., & Fahrenholtz, W. G. (2021). Measurement of the melting temperature of ZrB2 as determined by laser heating and spectrometric analysis. *Journal of the American Ceramic Society*, *104*(6), 2780–2787. doi:10.1111/jace.17634

Storr, B., Moore, L., Chakrabarty, K., Mohammed, Z., Rangari, V., Chen, C.-C., & Catledge, S. A. (2022). Properties of high entropy borides synthesized via microwave-induced plasma. *APL Materials*, *10*(6), 061109. doi:10.1063/5.0098276

Tallarita, G., Licheri, R., Garroni, S., Barbarossa, S., Orrù, R., & Cao, G. (2020). High-entropy transition metal diborides by reactive and non-reactive spark plasma sintering: A comparative investigation. *Journal of the European Ceramic Society*, *40*(4), 942–952. doi:10.1016/j.jeurceramsoc.2019.10.031

Tallarita, G., Licheri, R., Garroni, S., Orrù, R., & Cao, G. (2019). Novel processing route for the fabrication of bulk high-entropy metal diborides. *Scripta Materialia*, *158*, 100–104. doi:10.1016/j.scriptamat.2018.08.039

Tan, Y., Xia, Y., Teng, Z., Chen, C., Zhou, X., & Zhang, H. (2021). Synthesis and enhanced mechanical properties of compositionally complex MAX phases. *Journal of the European Ceramic Society*, *41*(8), 4658–4665. doi:10.1016/j.jeurceramsoc.2021.03.027

Teng, Z., Tan, Y., Zeng, S., Meng, Y., Chen, C., Han, X., & Zhang, H. (2021). Preparation and phase evolution of high-entropy oxides A2B2O7 with multiple elements at A and B sites. *Journal of the European Ceramic Society*, *41*(6), 3614–3620. doi:10.1016/j.jeurceramsoc.2021.01.013

Vega, H., Qin, M., & Luo, J. (2023). Thermodynamics of dual-phase compositionally complex ceramics: A case study of ultrahigh-entropy fluorite-bixbyite refractory oxides. *Journal of the European Ceramic Society*, *43*(5), 2104–2114. doi:10.1016/j.jeurceramsoc.2022.12.033

Wang, K., Chen, L., Xu, C., Zhang, W., Liu, Z., Wang, Y., Ouyang, J., Zhang, X., Fu, Y., & Zhou, Y. (2020). Microstructure and mechanical properties of (TiZrNbTaMo)C high-entropy ceramic. *Journal of Materials Science and Technology*, *39*, 99–105. doi:10.1016/j.jmst.2019.07.056

Wang, Q., Sarkar, A., Wang, D., Velasco, L., Azmi, R., Bhattacharya, S. S., Bergfeldt, T., Düvel, A., Heitjans, P., Brezesinski, T., Hahn, H., & Breitung, B. (2019). Multi-anionic and -cationic compounds: New high entropy materials for advanced Li-ion batteries. *Energy & Environmental Science, 12*(8), 2433–2442. doi:10.1039/C9EE00368A

Wei, X. F., Liu, J.-X., Li, F., Qin, Y., Liang, Y.-C., & Zhang, G.-J. (2019). High entropy carbide ceramics from different starting materials. *Journal of the European Ceramic Society, 39*(10), 2989–2994. doi:10.1016/j.jeurceramsoc.2019.04.006

Wright, A. J., & Luo, J. (2020). A step forward from high-entropy ceramics to compositionally complex ceramics: A new perspective. *Journal of Materials Science, 55*(23), 9812–9827. doi:10.1007/s10853-020-04583-w

Xiang, S., Du, C., Cao, X. & Chen, L. (2022). High-Entropy Ceramics Diborides for Hypersonic Applications. *Lect. Notes Electr. Eng., 821*, 247–256.

Xie, H. (2022). Research Progress of High-entropy Ceramic Materials. *Cailiao Daobao. Cailiao Daobao, 36*, 20070201–20070208.

Yahya, N. A., Rashid, T. N. I. T. A., & Hasmaliza, M. (2022). Effect of sintering temperature on physical properties of Trong clay and prefired material in low clay whiteware ceramic fabrication. *Materials Today: Proceedings, 66*, 2716–2719. doi:10.1016/j.matpr.2022.06.501

Ye, B., Wen, T., Huang, K., Wang, C. Z., & Chu, Y. (2019). First-principles study, fabrication, and characterization of (Hf 0.2 Zr 0.2 Ta 0.2 Nb 0.2 Ti 0.2)C high-entropy ceramic. *Journal of the American Ceramic Society, 102*(7), 4344–4352. doi:10.1111/jace.16295

Yeh, J. W. (2006). Recent progress in high-entropy alloys. *Ann. Chim. Sci. des Mater., 31*, 633–648.

Yeh, J. W., Yeh, A. C. & Chang, S. Y. (2016). Potential applications and prospects. *High-Entropy Alloy. Fundam. Appl.*, 493–512.. doi:10.1007/978-3-319-27013-5_15

Yuan, K., Tu, T., Shen, C., Zhou, L., Liu, J., Li, J., Xie, K., & Zhang, G. (2022). Self-ball milling strategy to construct high-entropy oxide coated LiNi0.8Co0.1Mn0.1O2 with enhanced electrochemical performance. *J. Adv. Ceram., 11*(6), 882–892. doi:10.1007/s40145-022-0582-6

Zhang, Y., Guo, W.-M., Jiang, Z.-B., Zhu, Q.-Q., Sun, S.-K., You, Y., Plucknett, K., & Lin, H.-T. (2019). Dense high-entropy boride ceramics with ultra-high hardness. *Scripta Materialia, 164*, 135–139. doi:10.1016/j.scriptamat.2019.01.021

Zhang, Y., Jiang, Z.-B., Sun, S.-K., Guo, W.-M., Chen, Q.-S., Qiu, J.-X., Plucknett, K., & Lin, H.-T. (2019). Microstructure and mechanical properties of high-entropy borides derived from boro/carbothermal reduction. *Journal of the European Ceramic Society*, *39*(13), 3920–3924. doi:10.1016/j.jeurceramsoc.2019.05.017

Zhang, Y., Sun, S.-K., Guo, W.-M., Xu, L., Zhang, W., & Lin, H.-T. (2021). Optimal preparation of high-entropy boride-silicon carbide ceramics. *J. Adv. Ceram.*, *10*(1), 173–180. doi:10.1007/s40145-020-0418-1

Zhang, Y., Zuo, T. T., Tang, Z., Gao, M. C., Dahmen, K. A., Liaw, P. K., & Lu, Z. P. (2014). Microstructures and properties of high-entropy alloys. *Progress in Materials Science*, *61*, 1–93. doi:10.1016/j.pmatsci.2013.10.001

Zhao, P., Zhu, J., Li, M., Shao, G., Wang, H., & He, J. (2023). Influence of molten-salt-synthesised high-entropy (V0.2Cr0.2Mo0.2W0.2Ni0.2)B powder characteristics on superhard mechanism during spark plasma sintering. *Journal of Alloys and Compounds*, *947*, 169618. doi:10.1016/j.jallcom.2023.169618

Zhong, Y., Sabarou, H., Yan, X., Yang, M., Gao, M. C., Liu, X., & Sisson, R. D. Jr. (2019). Exploration of high entropy ceramics (HECs) with computational thermodynamics - A case study with LaMnO3±δ. *Materials & Design*, *182*, 108060. doi:10.1016/j.matdes.2019.108060

Chapter 4
Mechanical Alloying of Aluminium Alloys

Rayappa Shrinivas Mahale
https://orcid.org/0000-0002-0508-9871
REVA University, India

P. C. Sharath
https://orcid.org/0000-0003-2939-5889
Jain University, India

Krishnamurthy P. Goggi
Jain College of Engineering and Research, India

Rajendrachari Shashanka
Bartin University, Turkey

Shamanth Vasanth
REVA University, India

Prashant Kakkamari
KLS Gogte Intstitute of Technology, India

Gangadhar M. Kanaginahal
https://orcid.org/0000-0002-0823-6507
Jain College of Engineering and Research, India

Adarsh Patil
https://orcid.org/0000-0002-6917-6245
KLE Technological University, India

Vaibhav S. Raibole
Jain College of Engineering and Research, India

ABSTRACT

Mechanical alloying is a solid-state powder processing technique that has revolutionized the development of advanced materials, including aluminium alloys. This chapter provides a concise overview of the key aspects of mechanical alloying applied to aluminium alloys, exploring its principles, advantages, and potential applications. The process of mechanical alloying involves subjecting aluminium powder and alloying elements to intense mechanical deformation using a high-energy ball mill. This results in repeated welding, fracturing, and cold welding of particles, leading to the creation of a homogenous and fine-grained microstructure. The alloying mechanisms encompass solid-state diffusion and deformation-induced mixing, facilitating the uniform dispersion of alloying elements within the aluminium matrix. Mechanical alloying has several distinct advantages over conventional alloying methods.

DOI: 10.4018/978-1-6684-9385-4.ch004

1. INTRODUCTION

1.1 Background and Significance

In the field of materials science and engineering, mechanical alloying has emerged as a pioneering technique that allows the synthesis of novel materials with exceptional properties. Specifically, the application of mechanical alloying to aluminium alloys has garnered substantial attention due to the versatile range of properties these materials can exhibit (Zhang, Guo, Chen et al, 2019). Conventional methods of alloying aluminium, such as melting and casting, often result in limited control over microstructures and may lead to the segregation of alloying elements. In contrast, mechanical alloying offers a solid-state processing route, enabling the creation of homogeneous and finely dispersed alloy compositions, which in turn, yield superior mechanical and chemical characteristics (Mao et al., 2019). The increasing demand for lightweight, high-performance materials in industries like aerospace, automotive, and electronics has spurred the interest in developing advanced aluminium alloys through mechanical alloying (Zheng et al., 2019).

1.2 Objectives of the Chapter

The primary objectives of this chapter are to provide a comprehensive overview of the principles, techniques, and applications of mechanical alloying in the context of aluminium alloys. The chapter aims to delve into the fundamental aspects of the mechanical alloying process, including the mechanisms behind alloy formation and microstructural evolution. Additionally, it seeks to explore the influence of process parameters on the properties of mechanically alloyed aluminium alloys. Through a detailed examination of mechanical properties, corrosion behavior, and potential applications, this chapter aims to showcase the superior performance of these materials in comparison to conventionally alloyed counterparts. Furthermore, the chapter intends to identify the challenges faced in the mechanical alloying process and potential solutions to overcome them, making it a valuable resource for researchers, engineers, and students working in the field of materials science.

1.3 Scope and Limitations

While mechanical alloying has demonstrated significant promise in enhancing the properties of aluminium alloys, it is essential to recognize the scope and limitations of this technique. The chapter will focus on the fundamental principles and theoretical aspects of mechanical alloying as applied to aluminium alloys (Affolter et al., 2019).

It will delve into the various process parameters that affect the microstructural evolution and properties of the resulting materials. Furthermore, the chapter will discuss the mechanical properties and corrosion behavior of mechanically alloyed aluminium alloys, exploring their potential applications in aerospace, automotive, and other engineering fields (Nacházel et al., 2019).

However, it is essential to acknowledge that mechanical alloying also has certain limitations, such as the potential for contamination from the milling media and the requirement for costly high-energy milling equipment (Cui et al., 2019). The chapter will address these challenges and discuss potential strategies to mitigate their impact. Additionally, the scope of the chapter may not cover every aspect of mechanical alloying of aluminium alloys comprehensively, as the field is constantly evolving, and ongoing research may yield new findings and advancements beyond the chapter's knowledge cutoff (Kumar et al., 2020).

2. FUNDAMENTALS OF MECHANICAL ALLOYING

2.1 Principles of Mechanical Alloying

Mechanical alloying is a solid-state powder processing technique that involves the synthesis of alloyed materials through repeated mechanical deformation, welding, and fracturing of powder particles (Gontijo et al., 2020). The process operates on the principles of cold welding and solid-state diffusion, where high-energy collisions between particles lead to localized plastic deformation and promote the intermixing of alloying elements (Park et al., 2020). The key to successful mechanical alloying lies in achieving a delicate balance between the kinetic energy of the milling balls and the adhesion forces between the powders, ensuring effective alloying without excessive contamination (Ahmed et al., 2020).

The initial stage of mechanical alloying involves the loading of aluminium powder and selected alloying elements, often in elemental or pre-alloyed form, into a milling container along with grinding balls (Siroky et al., 2020). The milling container undergoes high-energy oscillations or rotations, causing the grinding balls to impact the powders, leading to a series of mechanical processes. Cold welding occurs when powders are brought into intimate contact due to the impact and compression, while fracturing results from the stress induced during the collision, facilitating further alloying (Gao, Geng, Sun et al, 2022). The process continues until a homogenized, fine-grained structure is achieved, making it distinct from traditional melting and casting methods (Shrinivas Mahale, Shamanth, Sharath, Shashanka, & Hemanth, 2021).

2.2 Evolution of Mechanical Alloying

The history of mechanical alloying can be traced back to the 1960s, with significant advancements made in the 1970s and 1980s. The pioneering work by Benjamin and Volin in the 1960s laid the foundation for this technique, with early experiments demonstrating the feasibility of producing alloyed powders by mechanical means. In the following decades, researchers explored various milling equipment and process parameters, leading to refined techniques for producing nanocrystalline and amorphous materials (Camarano et al., 2019; Friák et al., 2019; Nithin et al., 2022; Phutela et al., 2019; Sun et al., 2019; Zhang, Mu, Wang et al, 2019).

The evolution of mechanical alloying also witnessed improvements in the understanding of alloying mechanisms, microstructural evolution, and the influence of milling conditions on material properties. Researchers identified the critical role of process parameters, such as milling time, rotational speed, and ball-to-powder ratio, in tailoring the final microstructure and properties of the mechanically alloyed materials (Kanaginahal et al., 2023; Shashanka, 2014; Shashanka & Chaira, 2015; Shashanka et al., 2020; Suryanarayana & Froes, 1990).

2.3 High-Energy Ball Milling Equipment

High-energy ball milling equipment is the core component of the mechanical alloying process. The design and construction of these mills have evolved over time to meet the demands of modern materials research. High-energy ball mills utilize a jar or vial containing the powders and grinding balls (Prabhu et al., 2006; Suryanarayana, 2001; Suryanarayana & Al-Aqeeli, 2013; Suryanarayana et al., 2001; Verma et al., 2020; Voltammetric Determination of Various Food Azo Dyes Using Different Modified Carbon Paste Electrodes, 2021). The mill is operated either in a batch or continuous mode, with the milling container subjected to intense mechanical vibrations or rotational movements.

Various types of high-energy mills are employed, such as planetary ball mills, attritor mills, and SPEX mills, each offering unique advantages in terms of milling efficiency, scalability, and versatility. These mills can achieve milling speeds of hundreds of revolutions per minute, resulting in high impact energies that promote efficient alloying. Additionally, the mills can be operated under controlled atmospheres, including inert gas environments, to mitigate the risk of contamination during the milling process. In recent years, advancements in high-energy ball milling equipment have further improved the control and precision of the mechanical alloying process, enabling researchers to explore a wide range of materials and optimize the properties of mechanically alloyed aluminium alloys for diverse engineering applications (Barro et al., 2020; Chaoyang et al., 2020; Li et al., 2020; Mahale et al., 2020; ##NO_NAME##, 2021; Richardson et al., 2020).

2.4 Mechanisms of Alloying in Aluminium Alloys

The alloying mechanisms in mechanical alloying of aluminium alloys involve a combination of diffusive processes and deformation-induced mixing. As the high-energy ball milling progresses, several key mechanisms contribute to the homogenization and dispersion of alloying elements within the aluminium matrix:-

1. Solid-State Diffusion: Solid-state diffusion is a fundamental mechanism in mechanical alloying. Under the influence of the high-energy collisions and localized plastic deformation, the alloying elements and aluminium atoms undergo diffusive mixing at the atomic scale. This enables the transfer of atoms between adjacent particles, promoting the formation of a uniform composition.
2. Cold Welding and Fracturing: During the milling process, particles experience intense plastic deformation due to the high-energy impacts. Cold welding occurs when freshly fractured surfaces of powder particles come into intimate contact, allowing them to bond together. Subsequent fracturing events lead to further mixing and breaking down of large particles, increasing the surface area available for diffusion and alloying (Cao et al., 2020; Chiu et al., 2020; Fan et al., 2022; Simoncini et al., 2020; Wang, Pan, Sun et al, 2022; Yang, Zhang, Chen et al, 2022).
3. Dislocation-Mediated Mixing: Deformation-induced dislocations generated during the mechanical alloying process facilitate the movement of atoms and enhance the diffusion of alloying elements. Dislocations act as diffusion paths, promoting the exchange of atoms between adjacent regions within the powders, thereby contributing to the formation of a homogeneous alloy.
4. Thermal and Strain-Induced Activation: The combination of mechanical deformation and localized heating during the milling process results in thermal and strain-induced activation of diffusion mechanisms. These activations facilitate the atomic rearrangement and promote alloying between the aluminium and alloying elements (Contuzzi et al., 2022; Imshinetskiy et al., 2022; Li et al., 2022; Saha & Roy, 2022; Zaara et al., 2022; Zhang, Zhang, Zhao et al, 2022).
5. Mechanical Activation of Reactants: Mechanical activation refers to the increase in chemical reactivity of powders caused by high-energy milling. The mechanical deformation of particles enhances their reactivity, leading to accelerated alloy formation and the reduction of reaction barriers.

The synergistic interplay of these mechanisms during mechanical alloying leads to the creation of a finely homogenized microstructure, characterized by a uniform distribution of alloying elements within the aluminium matrix. The prolonged milling time and controlled milling conditions further enhance the extent of alloying, resulting in the development of nanocrystalline or amorphous structures with superior mechanical and chemical properties (Mahale, V., K., C., & Shashanka, 2022; Wang, Luo, Kuang et al, 2022; Wu, Liu, Wei et al, 2022; Yang, Xue, Li et al, 2022; Zhao, Vdonin, Bezgans et al, 2022).

Understanding the underlying mechanisms of alloying in aluminium alloys during mechanical alloying is crucial for optimizing the process parameters and tailoring the microstructure and properties of the resulting materials for specific engineering applications. Researchers continue to investigate these mechanisms to further improve the efficiency and effectiveness of the mechanical alloying process for aluminium alloys (Kong et al., 2022; Marušáková et al., 2022; Shi et al., 2022; Tang et al., 2022; Wei et al., 2022; Wu, Lin, Qiao et al, 2022; Zhang, Liu, Liu et al, 2022).

3. PROCESS PARAMETERS AND OPTIMIZATION

3.1 Milling Time and Rotational Speed

Case Study 1: Milling Time and Rotational Speed

In this case study, researchers sought to optimize the milling time and rotational speed to synthesize an aluminium-based nanocomposite. They used a high-energy ball mill and varied the milling time from 2 hours to 20 hours with a constant rotational speed of 500 rpm. X-ray diffraction (XRD) analysis revealed that after 2 hours of milling, the alloy showed a partially alloyed structure with some unreacted aluminium. As the milling time increased up to 10 hours, the alloy fully alloyed, exhibiting a fine-grained structure with a uniform dispersion of reinforcing nanoparticles.

However, beyond 10 hours, excessive milling time led to grain growth and agglomeration of the nanoparticles, diminishing the desired properties. The researchers observed that the optimal milling time for achieving the desired microstructure and properties of the nanocomposite was around 10 hours. This study emphasizes the importance of carefully selecting the milling time to achieve efficient alloying while avoiding over-milling (Chausov et al., 2022; Du et al., 2022; Guan et al., 2022; Jiao et al., 2022; Yamazaki & Haga, 2022; Yang, Liang, Gao et al, 2022; Zhang, Wang, Chen et al, 2022).

3.2 Ball-to-Powder Ratio

Case Study 2: Ball-to-Powder Ratio

In this study, researchers investigated the effect of ball-to-powder ratio on the synthesis of aluminium-titanium carbide nanocomposites. They used ball-to-powder ratios of 5:1, 10:1, and 15:1 while keeping the milling time and rotational speed constant. XRD and transmission electron microscopy (TEM) analysis showed that a ball-to-powder ratio of 10:1 resulted in the best balance between efficient alloying and uniform dispersion of titanium carbide nanoparticles in the aluminium matrix.

The higher ball-to-powder ratio of 15:1 led to increased milling time without a significant improvement in nanoparticle dispersion, indicating that an optimal ball-to-powder ratio is critical to prevent over-milling and achieve effective alloying. This case study highlights the importance of optimizing the ball-to-powder ratio to maximize the efficiency of the mechanical alloying process (Arendarchuck et al., 2022; Ferreira et al., 2022; Gao, Yang, Xiao et al, 2022; Omarov et al., 2022; Zhao, Zhong, Zhai et al, 2022).

3.3 Milling Atmosphere

Case Study 3: Milling Atmosphere

To investigate the impact of milling atmosphere on the synthesis of mechanically alloyed aluminium-magnesium alloys, researchers employed two different atmospheres: inert argon gas and a controlled nitrogen atmosphere. The milling time and rotational speed were kept constant for both atmospheres. XRD and energy-dispersive X-ray spectroscopy (EDX) analysis revealed that milling under the inert argon atmosphere produced a well-alloyed structure with minimal oxidation, leading to improved mechanical properties.

In contrast, milling in the nitrogen atmosphere resulted in the formation of undesired nitride phases, compromising the alloy's performance. This case study demonstrates the significance of carefully selecting the milling atmosphere to avoid unwanted chemical reactions during mechanical alloying and highlights the impact of atmosphere on the final properties of the alloy (Aizawa et al., 2022; Di Egidio et al., 2022; Liu et al., 2022; Miao et al., 2022; Park et al., 2022; Yamanaka et al., 2022; Yasui et al., 2022).

3.4 Temperature Control During Milling

Case Study 4: Researchers aimed to optimize the mechanical properties of a mechanically alloyed aluminium-silicon-carbon nanocomposite. To control the temperature during milling, an intermittent milling approach was employed, with 10 minutes of milling followed by 5 minutes of cooling using a cooling jacket. The milling time and rotational speed were maintained constant. TEM analysis revealed that the intermittent milling approach successfully limited the temperature rise, preventing excessive grain growth and maintaining a fine-grained microstructure with uniform nanoparticle dispersion. This study demonstrates the importance of effective temperature control to preserve the desired microstructural characteristics in mechanically alloyed materials.

These case studies highlight the significance of process parameter optimization in mechanical alloying of aluminium alloys. Understanding the effects of milling time, rotational speed, ball-to-powder ratio, milling atmosphere, and temperature control is crucial for achieving desired microstructures and tailoring the properties of mechanically alloyed aluminium alloys for specific applications. The results from these studies provide valuable insights into the practical considerations and strategies employed to optimize the mechanical alloying process in real-time scenarios (Bedolla-Becerril et al., 2022; Ding et al., 2022; Ghosh et al., 2022; Gibbs et al., 2014; Jost & Schmidt, 1986; Jost & Schmidt, 1986; Latypov et al., 2016; Meng et al., 2022; Vishwakarma et al., 2022).

4. MICROSTRUCTURAL EVOLUTION

4.1 Nanocrystalline Structure in Mechanically Alloyed Aluminium Alloys

Nanocrystalline structures are characterized by extremely small grain sizes at the nanometer scale. In the context of aluminum alloys, the mechanical alloying process involves the synthesis of materials through mechanical means, typically ball milling, leading to the formation of finely dispersed alloying elements within the aluminum matrix. This process can result in the creation of nanocrystalline structures with unique mechanical, electrical, and thermal properties compared to conventional coarse-grained alloys. Understanding the formation, stability, and properties of nanocrystalline structures in mechanically alloyed aluminum alloys is crucial for tailoring these materials for specific applications, such as lightweight structural components and advanced electronic devices (Chen et al., 2014; Chen, Ye, & Lin, 2017; Ge et al., 2017; Rendón & Olsson, 2009).

4.2 Influence of Process Parameters on Microstructure

The microstructure of a material is strongly influenced by the processing parameters used during its fabrication. This includes factors such as temperature, pressure, alloy

Figure 1. SEM image of atomized aluminum alloy (Rendón & Olsson, 2009)

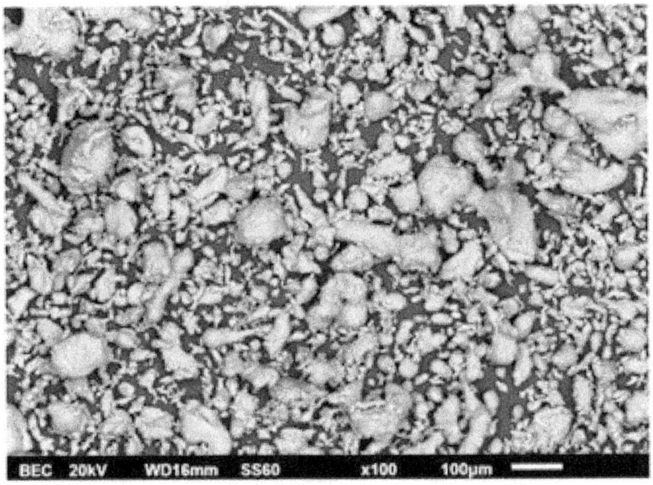

composition, cooling rates, and mechanical treatments. Researchers and engineers carefully study the effects of these parameters to optimize the resulting microstructure for desired properties. For example, in the context of aluminum alloys, the choice of heat treatment, solidification conditions, and deformation processes can significantly impact the final microstructure, affecting mechanical strength, corrosion resistance, and other material characteristics. Understanding the relationships between process parameters and microstructure is essential for producing materials with specific performance requirements (Chen, Zhao, Wang et al, 2017; Yan, Hu, Wang et al, 2021; Yan, Hu, Yu et al, 2021).

4.3 Grain Growth and Microstructure Stability

Grain growth is a phenomenon that occurs in polycrystalline materials, including aluminum alloys, where the individual crystalline grains increase in size over time under the influence of temperature and other factors. Controlling grain growth is crucial to maintaining the desired properties of a material. Excessive grain growth can lead to decreased mechanical strength and other undesirable effects. Studying the mechanisms of grain growth and developing strategies to inhibit it, such as grain boundary engineering or the addition of certain alloying elements, is essential for ensuring the long-term stability of microstructures in aluminum alloys and other materials.

Case Study 1: Nanocrystalline Aluminum Alloys for Lightweight Aerospace Components:

Figure 2. Images from the milled powder of AlCrFeMnNi HEA in (a) panoramic view (×50) and (b) higher magnification (×500) (Yan, Hu, Wang et al, 2021)

Aerospace industries are always seeking materials that provide high strength-to-weight ratios. In this case study, researchers aimed to develop nanocrystalline aluminum alloys for lightweight structural components. They used mechanical alloying to disperse fine ceramic particles (e.g., nanoparticles of carbides) within the aluminum matrix.

Key findings and considerations:

1. Nanocrystalline Structure: The mechanical alloying process led to the formation of a nanocrystalline structure in the aluminum matrix, with average grain sizes in the nanometer range. This significantly increased the material's strength while maintaining a reasonable level of ductility.
2. Thermal Stability: Researchers investigated the stability of the nanocrystalline structure at elevated temperatures, simulating the conditions experienced

Figure 3. Shown are panoramic images of the microstructures in composite materials with varying AlCrFeMnNi content: 1 vol.%, 3 vol.%, and 5 vol.%. The presence of black spots in all these images signifies the presence of some level of porosity (Chen, Zhao, Wang et al, 2017)

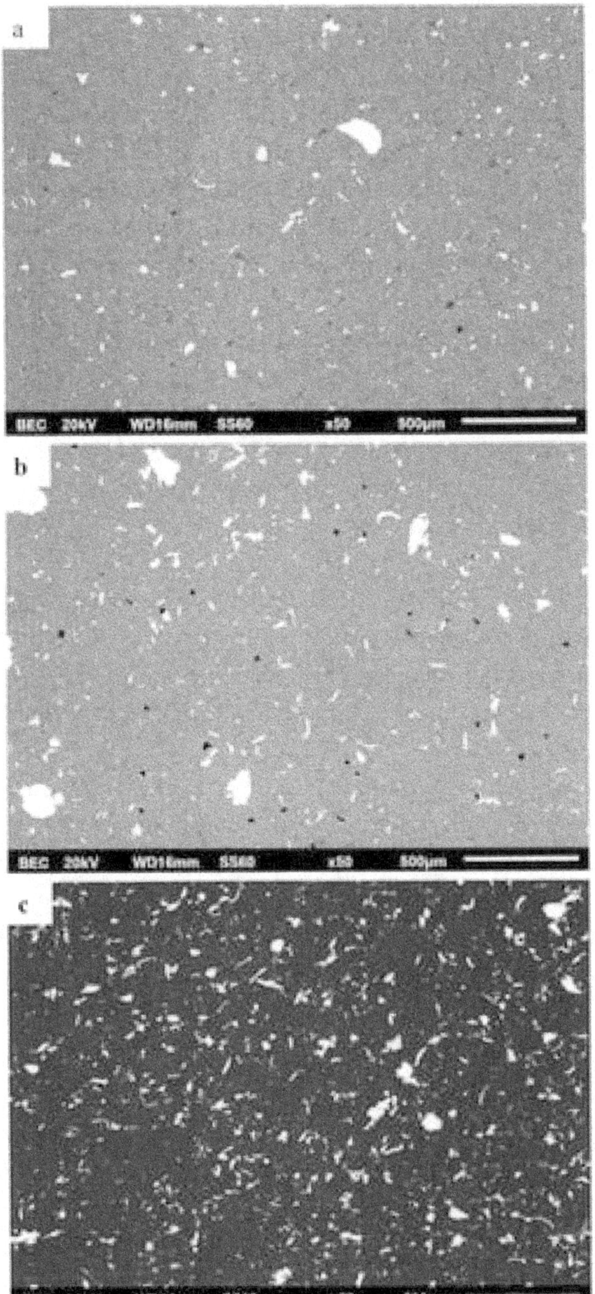

during aerospace operations. They found that the nanocrystalline structure remained stable up to a certain temperature range but started to coarsen at higher temperatures, affecting the mechanical properties.

3. Optimizing Alloy Composition: Further research focused on optimizing the alloy composition and processing parameters to enhance the high-temperature stability of the nanocrystalline structure. Small additions of certain elements (e.g., zirconium) were found to effectively retard grain growth at elevated temperatures.
4. Applications: The resulting nanocrystalline aluminum alloy showed promise for lightweight aerospace components, offering improved mechanical properties compared to conventional alloys. However, the selection of the right alloy composition and post-processing treatments was critical to ensure long-term microstructural stability under operational conditions.

Case Study 2: Control of Grain Growth in Aluminum Alloys for Automotive Applications:

In the automotive industry, aluminum alloys are used to produce lightweight components that contribute to fuel efficiency and reduce emissions. However, maintaining the desired microstructure during processing and service life is crucial.

Key findings and considerations:

1. Grain Growth Sensitivity: Researchers investigated the sensitivity of grain growth in different aluminum alloy compositions commonly used in automotive applications. They found that certain alloying elements, such as titanium and niobium, effectively pinned grain boundaries, inhibiting excessive grain growth.
2. Thermal Treatments: Researchers studied the effects of various heat treatment processes on the microstructure of aluminum alloys. Rapid quenching after solution heat treatment was found to be essential for preserving a fine-grained structure.
3. Hot Forming: In some automotive manufacturing processes, aluminum sheets are subjected to high-temperature forming operations. The study found that the microstructure's stability during hot forming depended on the initial grain size and the alloy's composition.
4. Durability and Performance: The optimized aluminum alloy with controlled grain growth exhibited excellent durability, impact resistance, and corrosion resistance while meeting the weight-saving requirements of automotive components.

5. MECHANICAL PROPERTIES OF MECHANICALLY ALLOYED ALUMINIUM ALLOYS

5.1 Strength and Hardness

Strength and hardness are key indicators of a material's ability to withstand external forces without deformation or failure. In mechanically alloyed aluminum alloys, the finely dispersed alloying elements and the nanocrystalline structure can lead to enhanced strength and hardness compared to conventionally processed alloys.

Case Study: High-Strength Aluminum-Alloy Composites for Structural Applications:

In this case study, researchers aimed to develop aluminum alloys with superior strength and hardness for use in structural components, such as load-bearing beams in construction.

Key findings and considerations:

1. Nanocrystalline Structure: The mechanical alloying process produced a nanocrystalline structure with a high density of grain boundaries and fine precipitates, resulting in significant strengthening.
2. Tensile Strength: The alloy exhibited higher tensile strength compared to traditional aluminum alloys, making it suitable for structural applications where high load-bearing capacity is required.
3. Hardness: The hardness of the mechanically alloyed aluminum alloy was considerably higher, indicating resistance to wear and deformation under applied loads.
4. Correlation with Microstructure: Researchers found a direct correlation between the nanocrystalline structure and the material's strength and hardness. The fine grain boundaries and homogeneously distributed alloying elements contributed to the enhanced mechanical properties (He et al., 1987).

Figure 4 displays the results of hardness measurements conducted on the three distinct composites. There is an observable rise in hardness with an increase in the content of the reinforcement phase. This increase in hardness can be attributed to the growing density of particles on the composite surface, which in turn makes mass movement more challenging.

Figure 4. The image represents the x-ray diffractograms of the produced composites (He et al., 1987)

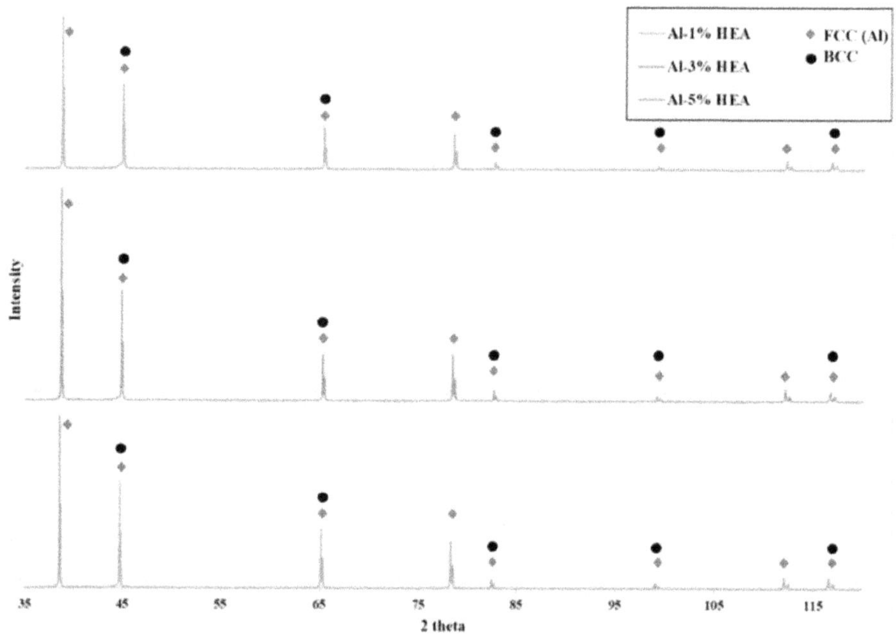

5.2 Wear Resistance

Wear resistance is crucial in applications were materials experience friction, abrasion, or sliding contact. Mechanically alloyed aluminum alloys can exhibit improved wear resistance due to their refined microstructure and enhanced hardness.

Case Study: Wear-Resistant Aluminum-Alloy Coatings:

In this case study, researchers focused on developing wear-resistant coatings based on mechanically alloyed aluminum alloys for applications such as industrial machinery components.

Key findings and considerations:

1. Surface Hardness: The mechanically alloyed aluminum alloy coatings displayed significantly higher surface hardness compared to conventional coatings. This property was attributed to the nanocrystalline structure and the presence of hard precipitates.

Figure 5. Graphs illustrating the relationship between mass loss and sliding distance have been plotted for each composite system (He et al., 1987)

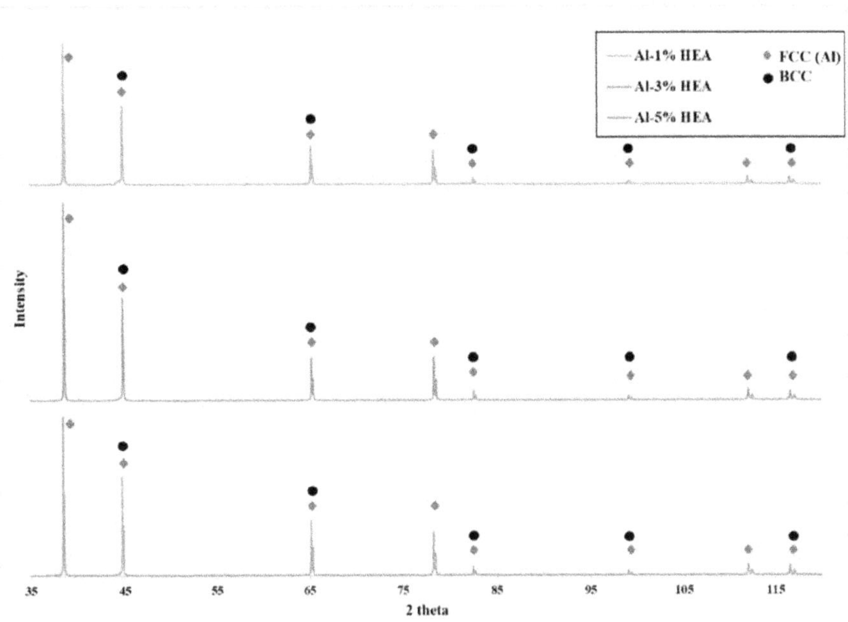

2. Wear Testing: Tribological tests, including sliding wear and abrasive wear tests, demonstrated that the alloyed coatings had improved wear resistance, leading to reduced material loss and extended component lifespan.
3. Application Considerations: The wear-resistant coatings showed promise for protecting critical components in industrial machinery that are prone to wear and require prolonged service life (Rodriguez et al., 2023).

5.3 Ductility and Toughness

Ductility and toughness are essential for materials that need to undergo plastic deformation without fracture, absorb energy before failure, and resist brittle behavior.

Case Study: Ductile Aluminum-Alloy Composites for Impact-Resistant Structures:

In this case study, researchers aimed to develop aluminum alloys with improved ductility and toughness for use in impact-resistant structures, such as automotive crash components.

Key findings and considerations:

Figure 6. Graphs depicting mass loss in relation to sliding distance have been generated for each composition, with data collected beyond the initial 200 meters (He et al., 1987)

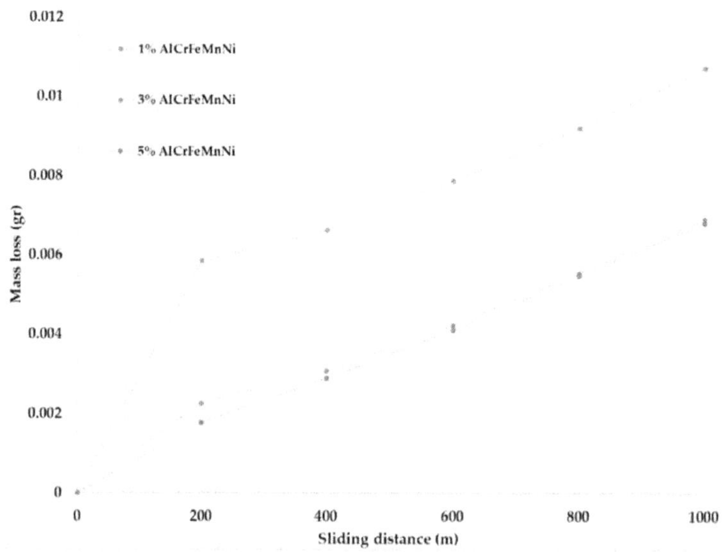

Figure 7. The wear rates have been calculated separately for three distinct intervals: the initial 200 meters, the range from 200 to 1000 meters, and the overall sliding distance, for each composition (He et al., 1987)

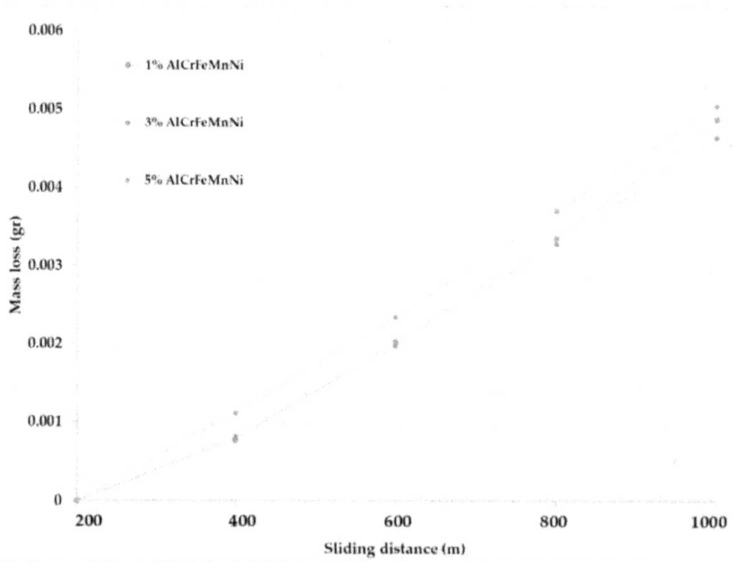

Figure 8. Panoramic images showcasing the worn surfaces of specimens are presented for three different compositions: (a) 1 vol.%, (b) 3 vol.%, and (c) 5 vol.% (He et al., 1987)

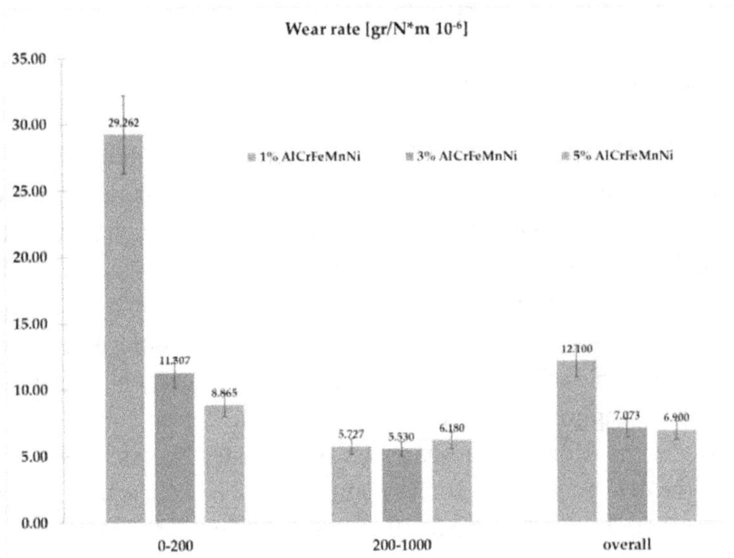

1. Ductility Enhancement: The carefully designed alloy composition, combined with the nanocrystalline structure, led to improved ductility compared to conventional aluminum alloys. This property was essential for absorbing impact energy.
2. Toughness: The alloy exhibited increased toughness, reducing the likelihood of catastrophic failure during impact events.
3. Crash-Worthiness: The impact behavior of the alloy was studied through crash simulations and testing, demonstrating its suitability for use in automotive structures that require high crash-worthiness (S et al., 2023).

6. CORROSION BEHAVIOR OF MECHANICALLY ALLOYED ALUMINIUM ALLOYS

6.1 Corrosion Resistance Mechanisms

Corrosion resistance mechanisms in aluminum alloys involve the protective passivation of the metal surface, the presence of alloying elements that form corrosion-resistant phases, and the impact of the microstructure on the corrosion process.

Case Study: Corrosion-Resistant Mechanically Alloyed Aluminum Alloy for Marine Applications:

In this case study, researchers aimed to develop an aluminum alloy with improved corrosion resistance for use in marine environments, where the material is exposed to seawater.

Key findings and considerations:

1. Passivation: The mechanically alloyed aluminum alloy exhibited a passivation layer that formed on the surface, providing a barrier against corrosion. This passivation was attributed to the presence of certain alloying elements, such as chromium or zinc, which enhanced the formation of protective oxide layers.
2. Alloy Composition: The selection of alloying elements played a crucial role in enhancing corrosion resistance. These elements could participate in the formation of stable and protective corrosion products, reducing the vulnerability of the alloy to aggressive environmental factors.
3. Localized Corrosion: The researchers investigated the susceptibility of the alloy to localized corrosion, such as pitting or crevice corrosion, which can be particularly problematic in marine environments. The alloy's microstructure was found to play a significant role in determining the initiation and propagation of localized corrosion (Farahani et al., 2023).

6.2 Influence of Microstructure on Corrosion Behavior

The microstructure of mechanically alloyed aluminum alloys can significantly impact their corrosion behavior. Factors such as grain size, distribution of alloying elements, and the presence of precipitates at grain boundaries can all influence the material's susceptibility to corrosion.

Case Study: Microstructural Effects on Corrosion of Mechanically Alloyed Aluminum Alloys:

In this case study, researchers aimed to understand how the microstructure of a mechanically alloyed aluminum alloy affects its corrosion behavior in aggressive chemical environments.

Key findings and considerations:

1. Grain Size: The nanocrystalline structure resulting from mechanical alloying can lead to increased corrosion resistance due to the presence of a high density of grain boundaries. These boundaries can act as barriers to corrosion propagation.
2. Precipitates: The presence of fine precipitates, especially those that form at grain boundaries, can influence the electrochemical behavior of the alloy. Properly engineered precipitates can reduce the material's susceptibility to corrosion.

3. Uniformity: A uniform microstructure with homogeneously distributed alloying elements was found to enhance overall corrosion resistance, as it reduced the likelihood of localized corrosion initiation (Ahuir-Torres et al., 2023).

6.3 Corrosion Testing Techniques

To assess the corrosion behavior of mechanically alloyed aluminum alloys accurately, various testing techniques are employed. These techniques range from simple electrochemical methods to more complex accelerated corrosion tests.

Case Study: Evaluating Corrosion Resistance of Mechanically Alloyed Aluminum Alloys:

In this case study, researchers aimed to systematically evaluate the corrosion resistance of mechanically alloyed aluminum alloys using a combination of corrosion testing techniques.

Key findings and considerations:

1. Electrochemical Methods: Techniques such as potentiodynamic polarization and electrochemical impedance spectroscopy (EIS) were used to measure the corrosion rates, polarization behavior, and impedance of the alloy in different electrolytes. These methods provided valuable insights into the alloy's corrosion behavior and its response to varying conditions.
2. Accelerated Corrosion Tests: Salt spray tests, cyclic corrosion tests, and exposure to aggressive environments were conducted to simulate real-world conditions. These tests allowed researchers to assess the alloy's performance over an accelerated timeframe.
3. Microstructural Analysis: Post-corrosion analysis of the alloy's microstructure provided a deeper understanding of how corrosion mechanisms were influenced by the material's structure. This information was used to refine the alloy design and processing parameters to enhance corrosion resistance (Kejić et al., 2023).

7. APPLICATIONS OF MECHANICALLY ALLOYED ALUMINIUM ALLOYS

7.1 Aerospace Applications

Aerospace engineering demands lightweight yet strong materials that can withstand the challenging conditions of flight. Mechanically alloyed aluminum alloys offer a promising solution for various aerospace components:

Case Study: High-Strength Lightweight Aircraft Components:

In this case study, researchers aimed to develop mechanically alloyed aluminum alloys for aerospace applications, particularly for structural components in aircraft.

Key findings and considerations:

Weight Reduction: The high strength-to-weight ratio of mechanically alloyed aluminum alloys made them ideal for aircraft components, leading to weight reduction without compromising structural integrity.

Corrosion Resistance: The corrosion-resistant properties of these alloys were crucial for withstanding the harsh environmental conditions experienced during flight, including exposure to atmospheric moisture and chemicals.

Improved Fatigue Resistance: The tailored microstructures and enhanced mechanical properties of the alloys contributed to improved fatigue resistance, extending the lifespan of aircraft components.

Manufacturability: The alloys were designed to be manufacturable using standard aerospace manufacturing techniques, ensuring a smooth transition to production.

Applications in the aerospace industry include aircraft structural components (such as fuselage frames, wing structures, and landing gear components) and aerospace fasteners (Balordi et al., 2023; Chatterjee et al., 2023; Tan & Liew, 2023; Trzepieciński et al., 2023).

7.2 Automotive Applications

In the automotive industry, the primary focus is on lightweight materials that contribute to fuel efficiency and reduce emissions while maintaining durability and safety. Mechanically alloyed aluminum alloys have gained attention for various automotive components:

Case Study: Lightweight High-Strength Automotive Components:

In this case study, researchers aimed to utilize mechanically alloyed aluminum alloys for lightweight automotive components to improve fuel efficiency and overall vehicle performance.

Key findings and considerations:

Fuel Efficiency: The lightweight nature of the alloys allowed for reduced vehicle weight, leading to improved fuel economy and reduced emissions.

Crash-Worthiness: The combination of high strength and tailored microstructures contributed to enhanced crash-worthiness, making these alloys suitable for impact-resistant structural components in vehicles.

Corrosion Resistance: The corrosion resistance of these alloys was essential for ensuring the durability and long-term performance of automotive components, especially in regions with salted roads or exposure to corrosive environments.

Applications in the automotive industry include lightweight structural components (such as chassis, body panels, and suspension parts) and corrosion-resistant components (such as exhaust systems) (Bogdanoff et al., 2023; Snopiński, 2023; Yu et al., 2023; Zupanič et al., 2023).

7.3 Other Engineering Applications

Beyond aerospace and automotive sectors, mechanically alloyed aluminum alloys find applications in various engineering fields where a combination of strength, corrosion resistance, and tailored properties is required:

Case Study: High-Performance Marine Components:

In this case study, researchers focused on the use of mechanically alloyed aluminum alloys for marine applications, where materials must withstand harsh seawater environments.

Key findings and considerations:

Corrosion Resistance: The alloys exhibited excellent corrosion resistance, making them suitable for marine components exposed to seawater, such as ship structures, offshore platforms, and marine equipment.

Durability: The combination of corrosion resistance, mechanical strength, and tailored microstructures contributed to the durability and longevity of marine components, reducing maintenance and replacement costs.

Salt Spray Testing: The alloys underwent extensive salt spray testing to simulate the effects of marine environments, confirming their suitability for such applications.

Other engineering applications for mechanically alloyed aluminum alloys include industrial machinery components, chemical processing equipment, and high-performance sports equipment (Ben Said et al., 2023; Mahale, V., K., C., & Shashanka, 2022; Mahale et al., 2023; Shrinivas Mahale, Shamanth, Sharath, Shashanka, & Hemanth, 2021; Technology and Challenges in Additive Manufacturing of Duplex Stainless Steels, 2021; Yan et al., 2023).

8.1 Contamination Control

Contamination during mechanical alloying can negatively impact the material's properties and performance. This can occur from wear of the milling equipment, as well as from impurities in the raw materials.

Challenge: During milling, the grinding media and the material being milled can wear, leading to contamination by small amounts of the milling media (such as steel) and other impurities.

Solutions:

1. Choice of Milling Media: Select milling media with a composition that minimizes contamination. Consider using ceramic or hardmetal (carbide) milling balls to reduce the risk of introducing unwanted impurities.
2. Protective Milling Atmosphere: Perform milling in a controlled atmosphere, such as inert gas (e.g., argon) to minimize oxidation and prevent contamination from the surrounding air.
3. Regular Equipment Maintenance:Proper maintenance of the milling equipment is essential to minimize wear and prevent cross-contamination between different materials (Pradeep et al., 2022; Rajendrachari, 2022; Shashanka et al., 2016).

8.2 Scale-Up and Industrial Production

Transferring successful laboratory-scale mechanical alloying processes to large-scale industrial production can be challenging due to issues related to scalability and consistency.

Challenge: The parameters that work on a small scale might not directly translate to large-scale production. Achieving the same material properties and microstructures at an industrial scale can be challenging.

Solutions:

1. Process Optimization: Conduct extensive process optimization studies when scaling up, considering factors such as milling time, rotation speed, and powder loading to maintain the desired microstructure and properties.
2. Continuous Monitoring: Implement robust quality control measures to ensure the consistency of the alloyed material, using advanced characterization techniques to verify microstructural properties.
3. Pilot-Scale Testing: Prior to full-scale production, conduct pilot-scale tests to validate the scalability and assess the feasibility of the process at an intermediate production level (Shashanka & Chaira, 2016).

8.3 Cost Considerations and Economic Viability

Mechanical alloying processes, especially those involving high-energy ball milling, can be resource-intensive, leading to concerns about the economic viability of producing alloys through this method.

Challenge: The cost of equipment, energy consumption, and raw materials can make the production of mechanically alloyed materials expensive, affecting their competitiveness.

Solutions:

1. Raw Material Selection: Choose raw materials with cost-efficient compositions that provide the desired alloying effect without unnecessary expense.
2. Energy Efficiency: Optimize milling conditions to reduce energy consumption while maintaining desired alloying effects. This can involve adjusting milling parameters and using advanced milling techniques.
3. Economies of Scale: Explore batch sizes and production volumes that maximize efficiency, as larger quantities may lead to economies of scale, reducing overall costs.
4. Value-Added Products: Consider the unique properties of mechanically alloyed materials and target applications where these properties provide significant value, justifying the production costs.

Addressing these challenges is essential for realizing the full potential of mechanical alloying in producing advanced materials with tailored properties, from laboratory-scale research to large-scale industrial production (Gupta et al., 2015).

9. FUTURE DIRECTIONS AND EMERGING TRENDS

9.1 Multi-Element Aluminium Alloys

Traditionally, mechanical alloying has been used to combine two or a few elements to create novel materials. The future of mechanical alloying is likely to involve the development of multi-element aluminum alloys with precise control over composition and microstructure.

Emerging Trend: Researchers are exploring the design and synthesis of aluminum alloys with multiple alloying elements to achieve specific combinations of properties. These alloys might exhibit improved strength, corrosion resistance, thermal stability, and other tailored properties for advanced applications.

Application Potential: Multi-element aluminum alloys could find applications in industries ranging from aerospace and automotive to electronics and energy, where a combination of high performance and lightweight properties is essential (Nayak et al., 2016).

9.2 Hybrid Mechanical Alloying Techniques

The future of mechanical alloying may involve the integration of hybrid techniques that combine mechanical alloying with other processing methods, such as severe plastic deformation, high-pressure torsion, or advanced powder consolidation techniques.

Emerging Trend: Researchers are exploring the synergistic effects of combining mechanical alloying with severe plastic deformation to achieve ultrafine-grained structures and enhanced properties.

Application Potential: Hybrid techniques could lead to the development of materials with superior mechanical properties, extended fatigue life, and improved formability, making them suitable for high-stress applications, such as aerospace components and advanced structural materials (R & Chaira, 2014; Shrinivas Mahale et al., 2022).

9.3 Sustainability and Environmental Impact

As environmental concerns grow, future trends in mechanical alloying will likely focus on more sustainable practices, reduced energy consumption, and environmentally friendly processing.

Emerging Trend: Researchers and industries are exploring greener synthesis routes, including the use of recyclable materials, reducing waste generation, and developing more energy-efficient milling processes.

Application Potential: Sustainable mechanical alloying practices can lead to materials with lower carbon footprints and contribute to the overall sustainability of industries that rely on advanced materials, including automotive, aerospace, and electronics.

10. CASE STUDIES AND PRACTICAL EXAMPLES

10.1 Mechanically Alloyed Aluminium Alloy Compositions

Case Study 1: High-Strength Lightweight Aluminum Alloy for Aerospace Fasteners

Alloy Composition: Mechanically alloyed Al-4.5Cu-0.5Mg with the addition of nanoparticles (e.g., Al2O3) for dispersion strengthening.

Processing Routes and Microstructural Analysis:

Processing Route: Mechanical alloying was performed using a high-energy ball mill under an argon atmosphere.

Microstructural Analysis: Scanning electron microscopy (SEM) and transmission electron microscopy (TEM) were used to analyze the microstructure, revealing a nanocrystalline structure with fine precipitates uniformly distributed in the aluminum matrix.

Performance Evaluation in Specific Applications:

Application: Aerospace fasteners (rivets, bolts).

Performance Evaluation: Tensile tests demonstrated improved strength compared to conventional aluminum alloys. Fatigue testing showed enhanced durability due to the dispersion-strengthening effect. Corrosion tests in simulated aerospace environments confirmed good corrosion resistance (R & Chaira, 2015).

Case Study 2: Mechanically Alloyed Al-Mg-Cr Alloy for Automotive Structural Components

Alloy Composition: Mechanically alloyed Al-5Mg-1Cr with additional alloying elements for improved corrosion resistance.

Processing Routes and Microstructural Analysis:

Processing Route: High-energy ball milling followed by hot extrusion.

Microstructural Analysis: X-ray diffraction (XRD) and electron backscatter diffraction (EBSD) analysis revealed a refined microstructure with uniform distribution of alloying elements.

Performance Evaluation in Specific Applications:

Application: Automotive structural components (chassis, body panels).

Performance Evaluation: Tensile tests demonstrated increased strength and ductility. Corrosion testing in a salt spray chamber showed excellent corrosion resistance. Finite element simulations showed improved crash-worthiness compared to traditional alloys, making it suitable for impact-prone automotive components (Shashanka & Chaira, 2015).

Case Study 3: Mechanically Alloyed Al-Ti-Si Powder for 3D Printing

Alloy Composition: Mechanically alloyed Al-4Ti-2Si powder for 3D printing.

Processing Routes and Microstructural Analysis:

Processing Route: High-energy ball milling followed by powder bed fusion 3D printing.

Microstructural Analysis: Metallography and electron microscopy were used to analyze the printed parts, revealing a fine-grained microstructure with uniform dispersion of Ti-Si intermetallic phases.

Performance Evaluation in Specific Applications:

Application: Additive manufacturing for lightweight structural components.

Performance Evaluation: Tensile tests of the 3D-printed parts showed good mechanical properties. The microstructural analysis confirmed uniformity. The 3D-printed components demonstrated high strength-to-weight ratios, making them suitable for aerospace and automotive lightweight applications.

11. CONCLUSION

I. In conclusion, mechanical alloying stands as a transformative method in the advancement of aluminum alloys, offering a precise and controlled approach to enhancing their properties.

II. This abstract highlights its significance by underscoring the formation of a homogeneous, fine-grained microstructure, providing aluminum alloys with exceptional mechanical properties, improved corrosion resistance, and promising applications in the aerospace and automotive industries.
III. While this technique demonstrates remarkable progress in materials engineering, challenges like contamination control and scaling up for industrial production remain to be addressed.
IV. The potential of developing complex, multi-element aluminum alloys through mechanical alloying represents a promising avenue for future research.
V. In essence, mechanical alloying has opened new doors in the world of aluminum alloy development, enabling the creation of innovative materials that cater to the ever-evolving demands of modern engineering applications.
VI. As we look ahead, the continued exploration and refinement of this technique promise to further revolutionize the field, shaping the future of aluminum alloy design and utilization across various industries.

REFERENCES

Affolter, C., Thorwarth, G., Arabi-Hashemi, A., Müller, U., & Weisse, B. (2019). Ductile Compressive Behavior of Biomedical Alloys. *Metals*, *10*(1), 60. doi:10.3390/met10010060

Ahmed, M. M. Z., Ataya, S., El-Sayed Seleman, M. M., Mahdy, A. M. A., Alsaleh, N. A., & Ahmed, E. (2020). Heat Input and Mechanical Properties Investigation of Friction Stir Welded AA5083/AA5754 and AA5083/AA7020. *Metals*, *11*(1), 68. doi:10.3390/met11010068

Ahuir-Torres, J. I., Kotadia, H. R., Öpoz, T. T., & Sharp, M. C. (2023, February 28). A Study on the Corrosion Behaviour of Laser Textured Pure Aluminium in Saltwater. *Processes (Basel, Switzerland)*, *11*(3), 721. doi:10.3390/pr11030721

Aizawa, T., Kurihara, T., & Sakayori, H. (2022). Manufacturing of Aluminum Alloy Parts from Recycled Feedstock by PIG Die-Casting and Hot Stamping. *Lubricants (Basel, Switzerland)*, *11*(1), 13. doi:10.3390/lubricants11010013

Arendarchuck, B. E., Mayer, A. R., de Oliveira, W. R., Pukasiewicz, A. G. M., Lourençato, L. A., Fals, H. D. C., & Martínez-Cámara, E. (2022). Effect of Ti-B Grain Refiners on Wear and Corrosion of the A332 Alloy with Sr Modification. *Applied Sciences (Basel, Switzerland)*, *13*(1), 430. doi:10.3390/app13010430

Balordi, M., Santucci de Magistris, G., Casali, A., Pini, F., Cammi, A., Lacavalla, M., & Rotella, V. (2023, March 14). The Anti-Snow Behaviour of Icephobic Coatings: Laboratory and In-Field Testing. *Coatings*, *13*(3), 616. doi:10.3390/coatings13030616

Barro, Ó., Arias-González, F., Lusquiños, F., Comesaña, R., del Val, J., Riveiro, A., Badaoui, A., Gómez-Baño, F., & Pou, J. (2020). Improved Commercially Pure Titanium Obtained by Laser Directed Energy Deposition for Dental Prosthetic Applications. *Metals*, *11*(1), 70. doi:10.3390/met11010070

Bedolla-Becerril, E., Garcia-Guerra, J., Lopez-Morelos, V. H., Garcia-Renteria, M. A., Falcon-Franco, L. A., Martinez-Landeros, V. H., García-Villarreal, S., & Flores-Villaseñor, S. E. (2022). Tribological Behaviour of Al-2024/TiC Metal Matrix Composites. *Coatings*, *13*(1), 77. doi:10.3390/coatings13010077

Ben Said, L., Chabchoub, A. K., & Wali, M. (2023, April 22). Mathematical Model Describing the Hardening and Failure Behaviour of Aluminium Alloys: Application in Metal Shear Cutting Process. *Mathematics*, *11*(9), 1980. doi:10.3390/math11091980

Bogdanoff, T., Tiryakioğlu, M., Liljenfors, T., Jarfors, A. E. W., Seifeddine, S., & Ghassemali, E. (2023, March 15). On the Effectiveness of Rotary Degassing of Recycled Al-Si Alloy Melts: The Effect on Melt Quality and Energy Consumption for Melt Preparation. *Sustainability (Basel)*, *15*(6), 5189. doi:10.3390/su15065189

Camarano, A., Giuranno, D., & Narciso, J. (2019). SiC-IrSi3 for High Oxidation Resistance. *Materials (Basel)*, *13*(1), 98. doi:10.3390/ma13010098 PMID:31878174

Cao, X., Zhu, J., Gao, F., & Gao, Z. (2020). Influence of Fine Grains on the Bending Fatigue Behavior of Two Implant Titanium Alloys. *Materials (Basel)*, *14*(1), 171. doi:10.3390/ma14010171 PMID:33396522

Chaoyang, C., Guangjie, G., Lingfei, L., Fei, F., & Lin, L. (2020). Effect of Two-Stage Homogenization Heat Treatment on Microstructure and Mechanical Properties of AA2060 Alloy. *Crystals*, *11*(1), 40. doi:10.3390/cryst11010040

Chatterjee, A., Sen, S., Paul, S., Roy, P., Seikh, A. H., Alnaser, I. A., Das, K., Sutradhar, G., & Ghosh, M. (2023, March 13). Fabrication and Characterization of SiC-reinforced Aluminium Matrix Composite for Brake Pad Applications. *Metals*, *13*(3), 584. doi:10.3390/met13030584

Chausov, M., Pylypenko, A., Maruschak, P., Zasimchuk, V., Brezinová, J., Brezina, J., & Viňáš, J. (2022). Impact of the Initial Phase Composition of Alloys on the Effects Manifested by Yield Sites That Occur on Sheet Aluminum Alloys Subjected to Impact-Oscillatory Loading. *Materials (Basel)*, *16*(1), 249. doi:10.3390/ma16010249 PMID:36614587

Chen, H., Guo, H., & Shang, C. J. (2014). Study on Wear Behaviours of 400 and 450 Grade Wear Resistant Steels. *Materials Research Innovations*, *18*(sup4), S4-270. . doi:10.1179/1432891714Z.000000000691

Chen, H., Zhao, D., Wang, Q., Qiang, Y., & Qi, J. (2017). Effects of Impact Energy on the Wear Resistance and Work Hardening Mechanism of Medium Manganese Austenitic Steel. *Friction*, *5*(4), 447–454. doi:10.1007/s40544-017-0158-6

Chen, S., Ye, H., & Lin, X. (2017). Effect of Rare Earth and Alloying Elements on the Thermal Conductivity of Austenitic Medium Manganese Steel. *International Journal of Minerals Metallurgy and Materials*, *24*(6), 670–674. doi:10.1007/s12613-017-1449-7

Chiu, Y.-C., Pan, T.-A., Chen, G.-M., Jiang, X.-C., Bor, H.-Y., Tzeng, Y.-C., & Lee, S.-L. (2020). The Effects of a Trace Amount of Manganese and the Homogenization on the Recrystallization of Al–7Mg–0.15Ti Alloys. *Molecules (Basel, Switzerland)*, *26*(1), 168. doi:10.3390/molecules26010168 PMID:33396531

Contuzzi, N., Casalino, G., Boccaccio, A., Ballini, A., Charitos, I. A., Bottalico, L., & Santacroce, L. (2022). Metals Biotribology and Oral Microbiota Biocorrosion Mechanisms. *Journal of Functional Biomaterials*, *14*(1), 14. doi:10.3390/jfb14010014 PMID:36662061

Cui, C., Uhlenwinkel, V., Schulz, A., & Zoch, H.-W. (2019). Austenitic Stainless Steel Powders with Increased Nitrogen Content for Laser Additive Manufacturing. *Metals*, *10*(1), 61. doi:10.3390/met10010061

Di Egidio, G., Tonelli, L., Morri, A., Boromei, I., Shashkov, P., & Martini, C. (2022). Influence of Anodizing by Electro-Chemical Oxidation on Fatigue and Wear Resistance of the EV31A-T6 Cast Magnesium Alloy. *Coatings*, *13*(1), 62. doi:10.3390/coatings13010062

Ding, Z., Long, W., Jiu, Y., Zhong, S., Jiang, D., Liu, C., Yang, J., & Qiao, J. (2022). Formation and Morphological Transition of Diversified Eutectic Structures in Ag-Based Brazing Filler Metals with Sn Addition. *Crystals*, *13*(1), 68. doi:10.3390/cryst13010068

Du, M., Liu, B., Liu, Y., & Yang, Y. (2022). Dynamic Behavior of Additively Manufactured FeCoCrNi High Entropy Alloy. *Metals*, *13*(1), 75. doi:10.3390/met13010075

Fan, X., Mi, Z., Yang, L., & Su, H. (2022). Application of DFT Simulation to the Investigation of Hydrogen Embrittlement Mechanism and Design of High Strength Low Alloy Steel. *Materials (Basel), 16*(1), 152. doi:10.3390/ma16010152 PMID:36614491

Farahani, E., Liberati, A. C., Moreau, C., Dolatabadi, A., & Stoyanov, P. (2023, February 27). Comparative Evaluation of the Shear Adhesion Strength of Ice on PTFE Solid Lubricant. *Lubricants (Basel, Switzerland), 11*(3), 105. doi:10.3390/lubricants11030105

Ferreira, P. P., Carvalho, F. M., Ariza-Echeverri, E. A., Delfino, P. M., Bauri, L. F., Ferreira, A. M., Braga, A. P., Eleno, L. T. F., Goldenstein, H., & Tschiptschin, A. P. (2022). Synergism between B and Nb Improves Fire Resistance in Microalloyed Steels. *Metals, 13*(1), 84. doi:10.3390/met13010084

Friák, M., Golian, M., Holec, D., Koutná, N., & Šob, M. (2019). An Ab Initio Study of Magnetism in Disordered Fe-Al Alloys with Thermal Antiphase Boundaries. *Nanomaterials (Basel, Switzerland), 10*(1), 44. doi:10.3390/nano10010044 PMID:31878105

Gao, X., Geng, K., Sun, C., Zhang, S., Zhou, J., Wu, J., Zhang, X., & Wang, X. (2022). Effects of Graphite Particle Content and Holding Time on the Microstructure and Mechanical Properties of the Graphite/AZ91D Composite. *Metals, 13*(1), 57. doi:10.3390/met13010057

Gao, Y., Yang, Z., Xiao, H., Lei, Q., Liu, B., & Liu, Y. (2022). Effect of Welding Current on Wear Behavior of PTA-Welded Cu35Ni25Co25Cr15 HEA Coating. *Coatings, 13*(1), 52. doi:10.3390/coatings13010052

Ge, S., Wang, Q., & Wang, J. (2017). The Impact Wear-Resistance Enhancement Mechanism of Medium Manganese Steel and Its Applications in Mining Machines. *Wear, 376–377*, 1097–1104. doi:10.1016/j.wear.2017.01.015

Ghosh, M., Hendy, M., Raush, J., & Momeni, K. (2022). A Phase-Field Model for In-Space Manufacturing of Binary Alloys. *Materials (Basel), 16*(1), 383. doi:10.3390/ma16010383 PMID:36614722

Gibbs, P. J., De Cooman, B. C., Brown, D. W., Clausen, B., Schroth, J. G., Merwin, M. J., & Matlock, D. K. (2014). Strain Partitioning in Ultra-Fine Grained Medium-Manganese Transformation Induced Plasticity Steel. *Materials Science and Engineering A, 609*, 323–333. doi:10.1016/j.msea.2014.03.120

Gontijo, M., Hoflehner, C., Ilie, S., Six, J., & Sommitsch, C. (2020). Holding Time Influence on the Hot Ductility Behavior of a Continuously Cast Low Alloy Steel. *Metals, 11*(1), 64. doi:10.3390/met11010064

Guan, C., Yu, T., Zhao, Y., Chen, L., & Chen, Y. (2022). Repair of Gear by Laser Cladding Ni60 Alloy Powder: Process, Microstructure and Mechanical Performance. *Applied Sciences (Basel, Switzerland), 13*(1), 319. doi:10.3390/app13010319

Gupta, S., Shashanka, R., & Chaira, D. (2015, February 19). Synthesis of nano-structured duplex and ferritic stainless steel powders by planetary milling: An experimental and simulation study. *IOP Conference Series. Materials Science and Engineering, 75*, 012033. doi:10.1088/1757-899X/75/1/012033

He, Z.-M., Jiang, Q.-C., Fu, S.-B., & Xie, J.-P. (1987). Improved Work-Hardening Ability and Wear Resistance of Austenitic Manganese Steel under Non-Severe Impact-Loading Conditions. *Wear, 120*(3), 305–319. doi:10.1016/0043-1648(87)90024-X

Imshinetskiy, I., Kashepa, V., Nadaraia, K., Mashtalyar, D., Suchkov, S., Zadorozhny, P., Ustinov, A., Sinebryukhov, S., & Gnedenkov, S. (2022). PEO Coatings Modified with Halloysite Nanotubes: Composition, Properties, and Release Performance. *International Journal of Molecular Sciences, 24*(1), 305. doi:10.3390/ijms24010305 PMID:36613748

Jiao, J., Xu, J., Jing, C., Cheng, X., Wu, D., Ru, H., Zeng, K., & Sheng, L. (2022). Effect of Laser Micro-Texturing on Laser Joining of Carbon Fiber Reinforced Thermosetting Composites to TC4 Alloy. *Materials (Basel), 16*(1), 270. doi:10.3390/ma16010270 PMID:36614608

Jost, N., & Schmidt, I. (1986). Friction-Induced Martensitic Transformation in Austenitic Manganese Steels. *Wear, 111*(4), 377–389. doi:10.1016/0043-1648(86)90134-1

Kanaginahal, G. M., Mahale, R. S., Vasanth, S., Sharath, P. C., & Shashanka, R. (2023). Studies on Dielectric, Super-Capacitive and Photoluminescence Properties of Metal Organic Frameworks Subjected to Varying Temperature and Frequencies. *Nano Hybrids and Composites, 39*, 17–25. doi:10.4028/p-si6964

Kejić, D., Žuvelek, V., & Valčić, A. (2023, March 9). Parametric Numerical Study of Welded Aluminium Beam-to-Column Joints. *Buildings, 13*(3), 718. doi:10.3390/buildings13030718

Kong, X., Dang, Z., Liu, X., Wang, M., & Hou, N. (2022). Simulation and Experimental Analysis of Tool Wear and Surface Roughness in Laser Assisted Machining of Titanium Alloy. *Crystals, 13*(1), 40. doi:10.3390/cryst13010040

Kumar, P., Skotnicova, K., Mallick, A., Gupta, M., Cegan, T., & Jurica, J. (2020). Mechanical Characterization of Graphene Nanoplatelets-Reinforced Mg-3Sn Alloy Synthesized by Powder Metallurgy. *Metals, 11*(1), 62. doi:10.3390/met11010062

Latypov, M. I., Shin, S., De Cooman, B. C., & Kim, H. S. (2016). Micromechanical Finite Element Analysis of Strain Partitioning in Multiphase Medium Manganese TWIP+TRIP Steel. *Acta Materialia, 108*, 219–228. doi:10.1016/j.actamat.2016.02.001

Li, H., Jiang, C., Yu, Z., Wan, Y., Ma, Y., & Yang, Z. (2022). Numerical Study on the Progressive Damage Behavior of the Interfacial Debonding between Shape Memory Alloy and Polymer Matrix. *Materials (Basel), 16*(1), 168. doi:10.3390/ma16010168 PMID:36614508

Li, J. C. M., Feng, C. R., & Rath, B. B. (2020). Emission of Dislocations from Grain Boundaries and Its Role in Nanomaterials. *Crystals, 11*(1), 41. doi:10.3390/cryst11010041

Liu, X., Su, Y., Zhang, G., Wang, R., & Cai, X. (2022). Effect of Longitudinal Magnetic Field on the Microstructure and Properties of A-TIG Welding with Different TiO2 Coating Amounts. *Crystals, 13*(1), 66. doi:10.3390/cryst13010066

Mahale, R. S., V., S., K., H., C., S. P., & Shashanka, R. (2022). Mechanical Testing of Spark Plasma Sintered Materials: A Review. *9th National Conference On Recent Developments In Mechanical Engineering*. . doi:10.1063/5.0080183

Mahale, R. S., V., S., K., H., C., S. P., & Shashanka, R. (2022). Mechanical testing of spark plasma sintered materials: A review. *9th National Conference On Recent Developments In Mechanical Engineering*. doi:10.1063/5.0080183

Mahale, R. S., Shamanth, V., & Sharath, P. C. (2020). *Review on Processing and Characterization of Duplex Stainless Steels*. Lecture Notes in Mechanical Engineering., doi:10.1007/978-981-15-7557-0_20

Mahale, R. S., Vasanth, S., Chikkegouda, S. P., Rajendrachari, S., Narsimhachary, D., & Basavegowda, N. (2023, August 9). Electrochemical Determination of Ascorbic Acid by Mechanically Alloyed Super Duplex Stainless Steel Powders. *Metals, 13*(8), 1430. doi:10.3390/met13081430

Mao, F., Wang, F., Shuang, Y., Hu, J., & Chen, J. (2019). Deformation Behavior and Experiments on a Light Alloy Seamless Tube via a Tandem Skew Rolling Process. *Metals, 10*(1), 59. doi:10.3390/met10010059

Marušáková, D., Corrêa, C. A., Aparicio, C., Libera, O., Berka, J., Vilémová, M., & Gávelová, P. (2022). The Influence of High-Temperature Helium and the Amount of Revert Material on the Material Properties of Inconel 738. *Coatings*, *13*(1), 45. doi:10.3390/coatings13010045

. (2021). Mechanical and Tribological Characteristics of Aluminium 2618 Matrix Composite Reinforced with Boron Carbide. *Biointerface Research in Applied Chemistry*, *12*(4), 4544–4556. doi:10.33263/BRIAC124.45444556

Meng, J., Chen, X., Jiang, J., & Liu, L. (2022). Investigation on Anisotropic Mechanical Behavior of Ti-6Al-4V Alloy via Schmid Factor and Kernel Average Misorientation Distribution. *Metals*, *13*(1), 89. doi:10.3390/met13010089

Miao, Q., Ding, W., Kuang, W., Zhou, B., Hao, T., Dai, C., & Yin, Z. (2022). Microstructure Evolution and Formation of Gradient Structures in Single Crystal Nickel-Based Superalloy by Surface Mechanical Creep-Feed Grinding Treatment. *Materials (Basel)*, *16*(1), 321. doi:10.3390/ma16010321 PMID:36614660

Nacházel, J., Palán, J., Dlouhý, J., Sláma, P., & Nový, Z. (2019). Effects of Rotary Swaging Parameters and Artificial Ageing on Mechanical Properties and Microstructure of 2024 Precipitation-Hardenable Aluminium Alloy. *Materials (Basel)*, *13*(1), 143. doi:10.3390/ma13010143 PMID:31905879

Nayak, A. K., Shashanka, R., & Chaira, D. (2016, February). Effect of Nanosize Yittria and Tungsten Addition to Duplex Stainless Steel During High Energy Planetary Milling. *IOP Conference Series. Materials Science and Engineering*, *115*, 012008. doi:10.1088/1757-899X/115/1/012008

Nithin, S. K., Hemanth, K., Shamanth, V., Shrinivas Mahale, R., Sharath, P. C., & Patil, A. (2022). Importance of Condition Monitoring in Mechanical Domain. *Materials Today: Proceedings*, *54*, 234–239. doi:10.1016/j.matpr.2021.08.299

Omarov, S., Nauryz, N., Talamona, D., & Perveen, A. (2022). Surface Modification Techniques for Metallic Biomedical Alloys: A Concise Review. *Metals*, *13*(1), 82. doi:10.3390/met13010082

Park, B. Y., Kim, Y., & Lee, K. (2020). Optimization Studies of AC4CH Material in the Cylinder Block of a Diesel Engine Application. *Processes (Basel, Switzerland)*, *9*(1), 70. doi:10.3390/pr9010070

Park, C., Hwang, T., Kim, G.-D., Nam, H., & Kang, N. (2022). Effect of the Initial Grain Size on Laser Beam Weldability for High-Entropy Alloys. *Crystals*, *13*(1), 65. doi:10.3390/cryst13010065

Phutela, C., Aboulkhair, N. T., Tuck, C. J., & Ashcroft, I. (2019). The Effects of Feature Sizes in Selectively Laser Melted Ti-6Al-4V Parts on the Validity of Optimised Process Parameters. *Materials (Basel), 13*(1), 117. doi:10.3390/ma13010117 PMID:31887981

Prabhu, B., Suryanarayana, C., An, L., & Vaidyanathan, R. (2006). Synthesis and Characterization of High Volume Fraction Al–Al2O3 Nanocomposite Powders by High-Energy Milling. *Materials Science and Engineering A, 425*(1–2), 192–200. doi:10.1016/j.msea.2006.03.066

Pradeep, N., Hegde, M. R., Rajendrachari, S., & Surendranathan, A. (2022, August). Investigation of microstructure and mechanical properties of microwave consolidated TiMgSr alloy prepared by high energy ball milling. *Powder Technology, 408*, 117715. doi:10.1016/j.powtec.2022.117715

R, S., & Chaira, D. (2014, June). Phase transformation and microstructure study of nano-structured austenitic and ferritic stainless steel powders prepared by planetary milling. *Powder Technology, 259*, 125–136. doi:10.1016/j.powtec.2014.03.061

R, S., & Chaira, D. (2015, January). Development of nano-structured duplex and ferritic stainless steels by pulverisette planetary milling followed by pressureless sintering. *Materials Characterization, 99*, 220–229. doi:10.1016/j.matchar.2014.11.030

Rajendrachari, S. (2022, June 21). An Overview of High-Entropy Alloys Prepared by Mechanical Alloying Followed by the Characterization of Their Microstructure and Various Properties. *Alloys, 1*(2), 116–134. doi:10.3390/alloys1020008

Rendón, J., & Olsson, M. (2009). Abrasive Wear Resistance of Some Commercial Abrasion Resistant Steels Evaluated by Laboratory Test Methods. *Wear, 267*(11), 2055–2061. doi:10.1016/j.wear.2009.08.005

Richardson, H., Bopp, C., Ha, B., Thomas, R., & Santhanam, K. S. V. (2020). Electrodeposition from a Graphene Bath: A Sustainable Copper Composite Alloy in a Graphene Matrix. *Journal of Composites Science, 5*(1), 9. doi:10.3390/jcs5010009

Rodriguez, L., Paris, J. Y., Denape, J., & Delbé, K. (2023, February 6). Micro-Arcs Oxidation Layer Formation on Aluminium and Coatings Tribological Properties—A Review. *Coatings, 13*(2), 373. doi:10.3390/coatings13020373

S, V., U, M. I., & Ghani, J. A. (2023, February 17). A Study on the Effects of Hybridized Metal Oxide and Carbonaceous Nano-Cutting Fluids in the End Milling of AA6082 Aluminum Alloy. *Lubricants, 11*(2), 87. https://doi.org/ doi:10.3390/lubricants11020087

Saha, S., & Roy, S. (2022). Metallic Dental Implants Wear Mechanisms, Materials, and Manufacturing Processes: A Literature Review. *Materials (Basel), 16*(1), 161. doi:10.3390/ma16010161 PMID:36614500

Shashanka, R. (2014). Chaira, D. Phase Transformation and Microstructure Study of Nano-Structured Austenitic and Ferritic Stainless Steel Powders Prepared by Planetary Milling. *Powder Technology, 259*, 125–136. doi:10.1016/j.powtec.2014.03.061

Shashanka, R., & Chaira, D. (2015). Optimization of Milling Parameters for the Synthesis of Nano-Structured Duplex and Ferritic Stainless Steel Powders by High Energy Planetary Milling. *Powder Technology, 278*, 35–45. doi:10.1016/j.powtec.2015.03.007

Shashanka, R., & Chaira, D. (2015, July). Optimization of milling parameters for the synthesis of nano-structured duplex and ferritic stainless steel powders by high energy planetary milling. *Powder Technology, 278*, 35–45. doi:10.1016/j.powtec.2015.03.007

Shashanka, R., & Chaira, D. (2016, January). Effects of Nano-Y2O3 and Sintering Parameters on the Fabrication of PM Duplex and Ferritic Stainless Steels. *Acta Metallurgica Sinica, 29*(1), 58–71. doi:10.1007/s40195-015-0362-1

Shashanka, R., Chaira, D., & Chakravarty, D. (2016, June 28). Fabrication of Nano-Yttria Dispersed Duplex and Ferritic Stainless Steels by Planetary Milling Followed by Spark Plasma Sintering and Non-Lubricated Sliding Wear Behaviour Study. *Journal of Materials Science and Engineering B, 6*(3). Advance online publication. doi:10.17265/2161-6221/2016.5-6.001

Shashanka, R., Esgin, H., Yilmaz, V. M., & Caglar, Y. (2020). Fabrication and Characterization of Green Synthesized ZnO Nanoparticle Based Dye-Sensitized Solar Cells. *Journal of Science. Advanced Materials and Devices, 5*(2), 185–191. doi:10.1016/j.jsamd.2020.04.005

Shi, Y., Li, R., & Lei, Z. (2022). Influences of Synthetic Parameters on Morphology and Growth of High Entropy Oxide Nanotube Arrays. *Coatings, 13*(1), 46. doi:10.3390/coatings13010046

Shrinivas Mahale, R., Shamanth, V., Hemanth, K., Nithin, S., Sharath, P., Shashanka, R., Patil, A., & Shetty, D. (2022). Processes and applications of metal additive manufacturing. *Materials Today: Proceedings, 54*, 228–233. doi:10.1016/j.matpr.2021.08.298

Shrinivas Mahale, R., Shamanth, V., Sharath, P., Shashanka, R., & Hemanth, K. (2021). A review on spark plasma sintering of duplex stainless steels. *Materials Today: Proceedings*, *45*, 138–144. doi:10.1016/j.matpr.2020.10.357

Shrinivas Mahale, R., Shamanth, V., Sharath, P. C., Shashanka, R., & Hemanth, K. (2021). A Review on Spark Plasma Sintering of Duplex Stainless Steels. *Materials Today: Proceedings*, *45*, 138–144. doi:10.1016/j.matpr.2020.10.357

Simoncini, M., Costa, A., Fichera, S., & Forcellese, A. (2020). Experimental Analysis and Optimization to Maximize Ultimate Tensile Strength and Ultimate Elongation of Friction Stir Welded AA6082 Aluminum Alloy. *Metals*, *11*(1), 69. doi:10.3390/met11010069

Siroky, G., Kraker, E., Rosc, J., Kieslinger, D., Brunner, R., van der Zwaag, S., Kozeschnik, E., & Ecker, W. (2020). Analysis of Sn-Bi Solders: X-Ray Micro Computed Tomography Imaging and Microstructure Characterization in Relation to Properties and Liquid Phase Healing Potential. *Materials (Basel)*, *14*(1), 153. doi:10.3390/ma14010153 PMID:33396361

Snopiński, P. (2023, April 4). Electron Microscopy Study of Structural Defects Formed in Additively Manufactured AlSi10Mg Alloy Processed by Equal Channel Angular Pressing. *Symmetry*, *15*(4), 860. doi:10.3390/sym15040860

Sun, X., Zhang, H., Li, W., Ding, X., Wang, Y., & Vitos, L. (2019). Generalized Stacking Fault Energy of Al-Doped CrMnFeCoNi High-Entropy Alloy. *Nanomaterials (Basel, Switzerland)*, *10*(1), 59. doi:10.3390/nano10010059 PMID:31887990

Suryanarayana, C. (2001). Mechanical Alloying and Milling. *Progress in Materials Science*, *46*(1–2), 1–184. doi:10.1016/S0079-6425(99)00010-9

Suryanarayana, C., & Al-Aqeeli, N. (2013). Mechanically Alloyed Nanocomposites. *Progress in Materials Science*, *58*(4), 383–502. doi:10.1016/j.pmatsci.2012.10.001

Suryanarayana, C., & Froes, F. H. (1990). Nanocrystalline Titanium-Magnesium Alloys through Mechanical Alloying. *Journal of Materials Research*, *5*(9), 1880–1886. doi:10.1557/JMR.1990.1880

Suryanarayana, C., Ivanov, E., & Boldyrev, V. V. (2001). The Science and Technology of Mechanical Alloying. *Materials Science and Engineering A*, *304–306*, 151–158. doi:10.1016/S0921-5093(00)01465-9

Tan, J. L., & Liew, K. W. (2023, March 10). Tribological Behaviour and Microstructure of an Aluminium Alloy-Based g-SiC Hybrid Surface Composite Produced by FSP. *Lubricants (Basel, Switzerland)*, *11*(3), 124. doi:10.3390/lubricants11030124

Tang, L., Qian, H., Li, C., & Wu, X. (2022). Fretting Fatigue Test and Simulation Analysis of Steam Generator Heat Transfer Tube. *Metals, 13*(1), 67. doi:10.3390/met13010067

Technology and Challenges in Additive Manufacturing of Duplex Stainless Steels. (2021, April 26). *Biointerface Research in Applied Chemistry, 12*(1), 1110–1119. doi:10.33263/BRIAC121.11101119

Trzepieciński, T., Slota, J., Kaščák, U., Gajdoš, I., & Vojtko, M. (2023, March 14). Friction Behaviour of 6082-T6 Aluminium Alloy Sheets in a Strip Draw Tribological Test. *Materials (Basel), 16*(6), 2338. doi:10.3390/ma16062338 PMID:36984219

Verma, L., Andrew, J., Sivakumar, S. M., Balaganesan, G., Vedantam, S., & Dhakal, H. N. (2020). Ballistic Impact Behaviour of Glass/Epoxy Composite Laminates Embedded with Shape Memory Alloy (SMA) Wires. *Molecules (Basel, Switzerland), 26*(1), 138. doi:10.3390/molecules26010138 PMID:33396838

Vishwakarma, V., Kaliaraj, G. S., & Amirtharaj Mosas, K. K. (2022). Multifunctional Coatings on Implant Materials—A Systematic Review of the Current Scenario. *Coatings, 13*(1), 69. doi:10.3390/coatings13010069

Voltammetric Determination of Various Food Azo Dyes Using Different Modified Carbon Paste Electrodes. (2021). *Biointerface Research in Applied Chemistry, 12*(4), 4557–4566. doi:10.33263/BRIAC124.45574566

Wang, X., Pan, F., Sun, X., Li, Y., Zhou, J., Wang, Z., & Qin, C. (2022). Performance and Mechanism of Nanoporous Ni@NiO Composites for RhB Ultrahigh Electro-Catalytic Degradation. *Metals, 13*(1), 38. doi:10.3390/met13010038

Wang, Z., Luo, J., Kuang, W., Jin, M., Liu, G., Jin, X., & Shen, Y. (2022). Strain Rate Effect on the Thermomechanical Behavior of NiTi Shape Memory Alloys: A Literature Review. *Metals, 13*(1), 58. doi:10.3390/met13010058

Wei, Y., Niu, B., Liu, Q., Liu, Z., & Jiang, C. (2022). First-Principles Calculations of Structural and Mechanical Properties of Cu–Ni Alloys. *Crystals, 13*(1), 43. doi:10.3390/cryst13010043

Wu, J., Lin, X., Qiao, H., Zhao, J., Ding, W., & Zhu, R. (2022). Microstructural Evolution and Surface Mechanical Properties of the Titanium Alloy Ti-13Nb-13Zr Subjected to Laser Shock Processing. *Materials (Basel), 16*(1), 238. doi:10.3390/ma16010238 PMID:36614576

Wu, Z., Liu, B., Wei, J., Yang, Y., Zhang, X., & Deng, J. (2022). Design of Porous Shape Memory Alloys with Small Mechanical Hysteresis. *Crystals, 13*(1), 34. doi:10.3390/cryst13010034

Yamanaka, K., Mori, M., Onuki, Y., Sato, S., & Chiba, A. (2022). Dislocation Density of Electron Beam Powder Bed Fusion Ti–6Al–4V Alloys Determined via Time-Of-Flight Neutron Diffraction Line-Profile Analysis. *Metals*, *13*(1), 86. doi:10.3390/met13010086

Yamazaki, K., & Haga, T. (2022). High-Speed and Low-Load Twin-Roll Casting of Al–5%Mg Strip. *Metals*, *13*(1), 72. doi:10.3390/met13010072

Yan, X., Hu, J., Wang, L., Chai, Z., Sun, W., & Xu, W. (2021). The Coupled Effect of Thermal and Mechanical Stabilities of Austenite on the Wear Resistance in a 0.2C–5Mn-1.6Si Steel down to Cryogenic Temperatures. *Wear*, *486–487*, 204116. doi:10.1016/j.wear.2021.204116

Yan, X., Hu, J., Yu, H., Wang, C., & Xu, W. (2021). Unraveling the Significant Role of Retained Austenite on the Dry Sliding Wear Behavior of Medium Manganese Steel. *Wear*, *476*, 203745. doi:10.1016/j.wear.2021.203745

Yan, Y., Lu, C., Chen, Z., Zhuo, Y., Wang, C., & Jia, Q. (2023, April 17). Microstructures and High Temperature Tensile Behaviours of Laser Powder Bed Fusion Fabricated Al-Mn-Sc Alloy. *Metals*, *13*(4), 788. doi:10.3390/met13040788

Yang, G., Xue, B., Li, Z., Zhou, G., Zhang, S., Lu, N., Wen, L., & Zhang, D. (2022). Influence of Stress Field and Temperature Field on Residual Stress of 2A14 Aluminum Alloy Based on In Situ SAXS Method. *Materials (Basel)*, *16*(1), 170. doi:10.3390/ma16010170 PMID:36614509

Yang, S., Liang, P., Gao, F., Song, D., Jiang, P., Zhao, M., & Kong, N. (2022). The Comparation of Arrhenius-Type and Modified Johnson–Cook Constitutive Models at Elevated Temperature for Annealed TA31 Titanium Alloy. *Materials (Basel)*, *16*(1), 280. doi:10.3390/ma16010280 PMID:36614619

Yang, X., Zhang, X., Chen, J., Wang, Z., Li, X., Zhang, X., & Gao, L. (2022). Study of Hydrogen-Induced Plastic Damage Response of 7085-T7651 High-Strength Aluminum Alloy. *Metals*, *13*(1), 40. doi:10.3390/met13010040

Yasui, T., Hayashi, K., & Fukumoto, M. (2022). Behaviors of Micro-Arcs, Bubbles, and Coating Growth during Plasma Electrolytic Oxidation of β-Titanium Alloy. *Materials (Basel)*, *16*(1), 360. doi:10.3390/ma16010360 PMID:36614699

Yu, K., Su, J., Liu, M., Li, C., Lin, Y., Zhao, M., & Hao, C. (2023, March 17). Facile Synthesis of Al-20Si@Al2O3 Nanosheets Composite Powders and Its Refinement Performance on Primary Silicon in Al-20Si Alloy. *Crystals*, *13*(3), 514. doi:10.3390/cryst13030514

Zaara, K., Daza, J., Ben Mbarek, W., & Suñol, J.-J. (2022). Structural, Thermal and Magnetic Analysis of Two Fe-X-B (X = Nb, NiZr) Nanocrystalline Alloy. *Materials (Basel), 16*(1), 155. doi:10.3390/ma16010155 PMID:36614494

Zhang, C., Mu, A., Wang, Y., & Zhang, H. (2019). Study on Dynamic Mechanical Properties and Constitutive Model Construction of TC18 Titanium Alloy. *Metals, 10*(1), 44. doi:10.3390/met10010044

Zhang, W., Wang, D., Chen, J., Ghassemi-Armaki, H., Carlson, B., & Feng, Z. (2022). A New Perspective of Post-Weld Baking Effect on Al-Steel Resistance Spot Weld Properties through Machine Learning and Finite Element Modeling. *Journal of Manufacturing and Materials Processing, 7*(1), 6. doi:10.3390/jmmp7010006

Zhang, Y., Guo, Y., Chen, Y., Cao, Y., Qi, H., & Yang, S. (2019). Microstructure and Mechanical Properties of Al-12Si Alloys Fabricated by Ultrasonic-Assisted Laser Metal Deposition. *Materials (Basel), 13*(1), 126. doi:10.3390/ma13010126 PMID:31888072

Zhang, Y., Liu, W., Liu, Y., Zhang, M., Tian, Y., & Chen, L. (2022). Research Progress on Corrosion Behaviors and Improvement Methods of Medical Degradable Mg–Based Alloys. *Metals, 13*(1), 71. doi:10.3390/met13010071

Zhang, Z., Zhang, J., Zhao, X., Cheng, X., Jiang, S., & Zhang, Q. (2022). Effects of Al-Mg on the Microstructure and Phase Distribution of Zn-Al-Mg Coatings. *Metals, 13*(1), 46. doi:10.3390/met13010046

Zhao, D., Vdonin, N., Bezgans, Y., Radionova, L., & Glebov, L. (2022). Correlating Electrode Degradation with Weldability of Galvanized BH 220 Steel during the Electrode Failure Process of Resistance Spot Welding. *Crystals, 13*(1), 39. doi:10.3390/cryst13010039

Zhao, X., Zhong, Q., Zhai, P., Fan, P., Wu, R., Fang, J., Xiao, Y., Jiang, Y., Liu, S., & Li, W. (2022). Microstructure and Wear Resistance of Laser-Clad Ni–Cu–Mo–W–Si Coatings on a Cu–Cr–Zr Alloy. *Materials (Basel), 16*(1), 284. doi:10.3390/ma16010284 PMID:36614623

Zheng, X., Zhuang, X., Lei, Y., Chu, Z., Xu, J., Gao, L., & Sun, X. (2019). Corrosion Behavior of the Ti–6Al–4V Alloy in Sulfate-Reducing Bacteria Solution. *Coatings, 10*(1), 24. doi:10.3390/coatings10010024

Zupanič, F., Žist, S., Albu, M., Letofsky-Papst, I., Burja, J., Vončina, M., & Bončina, T. (2023, April 7). Dispersoids in Al-Mg-Si Alloy AA 6086 Modified by Sc and Y. *Materials (Basel), 16*(8), 2949. doi:10.3390/ma16082949 PMID:37109786

Chapter 5
Mechanical Alloying of Magnesium and Titanium Alloys

Sateeshkumar Kanakannavar
https://orcid.org/0000-0002-1169-1367
IIIT Bhagalpur, India

Kiran Shahapurkar
Adama Science and Technology University, Ethiopia

Gangadhar M. Kanaginahal
https://orcid.org/0000-0002-0823-6507
Jain College of Engineering and Research, India

Rayappa S. Mahale
https://orcid.org/0000-0002-0508-9871
Jain College of Engineering and Research, India

Prashant P. Kakkamari
KLS Gogte Institute of Tehnology, India

ABSTRACT

Mechanical alloying (MA) creates advanced Mg-Ti alloys via controlled microstructural changes and uniform elemental distribution. High-energy ball milling induces solid-state reactions, refining alloys. Parameters like duration, ball-to-powder ratio, and speed impact microstructure. Analytical tools reveal phase changes, microstructural evolution, and element distribution. Mg-Ti alloys exhibit enhanced mechanical properties. MA shows promise for aerospace and automotive applications, with challenges to address for full potential realization. This chapter discusses the different approaches opted for mechanical alloying of magnesium and titanium alloys.

DOI: 10.4018/978-1-6684-9385-4.ch005

INTRODUCTION TO MAGNESIUM ALLOYS

Magnesium, discovered in 1774, ranks as the sixth most abundant element on Earth, representing approximately 2% of the Earth's crust. Due to its classification in the alkaline earth metals group, magnesium is naturally found in compound forms rather than as a pure element. Notable sources of magnesium extraction include magnesite, dolomite, carnallite, and seawater. With an approximate density of 1.738 g/cm3, magnesium stands as the lightest structural metal, yet it boasts remarkable strength and stiffness. To enhance its properties, magnesium is alloyed with various elements, and these alloys are identified using an ASTM-based naming convention. In the realm of material processing, powder metallurgy plays a vital role in enhancing magnesium's mechanical characteristics and corrosion resistance. Subsequently, sintering is employed to promote uniformity and facilitate interdiffusion between particles, thereby improving the material's performance (Giridasappa et al., 2023; Guan, 2015). The realm of cast magnesium (Mg) alloys can be divided into two groups based on chemical composition: Mg- Aluminium (Al) alloys with controlled Al content and additional elements for cost-effective manufacturing, and Al-free alloys with Zn, RE, Y, and Zr for high-temperature applications. Wrought Mg alloys have heat treatable and non-heat treatable variants, limited by the hexagonal close-packed crystal structure of Mg. Advanced processing techniques offer potential for microstructure refinement. Cast Mg alloys dominate with around 90% usage. Incorporating 1-9% Al in Mg alloys enhances tensile strength, ductility, and castability below 120 °C, but beyond 2% Al, corrosion resistance improves while mechanical properties decline at higher temperatures. Adding Zinc (Zn) improves tensile properties but affects weldability. Mg-Zn based alloys exhibit strength and ductility, with research focusing on optimizing properties through alloying elements like the Mg-Zn- Zirconium (Zr) system. Mn enhances corrosion resistance but has limited strengthening effects. Adding Mn to Mg-Gd alloys increases strength but reduces ductility. Ca enhances mechanical properties and corrosion behavior in Mg alloys for biomedical applications, but excessive Ca content can render the alloy brittle. Mg-Zn- Calcium (Ca) alloys are considered for biodegradable implants. The addition of Yttrium (Y) in Mg alloys aims to enhance elevated temperature plasticity and creep resistance, while RE elements refine crystals and improve creep and corrosion resistance in Mg alloys for automobile engineering. Gd, in particular, shows high solubility in Mg, enabling solid solution and precipitation strengthening. Mg alloy systems containing large amounts of rare-earth elements result in high costs for practical applications (Cheng et al., 2016; Rajendrachari, 2023).

Researchers have developed magnesium alloys with the Mg2Si phase, exhibiting excellent mechanical properties. Al-Mg- Silicon (Si) alloys, including Mg2Si, show improved characteristics, making them attractive in automotive and aerospace

Figure 1. Classification of Mg alloys (Sharma et al., 2022)

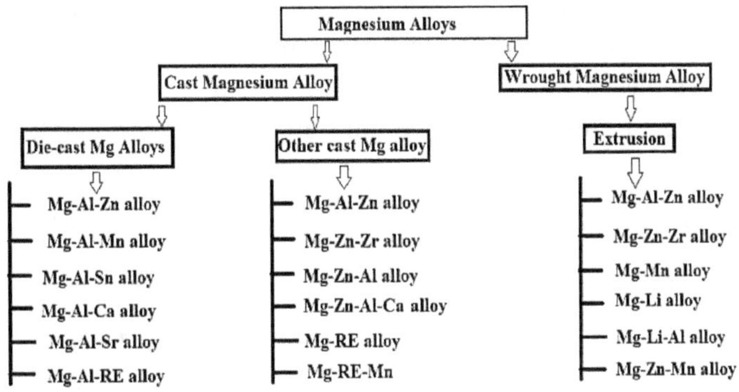

industries. The abundance and low cost of magnesium, aluminum, and silicon further enhance their appeal. Structural modifications of Mg2Si through various techniques hold promise for enhancing mechanical performance. AS series alloys, like AS21, developed by Volkswagen, demonstrate stress and creep resistance up to 150°C. These alloys feature Mg-rich solid solution, Mg17Al12, and beneficial Mg2Si phase, making them desirable for critical applications at elevated temperatures (Kanaginahal, Mahale, & Vasanth, 2023; Seth et al., 2020). A novel method using mechanical alloying (MA) to produce a biodegradable Mg-Zn alloy and investigates the influence of milling time on its properties. Prolonged milling time enhances the density and microhardness of the sintered Mg-Zn alloy. The study highlights the potential of Mg-based materials as loadbearing implants in the medical industry due to their excellent biocompatibility and biodegradability. Mechanical alloying improves the mechanical properties and corrosion resistance of Mg-based materials, emphasizing the importance of controlling milling time to achieve desired alloy properties (Keshavamurthy et al., 2023; Salleh et al., 2016). A comprehensive investigation into the impact of various mechanical alloying (MA) parameters on the composition of Zn-1Mg powder and the subsequent influence of MA preparation on Zn-6Mg and Zn-16Mg alloys. The utilization of the spark plasma sintering (SPS) technique ensures the preservation of a refined and uniform microstructure from the precursor powder, resulting in the emergence of new intermetallic phases. Notably, the compressive properties of the sintered samples are primarily governed by the concentration of alloying elements, with increased concentration leading to improved strength but reduced ductility. Among the examined alloys, the Zn-1Mg alloy demonstrates the most favorable properties. These findings provide valuable insights into the advanced preparation of zinc-magnesium alloys, employing the synergistic combination of mechanical alloying and spark plasma sintering techniques. The remarkable fine-grained and homogenous microstructure exhibited by the Zn-1Mg alloy holds great promise for enhancing the mechanical attributes of biodegradable medical devices, particularly in applications requiring exceptional compression-loaded performance. Furthermore, the high tensile strength values attained by the Zn-1Mg alloy underscore its potential suitability for engineering applications demanding robust mechanical performance (Nečas et al., 2022; Rajendrachari, Adimule, Jayaprakash et al, 2023).

Figure 2. Zn-1Mg alloy compacted by spark plasma sintering and their distribution (Nečas et al., 2022)

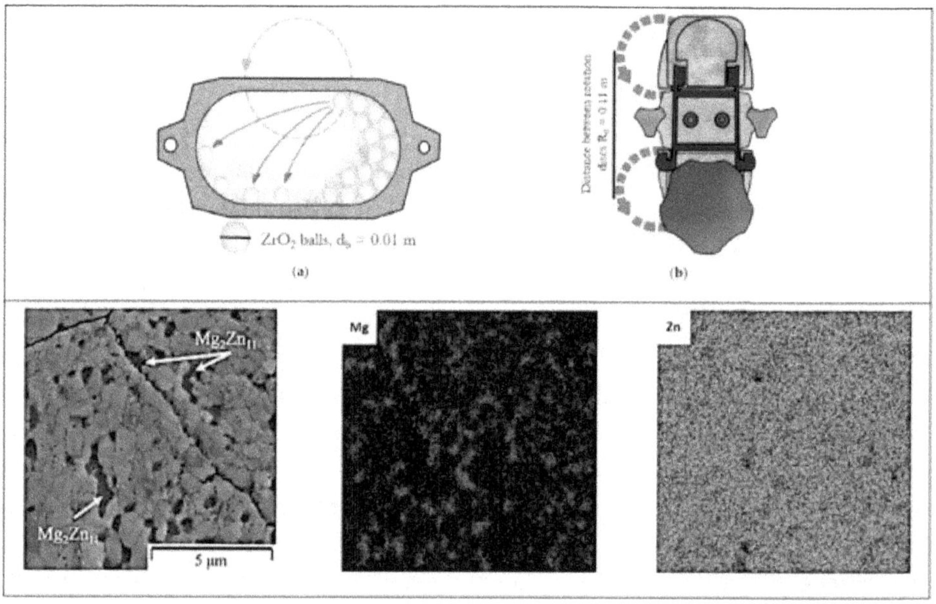

Mechanical alloying holds paramount significance in the realm of biomedical applications as it enables the fabrication of sophisticated materials with precisely tailored properties. This technique empowers the production of nanocrystalline or ultrafine-grained biomaterials, elevating their mechanical robustness, hardness, and wear resistance. Biodegradable constructs can be ingeniously devised through mechanical alloying, offering tunable degradation rates in alignment with tissue healing kinetics, thereby mitigating complications associated with implants. Integration of bioactive constituents, such as calcium and phosphorus, augments osteointegration and fosters osseous bonding. Moreover, antibacterial agents can be judiciously dispersed within the alloy, mitigating infection risks. The versatility and precise control afforded by mechanical alloying bestow paramount value in crafting biocompatible and high-performance biomaterials, propelling medical devices and implants towards enhanced patient outcomes in the biomedical domain.

An extensive investigation into the influence of various alloying elements on the mechanical attributes of Mg-based materials tailored for biomedical applications encompassed diverse binary and ternary Mg alloys, wherein ZEK100, WE43, and EW62 emerge as highly promising candidates, showcasing commendable biodegradability, biocompatibility, and reduced corrosion rates. Additionally, the study explores the implementation of surface modifications to effectively govern degradation rates, rendering lightweight Mg-based materials exceptionally suitable for biomedical implants. The noteworthy findings furnish invaluable insights for researchers and manufacturers in the biomedical industry, facilitating judicious material selection and fostering the development of cutting-edge implants meeting critical biocompatibility and mechanical prerequisites (Basavarajappa et al., 2023; Sharma et al., 2022). The work on ternary Mg-Zn-Ca alloys, with a particular emphasis on their mechanical properties, corrosion resistance, and biocompatibility results show that the MA process yields alloys with a fine microstructure, enhancing their mechanical characteristics. Additionally, the synthesized alloys exhibit improved corrosion resistance compared to pure magnesium. The biocompatibility of these alloys makes them promising candidates as temporary implants for bone regeneration applications. Notably, cytotoxicity tests reveal low adverse effects on human fibroblast cells, supporting their potential for safe clinical use (Gonzaga et al., 2023; Rajendrachari, Adimule, Gulen et al, 2022).

Figure 3. Cytotoxicity and potentiodynamic polarization results of pure and alloyed Mg materials (Gonzaga et al., 2023)

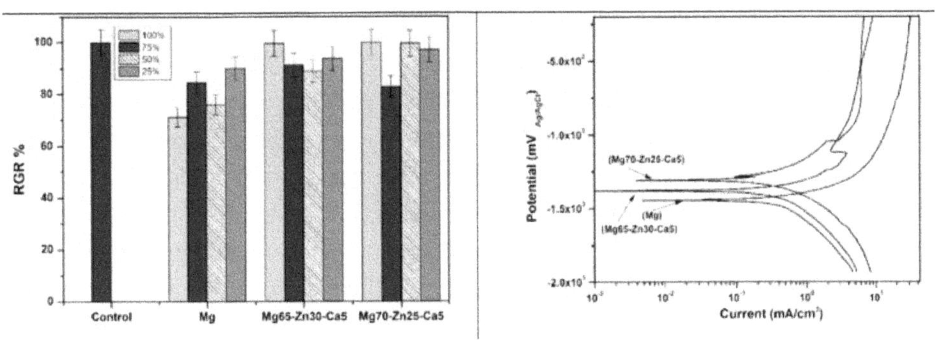

Significantly, the utilization of magnesium alloys as medical implants presents a promising prospect to counteract the limitations associated with conventional implant materials, including the potential release of toxic metallic ions and particles through corrosion, necessitating secondary surgical procedures for permanent fixture removal, resulting in escalated costs and patient morbidity.

The insights into the comprehensive investigation of the microstructure and mechanical properties of as-cast Mg-Zn-Mn-Ca alloys, specifically exploring their potential suitability as medical implants found that incorporation of Zn significantly refines the grain size of the alloys, thereby resulting in substantial enhancements in their mechanical characteristics. Notably, the main secondary phase observed in these alloys is identified as the Mg-Zn-Ca phase, with its volume fraction exhibiting an incremental trend alongside the Zn content. Furthermore, the application of solution treatment facilitates the dissolution of select secondary phases into the matrix, culminating in a marked augmentation of the alloy's plasticity. The findings of this investigation offer valuable avenues for the development of novel biodegradable magnesium alloys for medical implantation, catering to bone fractures and vascular tissue treatments due to their inherent degradability, biocompatibility, and well-suited elastic modulus (40-45GPa) (Kane et al., 2016; Ramakrishna et al., 2022). The synthesis of magnesium-based alloys utilizing mechanical alloying for prospective implant applications, particularly focusing on the Mg65-Zn30-Ca5 and Mg70-Zn25-Ca5 alloys, demonstrate heightened corrosion resistance and cytotoxicity compared to pure Mg, rendering them auspicious contenders for biomedical use. X-ray diffraction (XRD) analysis affirms the genesis of a MgZn intermetallic phase via the mechanical alloying process. The adoption of a low-energy regimen during mechanical alloying showcases its cost-effectiveness and feasibility in fabricating these alloys. Potentiodynamic polarization curves underscore the contributions of increased Zn content in engendering a protective film, thereby bolstering corrosion resistance. Cell viability tests, employing the indirect Sulforhodamine B (SRB) assay, evince the lack of cytotoxicity toward human fibroblast cells by the sintered Mg65-Zn30-Ca5 and Mg70-Zn25-Ca5 alloys (Gonzaga et al., 2023; Rajendrachari, 2022). The biodegradation behavior and mechanical properties of Mg-Sr-Y alloys intended for medical applications, specifically, the Mg-0.3Sr-0.5Y alloy, which exhibits a homogeneous degradation pattern characterized by a subdued corrosion rate and a well-balanced interplay between mechanical strength and bio-degradation properties, making it a compelling prospect for bio-applications. The study also sheds light on the osteogenic prowess of Sr, as it exerts a positive influence on bone growth while hindering bone resorption, rendering Mg-Sr-Y alloys a promising material choice for bone implants in the medical realm. The incorporation of Y in both the α-Mg phase and Mg17Sr2 phase results in a mitigated micro-galvanic corrosion effect, favoring a uniform degradation profile for the alloys. The formation of a compact composite film comprising Mg(OH)2, Y2O3, and Sr-substituted phosphate (Sr-HA) further bolsters matrix protection, ensuring consistent degradation (Li et al., 2023).

The impact of various alloying elements on these alloys, revealing the potential to enhance their mechanical characteristics shows use of powder metallurgy technique for fabricating magnesium-based alloys, with a detailed analysis of parameters and reinforcing materials that significantly influence mechanical properties. Notably, the selection of appropriate parameters and reinforcement materials emerges as a crucial factor in achieving desired mechanical attributes. The promising potential of alumina particles, silicon carbide, fullerene, and yttria as alloying elements to elevate the microhardness and mechanical performance of these alloys. As a result, magnesium-based alloys demonstrate considerable applicability in aerospace, automotive, and biomedical industries, owing to their favorable combination of lightweight and high strength properties. Researchers and engineers can utilize these insights to advance the development of lightweight, high-strength materials with diverse applications (Adimule et al., 2023; Esamael & Fatalla, n.d.).

This potentiality of Mechanical Alloying (MA) methodology to attain a refined microstructure in magnesium alloys accentuates the paramount significance of magnesium alloys in multifarious domains, encompassing automotive, aerospace, and medical applications. Augmenting the mechanical potency and corrosion resistance of magnesium alloys, the incorporation of Zn, Ca, and rare earth elements (REE) garners particular attention. Elucidating the unique applicability of Erbium (Er), in biomaterials, this research showcases how MA expedites the preparation of bespoke alloys, proving more expeditious, cost-effective, and specialized vis-à-vis conventional techniques. Delving into the interactions between erbium and magnesium alloys under varied milling durations, scrutinizing phase composition, microstructure, and hardness, the findings reveal an upsurge in hardness and a homogeneous dispersion of fine particulates with prolonged milling. Paving the way for future considerations, the Mg 65 Zn 30 Ca 4 Er 1 alloy powder emerges as a promising candidate for sintering materials (Adimule & Shashanka, 2023; Hrapkowiczk et al., 2021). Furthermore, the investigation delves into the outcomes of fabricating Al-Mg alloys through direct current and pulse electrodeposition within the AlC13-LiAlH4-benzene-tetrahydrofuran (THF) system. Notably, the pulse electrodeposition method exhibits superior efficiency and efficacy in yielding Al-Mg alloy coatings characterized by finer grain sizes and elevated magnesium contents. This breakthrough paves the way for the production of high-purity, uniform-quality nanometer alloys, signifying their potential applicability in diverse industries like aerospace, automotive, and electronics. Conversely, the mixing and mechanical alloying methods are beset by certain limitations, encompassing heightened heat consumption, metal oxidation loss, and inadvertent introduction of impurities. In stark contrast, the electrodeposition method emerges as a more promising alternative, proficiently generating high-quality Al-Mg alloy coatings (Zhu et al., 2016).

The exhaustive and comprehensive literature survey, elucidating the profound potential of biodegradable magnesium alloys as cutting-edge bone repair materials meticulously summarizes recent investigations, with a particular focus on alloy design, surface modification strategies, and the intricate biological performance of magnesium alloys. Distinguishing itself from broader reviews, this study furnishes targeted insights, catering to the specific niche of magnesium alloys for bone repair applications. The practical implications of this seminal work are substantial, heralding a promising path for the advancement of biodegradable magnesium alloys in orthopedic surgery. The inherent biocompatibility, osteoconductivity, and bone-inducing properties of these alloys elevate their standing as promising contenders for regenerative therapies (Liu et al., 2018).

The use of mechanical milling to create nanocrystalline AZ31 magnesium alloy with titanium (Ti) addition revealed that Ti addition refines the Mg matrix grains, leading to improved mechanical properties. This has practical implications for the development of nanocrystalline AZ31 Mg alloys with enhanced performance, making them potential candidates for aerospace, automotive, and biomedical applications. Mechanical milling effectively reduces crystallite size, achieving nanocrystalline dimensions for Ti additions of 18 wt% and 27 wt%. The alloy exhibits increased microhardness and yield strength, reaching 147 Hv and 293 MPa at room temperature, respectively (Zhou et al., 2016).

Fabricating an advanced Mg-based nanocomposite with exceptional strength was accomplished through the incorporation of thermally stable and coherent second phases. Employing the powder metallurgy technique, the Mg-Ni-Gd alloy demonstrated the formation of Mg5Gd, Mg2Ni, Gd2O3, and MgO phases, accompanied by the emergence of ultra-fine grains and uniform dispersion of second-phase particles in the 100-200 nm range, ultimately yielding an impressive strength of 804 MPa. The attainment of fine crystallite size following milling and the establishment of solid solution contribute to the elevated mechanical properties, resilient even under high-temperature conditions (Ravikanth Reddy et al., 2023). A novel lightweight Mg-based multicomponent alloy was fabricated, surpassing the mechanical prowess of conventional Mg and Al alloys. Employing the disintegrated melt deposition technique, followed by hot extrusion, the alloy manifests remarkable attributes of hardness, compressive strength, and ductility. This innovative multicomponent alloying strategy, incorporating Al, Cu, Mn, and Zn (each 5 at %), culminates in the integration of Al6Mn and Al2CuMg dispersed hard phases within the pliant Mg base hcp solid solution phase, culminating in extraordinary strength and ductility. The Mg 80 Al 5 Cu 5 Mn 5 Zn 5 alloy remarkably exhibits a substantial hardness of 196 HV, a significant advancement over commercially utilized alloys. It demonstrates a remarkable combination of compressive strength (616 MPa) and ductility (16.9%), outperforming conventional Mg alloys. Additionally, in its as-extruded state, the alloy showcases higher ultimate tensile strength (318 MPa) and commendable ductility (8.2%), obviating the need for supplementary heat treatment while maintaining symmetrical tensile/compression yielding behavior, underscoring its suitability for weight-critical industrial applications (Tun et al., 2019).

TITANIUM ALLOYS

Introduction

Around 1966, Mechanical alloying emerged as a development at The International Nickel Company's (INCO's) Paul D Merica Research Laboratory. This advancement was a component of a program aimed at creating an alloy that combined oxide dispersion strengthening with gamma prime precipitation hardening. The primary purpose was to design a nickel-based super-alloy suitable for use in gas turbine applications (Benjamin & Co, 1990). The INCO's extensive search to combine the high-temperature strength of oxide dispersion with the intermediate temperature strength of a gamma prime precipitate in a nickel-based superalloy served as the driving force behind the invention of mechanical alloying. Additionally, it was intended to include the essential corrosion and oxidation resistance. Modern mechanical alloying (MA) techniques have enabled sophisticated turbine blading materials like the INCONEL alloy MA 6000 to accomplish this desired result (Gilman & Benjamin, 1983; Kim & Merrick, 1979). Beyond the creation of complicated oxide dispersion-strengthened (ODS) alloys, mechanical alloying is a procedure. It is a technique for producing composite metal powders with carefully regulated, extraordinarily fine microstructures. As a result, it turns into a useful tool for creating alloys that are difficult to make or even impossible to make using conventional melting and casting methods (Gilman & Benjamin, 1983).

Over 200 years ago, in 1791, titanium was initially isolated and named by Reverend William Gregor (1761-1817), an amateur chemist and the Bishop of Creed in Cornwall, UK. In Stockholm, Sweden, Jöns Jacob Berzelius first isolated titanium metal in an impure form in 1825. Subsequently, in 1887, Nilson and Pettersson also accomplished the isolation of titanium metal. However, the first large quantity of pure titanium metal was first produced at General Electric by Mathew A. Hunter in 1910. The US Bureau of Mines plant in Albany, Oregon, showcased the first commercial production of titanium in 1946. Since then, the process has remained largely unchanged from the method developed by DuPont in 1948 to produce titanium sponge commercially (Cardarelli, 2008). Due to its remarkable qualities, titanium and titanium alloys have been widely used in industry since the early 1950s. High strength-to-weight ratios, high melting points, and exceptional corrosion resistance are a few of these, especially in chloride-containing settings (). Furthermore, titanium alloys find extensive application in chemical processing industries, encompassing heat exchangers, electrochemical reactors, desalination plants, the paper industry, and hydrocarbon processing. Moreover, these alloys are also utilized in various consumer goods, including sports equipment, jewelry, architecture, decoration, and eyeglasses (Pio et al., 2021). Since the 1950s, pure titanium has been extensively utilized as a material for surgical implants and medical instruments (). Brånemark made the ground breaking discovery that titanium could achieve permanent integration with bone, coining the term "osseointegration" to describe the stable bonding between titanium and bone tissues (Alipal et al., 2021). In addition to its exceptional corrosion resistance in biological settings, titanium exhibits non-cytotoxic properties, making it highly compatible with living tissues (Rajendrachari, Basavegowda, Vinaykumar et al, 2023; Sarraf et al., 2022).

An alloy refers to a material comprising two or more elements (which can be either metals or nonmetals) that are thoroughly blended through processes like fusion or electro-deposition. Utilizing this principle, titanium alloys are created by introducing elements like aluminum, vanadium, molybdenum, niobium, zirconium, and various others. This combination results in the formation of diverse alloys (Hamweendo & Botef, 2019; Rajendrachari, Basavegowda, Vinaykumar et al, 2023). These alloys exhibit remarkable characteristics, as demonstrated in the description below. Based on their impact on heat treating temperature and the alloying components, titanium alloys can be categorized into the subsequent three types:

1. Alpha (α) alloys
2. Alpha-beta (α-β) alloys
3. Beta (β) alloys

Alpha (α) Alloys

Unalloyed titanium, alpha alloys, and near-alpha alloys are the three main classes of alloys that include alpha titanium alloys. Under usual temperatures, unalloyed titanium and alpha alloys that contain alpha stabilisers like aluminium and tin exhibit a solitary-phase, hexagonal close-packed (hcp) crystal structure. These alloys have only moderate tensile strengths due to their monophase composition. Nevertheless, their notable thermal stability contributes to acceptable resistance to creep and toughness. Near-alpha alloys can have up to 0.5% Si added to create silicide precipitates and up to 2% beta stabilisers, which add trace amounts of the beta phase to the microstructure. These alloys, as opposed to unalloyed titanium and alpha alloys, exhibit the highest tensile strength and the best creep resistance of all titanium alloys when subjected to temperatures above 400 °C (Weiss & Semiatin, 1999). The frequently employed α-phase alloys encompass Ti-3Al-2.5V (designated as Ti-3-2.5) and Ti-5Al-2.5Sn (referred to as Ti-5-2.5) (Somashekharappa & Halappa, 2023; Williams & Boyer, 2020). High strength to weight ratio, corrosion resistance, biocompatibility, good ductility and toughness, high temperature performance, ease of fabrication, creep resistance, low thermal expansion, high fracture toughness and non-magnetic are the advantages of the alpha titanium alloys (Been et al., 1999; Tabie et al., 2020).

Alpha-Beta (α-β) Alloys

The α-β alloys stand out as the most extensively utilized titanium alloys for structural purposes. These alloys contain higher quantities of β-stabilizing elements, typically ranging between 4% and 6% (by weight), including substances like molybdenum, vanadium, tungsten, tantalum, and silicon. As a result, their β-phase content surpasses that of near-α alloys, allowing for processing methods such as heat treatment to achieve enhanced strengths. The primary mechanisms for bolstering involve maintaining some metastable β-phase at room temperature or the formation of martensite from the preceding β-phase upon rapid cooling to ambient conditions. Approximately fifty percent of all titanium alloys manufactured fall into this category. Within these alloys, aluminum serves as both an α-phase stabilizer and a hardening agent due to its capacity for reinforcing solutions. An illustrative instance of an α-β titanium alloy is Ti-6Al-4V, featuring 6% aluminum and 4% vanadium by weight (Hamweendo & Botef, 2019; Williams & Boyer, 2020). The α-β titanium alloys - display important properties such as notable creep resistance, excellent corrosion resistance, strong fatigue strength, and favourable hot formability (Elahinia et al., 2012; Kanaginahal, Suresh Hebbar, & Shahapurkar, 2023).

Beta (β) Alloys

Beta alloys are characterized as alloys that maintain an entirely beta structure when rapidly cooled from the beta phase range. These alloys include enough beta-stabilizing components to avoid martensite formation by keeping them from crossing the martensite start line (Ms). The composition threshold for what are referred to as metastable alloys is determined by the minimal content of beta stabilisers (βs) that identifies stable beta alloys (Rajendrachari, Somashekharappa, & Mahale, 2022; Weiss & Semiatin, 1998). These alloys serve as examples of the strongest titanium alloys and are referred to as "β -alloys" because they can quickly cool from the β -phase range while maintaining a totally metastable β -phase at ambient temperature. Later ageing causes the α-phase to precipitate, which then triggers reinforcing. For β -alloys, the ageing temperature is normally high enough to relieve any remaining stresses brought on by the solution treatment's quenching process. Conversely, stable beta alloys lack the capability for age hardening (Weiss & Semiatin, 1998; Williams & Boyer, 2020). These β -titanium alloys have great biological properties, including amazing characteristics like an extraordinarily high strength-to-weight ratio, a lowered elastic modulus, super-elasticity, increased elastic deflections, and less toxicity (Elahinia et al., 2012; Sidambe, 2014; Yang et al., 2014).

TITANIUM ALLOYS PRODUCTION PROCESSES

The titanium alloys can be produced from different methods:

1. Powder metallurgy
2. Atomisation
    2.1. Gas atomisation process
    2.2. Plasma atomisation process
3. Cold spray process
4. Electron beam melting
5. Selective laser melting
6. 3D printing

Powder Metallurgy

The creation, processing, and compacting of finely split metal particles to produce useful engineering components is known as powder metallurgy (PM) (Froes & Eylon, 1990; Kanaginahal et al., 2018). Powder metallurgy (PM) techniques have been used to produce a wide variety of structural parts, self-lubricating bearings, and cutting tools since the 1920s. Compared to competing metalworking methods, PM processing has advantages such as form and material flexibility, application versatility, near-net shape production capability, and recognition as an ecologically benign technology (Alexandra et al., 2017; Mahale et al., 2022).

Figure 4. Powder metallurgy process (Hamweendo et al., 2016)

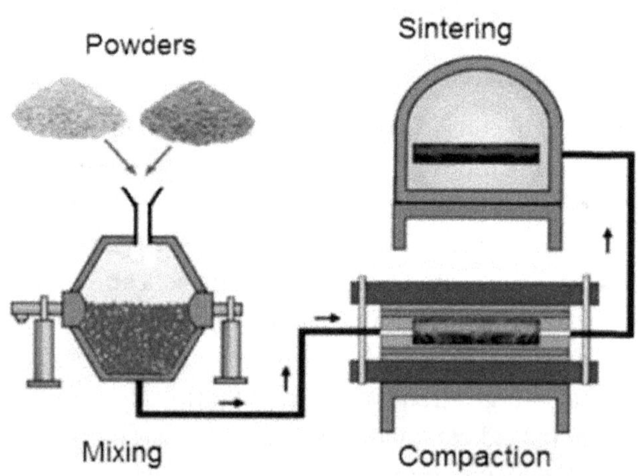

Depicted in Figure, conventional sintering is a prevalent technique within the realm of powder metallurgy (PM) for producing bone implants. In this method, feedstock powders are initially blended, then compacted under significant pressure, and ultimately subjected to sintering. The sintering step involves subjecting the materials to high temperature and pressure, inducing particle cohesion with minimal alteration to their shape. This process also enables controlled temperature regulation to facilitate the creation of porosity within the final product (Hamweendo et al., 2016; Shashanka, 2022).

Atomization

In the present market, nearly all commercially available spherical titanium powder is generated through atomization techniques or plasma spheroidization. Numerous atomization methods are accessible, with gas atomization (GA), plasma atomization (PA), and plasma rotating electrode process (PREP) being employed in commercial applications for producing spherical titanium alloy powders. All atomization procedures encompass three primary interconnected stages: melting, atomization, and solidification. The melting phase can be achieved using methodologies like vacuum induction melting, plasma arc melting, induction drip melting, or direct plasma heating. Atomization is the procedure where liquid metal is fragmented into droplets that solidify while airborne within a cooling chamber, safeguarded by inert gas. Typically, atomization involves employing high-pressure gas to disintegrate a liquid stream via a nozzle. Another approach involves the rotation of a liquid stream off a disc, leading to the creation of molten droplets that experience centrifugal force away from the center of rotation. Subsequently, these droplets solidify during their flight. Commercial atomization processes are usually conducted within ultra-high-purity inert gases like Argon (Ar) or Helium (He) to minimize the absorption of oxygen. Oxygen pickup is generally controlled to range between 100 and 500 parts per million (ppm), contingent on the specific process and powder size. While each atomization technique varies in certain aspects, they all share the fundamental three steps described previously. The subsequent section presents these distinct techniques (Kanaginahal & Murthy, 2017; Sun et al., 2017).

Gas Atomization (GA)

The Crucible Materials Corporation invented the titanium gas atomization process in the 1980s. In this process, pre-alloyed titanium alloy ingots or bars are induction skull-melted in a copper crucible that has been water-cooled while being conducted in a vacuum. The molten metal is poured into a tundish through a refractory metal nozzle once the alloy composition has become homogeneous after spending enough time in the molten state. It is then atomized by streams of high-pressure input gas (typically, Argon is preferred over Helium due to cost savings). The molten metal stream descends under the influence of gravity for a predetermined distance prior to atomization, and this process is known as free-fall gas atomization (FFGA). The supplied Figure depicts the basic design of the FFGA procedure (Sun et al., 2017).

Figure 5. Gas atomization process (Sun et al., 2017)

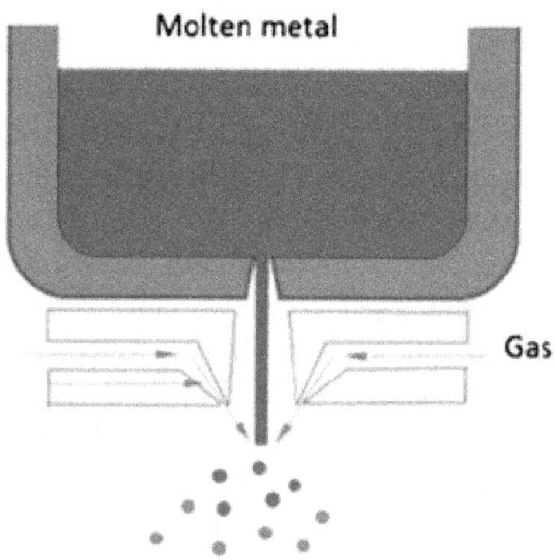

Plasma Atomization (PA)

The Plasma Atomization (PA) technique was developed in 1996 with the aim of producing finely spherical powders. Illustrated in Figure, the plasma atomization process involves introducing a pre-alloyed wire (for instance, 1/16" or 1/8") into a high-temperature region (approximately 10,000 K) that is heated by plasma torches. This causes the wire to melt and break into droplets that cool rapidly. The cooling rate typically falls within the range of 10 to 1000 degrees Celsius per second (10 - 1000 °C/s). A notable distinction between plasma atomization and gas atomization lies in the method by which the wire is handled. In PA, the wire is simultaneously melted and atomized by an extremely high-temperature plasma, whereas in GA, the metal is melted using an induction coil or another heat source before being atomized by cold, high-pressure gas. Because the liquid metal is kept separate from any refractory metals or solid materials that could contaminate the powder before it hardens, plasma-atomized titanium powder boasts exceptional purity (Mahale & Shashanka, 2022; Sun et al., 2017).

Figure 6. Plasma atomization (Sun et al., 2017)

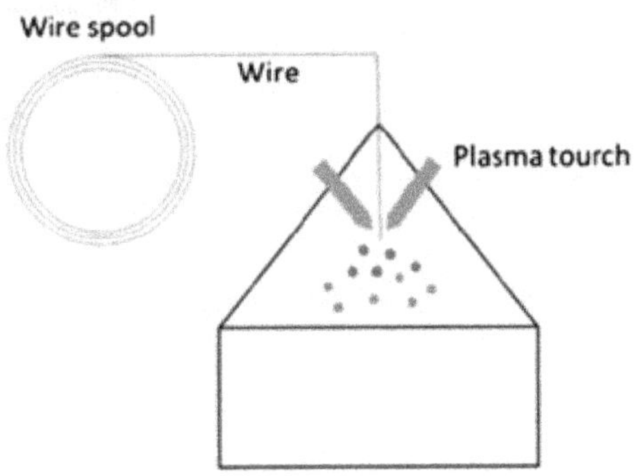

Plasma Rotating Electrode Process (PREP)

Nuclear Metals/Starmet (now a division of TIMET) invented the plasma rotating electrode process (PREP), a centrifugal atomization method for producing titanium pre-alloyed powder. This technique involves using helium plasma to liquefy the tip of a quickly spinning centerless ground bar. The molten droplets are hurled away and solidify in the air, surrounded by helium. Figure 2.6 shows the procedure schematically. A stainless steel chamber with a diameter of 2.4 metres (96 inches) and positive helium pressure is used for the melting and atomization procedures. Pre-alloyed bars with a diameter of around 64 mm (2.5 inches) are used as the initial electrodes. These bars can rotate at up to 15,000 revolutions per minute. The resulting PREP powder has a spherical shape and superior flow and packing characteristics. Depending on the particular alloy, electrode diameter, and rotational speed, the particle size distribution changes. For instance, Ti-6Al-4V powder typically has a size range of 100 to 300 m, with a median particle size (d50) of roughly 175 m. Less than 1% of the powder produced during this manufacturing method has a particle size less than 50 m (Yolton & Froes, 2015).

Cold Spray Process

The most recent innovation in surface coating techniques is cold spray (CS) process. It involves coating metallic or dielectric substrates with metal, alloy, or composite

Figure 7. Plasma rotating electrode process (Sun et al., 2017)

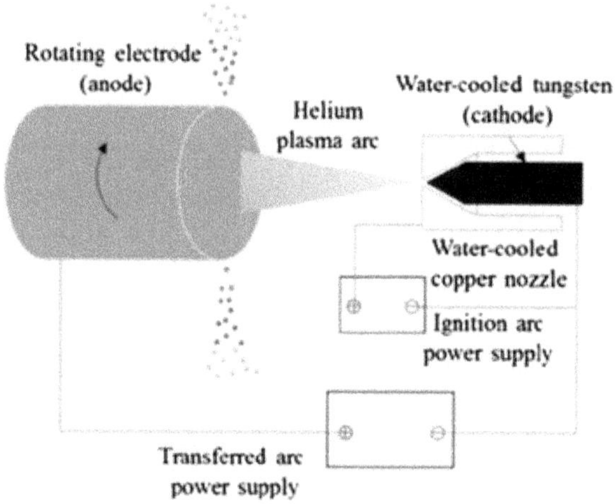

powders. This is accomplished by launching a high-velocity jet of small particles—sized between 5 m and 50 m at speeds between 300 and 1200 m/s through a specially designed nozzle into a stream of compressed and preheated gas. A classic cold-spray system is schematically shown in the Figure . Initially, the compressed gas is heated by a gas heater; then powder is added to the gas stream; and the gas-powder mixture is accelerated through a nozzle of the de Laval type and applied to a substrate. To improve the core system, additional machinery for in-process tasks like milling, laser treatment, and helium recycling has been developed. Gas conditions, nozzle location and movement, and powder feed rates are the key process variables. It is possible to create compact coatings with deposition efficiency greater than 95% (for example, for CP-Ti) when these parameters are fine-tuned. This deposition technique produces coatings that have a number of benefits while operating at temperatures that are much below the melting point of the majority of feedstock powders. These benefits include low oxidation, outstanding repeatability, great flexibility, rapid deposition rates, and cost savings (Hamweendo et al., 2016; Vo et al., 2015).

Electron Beam Melting

One method for producing titanium coatings using additive manufacturing is electron beam melting (EBM), as seen in Figure . This process includes using an electron beam that is magnetically directed to melt and deposit metal particles one layer at a time. Although this method has been successful, it has higher initial setup costs because a high vacuum environment is required (Hamweendo & Botef, 2019).

Figure 8. Schematic of cold spray process (Hamweendo et al., 2016)

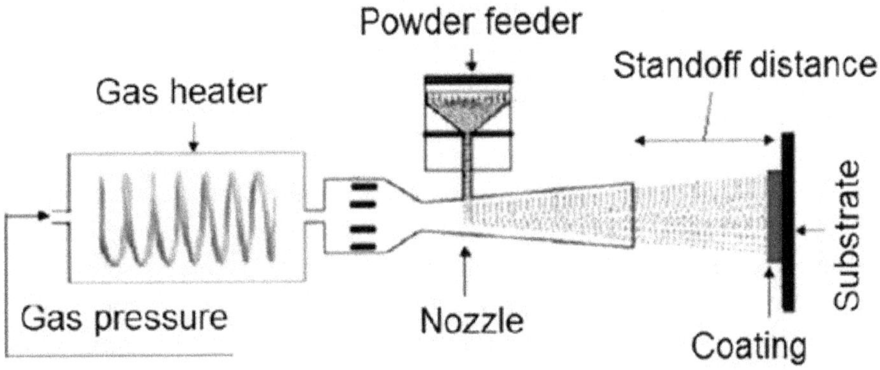

Selective Laser Melting

Another additive manufacturing method used to produce titanium alloy coatings is selective laser melting (SLM), as seen in Figure . In this procedure, a strong laser beam is used to completely melt the powder. Due to the requirement for complex quick cooling systems. Furthermore, the quality of the finished products may suffer

Figure 9. Schematic of selective laser melting (Hamweendo et al., 2016)

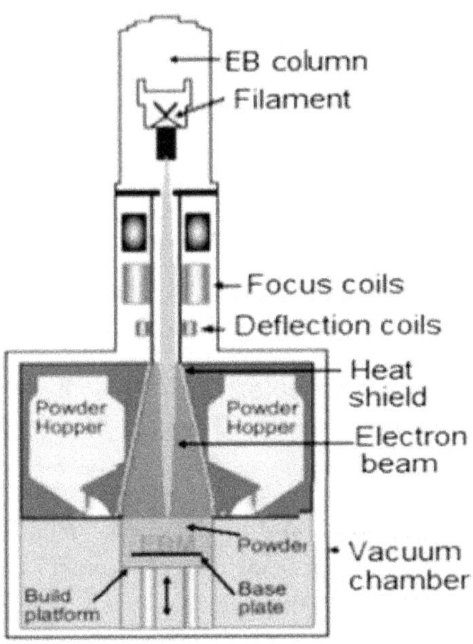

as a result of temperature changes that occur during processing (Hamweendo & Botef, 2019; Hamweendo et al., 2016).

CONCLUSION

The investigation into mechanical alloying of magnesium and titanium alloys highlights its potential in producing advanced lightweight materials with enhanced mechanical properties. High-energy ball milling induces solid-state reactions, resulting in refined microstructures and improved mechanical attributes. The optimization of milling parameters, such as duration, ball-to-powder ratio, and rotational velocity, plays a pivotal role in achieving desirable alloy properties. Analytical techniques like X-ray diffraction, scanning electron microscopy, and energy-dispersive X-ray spectroscopy provide insights into phase changes, microstructural evolution, and elemental distribution. Mechanical testing confirms the superior performance of the synthesized Mg-Ti alloys, positioning them for applications in aerospace and automotive sectors. Overcoming challenges of contamination and phase stability during milling remains a focus for further advancement. This study underscores the potential of mechanical alloying in tailoring materials for specific applications through controlled microstructural refinement and property enhancement.

REFERENCES

Adimule, V., Keri, R., Rajendrachari, S., Kendrekar, P., & Kulkarni, C. V. (2023). Highly photoluminescence and wide band gap insulating metal hybrid nanoparticles array of samarium-doped SrO:CoO: synthesis, characterizations and sensor characteristics. *Journal of Materials Science Materials in Electronics*, *34*(5), 34. doi:10.1007/s10854-023-09899-2

Adimule, V., & Shashanka, R. (2023). Dielectric, Supercapacitive, Photoluminescent Properties of Hybrid Metal Organic Frameworks. Academic Press.

Alexandra, B., Hidalgo, A., & Frykholm, R. (2017). *Powder Metallurgy Strategies to Improve Properties and Processing of Titanium Alloys: A Review*. doi:10.1002/adem.201600743

Alipal, J., Lee, T. C., Koshy, P., Abdullah, H. Z., & Idris, M. I. (2021). Evolution of anodised titanium for implant applications. *Heliyon*, *7*(7), e07408. doi:10.1016/j.heliyon.2021.e07408 PMID:34296002

Balazic, M., Kopac, J., Jackson, M. J., & Ahmed, W. (2007). Review: Titanium and titanium alloy applications in medicine. *Int J Nano Biomater*, *1*(1), 3–34. doi:10.1504/IJNBM.2007.016517

Basavarajappa, P. N., Hegde, M. M. R., Rajendrachari, S., & Surendranathan, A. O. (2023). Investigation of Structural and Mechanical Properties of Nanostructured TiMgSr Alloy for Biomedical applications. *Biointerface Research in Applied Chemistry*, *13*, 1–12.

Been, J., Faller, K., & Eli, T. (1999). Using Ti-5111 for Marine Fastener Applications. *Journal of the Minerals Metals & Materials Society*, *51*(6), 21–24. doi:10.1007/s11837-999-0088-5

Benjamin, J. S., & Co, A. (1990). Mechanical Alloying - A Perspective. *Metal Powder Report*, *45*(2), 122–127. doi:10.1016/S0026-0657(10)80124-9

Cardarelli, F. (2008). *Less Common Nonferrous Metals*. Springer.

Cheng, Y.-L., Lee, C.-Y., & Huang, Y.-L. (2016). Alloying Elements of Magnesium Alloys: A Literature Review. Intech.

Elahinia, M. H., Hashemi, M., Tabesh, M., & Bhaduri, S. B. (2012). Progress in Materials Science Manufacturing and processing of NiTi implants : A review. *Progress in Materials Science*, *57*(5), 911–946. doi:10.1016/j.pmatsci.2011.11.001

Esamael, S.K., & Fatalla, A.A. (n.d.). *Analysis of Effect of Alloying Magnesium by Powder Metallurgy on Mechanical Properties: Review*. Academic Press.

Eylon, D., Fujishiro, S., Postans, P. J., & Froes, F. H. (1984). High-Temperature Titanium Alloys—A Review. *JOM*, *36*(11), 55–62. doi:10.1007/BF03338617

Froes, F. H., & Eylon, D. (1990). Powder -metallurgy of titanium alloys. *International Materials Reviews*, *35*(1), 162–184. doi:10.1179/095066090790323984

Geetha, M., Singh, A. K., Asokamani, R., & Gogia, A. K. (2009). Ti based biomaterials, the ultimate choice for orthopaedic implants – A review. *Progress in Materials Science*, *54*(3), 397–425. doi:10.1016/j.pmatsci.2008.06.004

Gilman, P. S., & Benjamin, J. S. (1983). Mechanical Alloying. *Annual Review of Materials Science*, *13*(1), 279–300. doi:10.1146/annurev.ms.13.080183.001431

Giridasappa, A., Kiran, K. K., & Rajendrachari, S. (2023). Nanotechnology and their Evaluation of Bi-Functional Applications. *J Metastable Nanocrystalline Mater*, *37*, 33–43. doi:10.4028/p-dKORv2

Gonzaga, S., Molina, A., Guardián, R., Martínez, H., Vélez, E. V., & Tapia, J. S.-O. (2023). Synthesis of Magnesium-Based Alloys by Mechanical Alloying for Implant Applications. *Coatings*, *13*(2), 1–15. doi:10.3390/coatings13020260

Guan, D. (2015). *Processing, microstructure and mechanical behavior of bulk nanostructured Mg alloy*. Academic Press.

Hamweendo, A., & Botef, I. (2019). Modern Production Methods for Titanium Alloys: A review. Titanium Alloys-Novel Aspects of Their Manufacturing and Processing, 1–17.

Hamweendo, A., Malama, T., & Botef, I. (2016). Titanium-Nickel Alloys for Bone Tissue Engineering Application via Cold Titanium-Nickel Alloys for Bone Tissue Engineering Application. *International Conference on Competitive Manufacturing COMA16*, 273–279.

Hrapkowiczk, B., Lesz, S., & Kremzer, M. (2021). Mechanical alloying of Mg-Zn-Ca-Er alloy. *Bulletin of the Polish Academy of Sciences. Technical Sciences*, *69*, 1–8. doi:10.24425/bpasts.2021.137587

Kanaginahal, G. M., Kumar Muniraju, A., & Madhav Murthy, M. (2018). Coatings for Enhancement of Properties of Polymer Matrix Composites: A review. *Materials Today: Proceedings*, *5*(1), 2462–2465. doi:10.1016/j.matpr.2017.11.026

Kanaginahal, G. M., Mahale, R. S., & Vasanth, S. (2023) Studies on Dielectric, Super-Capacitive and Photoluminescence Properties of Metal Organic Frameworks Subjected to Varying Temperature and Frequencies. *Nano Hybrids Compos, 39*. https://doi.org/https://doi.org/10.4028/p-si6964

Kanaginahal, G. M., & Murthy, M. (2017). Microstructural Study and Evaluation of Few Mechanical Properties of Hybrid Composites. *Adv Mater Manuf Charact, 7*, 7–14. https://doi.org/http://dx.doi.org/10.11127/ijammc2017.04.02C

Kanaginahal, G. M., Suresh Hebbar, H., & Shahapurkar, K. (2023). Leverage of Weave Pattern and Composite Thickness on Dynamic Mechanical Analysis, Water Absorption and Flammability Response of Bamboo Fabric/Epoxy Composites. *Heliyon*, *9*(1), 1–13. doi:10.1016/j.heliyon.2023.e12950 PMID:36820186

Kane, S. N., Mishra, A., & Dutta, A. K. (2016). Microstructure and mechanical properties of the as-cast Mg-Zn-Mn-Ca alloys. *Journal of Physics: Conference Series*, *755*. Advance online publication. doi:10.1088/1742-6596/755/1/011001

Kim, Y.G., & Merrick, H.F. (1979). *Characterization of an oxide dispersion strengthened superalloy, MA-6000E, for turbine blade applications*. Academic Press.

Li, Y., Yuan, Y., & Wang, J. (2023). Controllable degradation behavior of Mg-Sr-Y alloys for the bio-applications. *NPJ Mater Degrad, 7*. doi:10.1038/s41529-023-00366-1

Liu, C., Ren, Z., Xu, Y., Pang, S., Zhao, X., & Zhao, Y. (2018). Biodegradable Magnesium Alloys Developed as Bone Repair Materials: A Review. *Scanning, 2018*, 1–15. Advance online publication. doi:10.1155/2018/9216314 PMID:29725492

Mahale, R. S., Kanaginahal, G. M., & Vasanth, S. (2023). Applications of Fused Deposition Modeling in Dentistry. In R. Keshavamurthy, V. Tambrallimath, & J. P. Davim (Eds.), *Development, Properties, and Industrial Applications of 3D Printed Polymer Composites* (pp. 211–219). IGI Global. doi:10.4018/978-1-6684-6009-2.ch012

Mahale, R. S., & Shashanka, R. (2022). Mechanical testing of spark plasma sintered materials: A review. AIP Conference Proceedings, 1–10. doi:10.1063/5.0080183

Mahale, R. S., Vasanth, S., & Krishna, H. (2022). Electrochemical sensor applications of nanoparticle modified carbon paste electrodes to detect various neurotransmitters: A review. *Applied Mechanics and Materials*, *908*, 908. doi:10.4028/p-mizm85

Nečas, D., Marek, I., Pinc, J., Vojtěch, D., & Kubásek, J. (2022). Advanced Zinc–Magnesium Alloys Prepared by Mechanical Alloying and Spark Plasma Sintering. *Materials (Basel)*, *15*(15), 5272. Advance online publication. doi:10.3390/ma15155272 PMID:35955207

Peters, B. M., Kumpfert, J., Ward, C. H., & Leyens, C. (2003). Titanium Alloys for Aerospace Applications. *Advanced Engineering Materials*, *5*(6), 419–427. doi:10.1002/adem.200310095

Pio, E., Medina, A., & Mart, C. (2021). Evolution of Face-Centered Cubic Ti Alloys Transformation by X-ray Diffraction Profile Analysis in Mechanical Alloying. *Metals*, *11*(11), 1–17. doi:10.3390/met11111841

Rajendrachari, S. (2022). An Overview of High-Entropy Alloys Prepared by Mechanical Alloying Followed by the Characterization of Their Microstructure and Various Properties. *Alloys*, *1*(2), 116–132. doi:10.3390/alloys1020008

Rajendrachari, S. (2023). New Advances in Powder Technology. Academic Press.

Rajendrachari, S., Adimule, V., Gulen, M., Khosravi, F., & Somashekharappa, K. K. (2022). Synthesis and Characterization of High Entropy Alloy 23Fe-21Cr-18Ni-20Ti-18Mn for Electrochemical Sensor Applications. *Materials (Basel)*, *15*(21), 1–14. doi:10.3390/ma15217591 PMID:36363181

Rajendrachari, S., Adimule, V. M., Jayaprakash, G. K., & Pandith, A. (2023). Electrochemical oxidation of methylene blue dye in wastewater using mechanically alloyed high entropy alloy modified carbon paste electrode using cyclic voltammetry. *Materials Research Express*, *10*(5), 1–13. doi:10.1088/2053-1591/acd3d7

Rajendrachari, S., Basavegowda, N., Vinaykumar, R., Narsimhachary, D., Somu, P., & Lee, M.-J. (2023). Electrocatalytic determination of methyl orange dye using mechanically alloyed novel metallic glass modified carbon paste electrode by cyclic voltammetry. *Inorganic Chemistry Communications*, *155*, 1–10. doi:10.1016/j.inoche.2023.111010

Rajendrachari, S., Somashekharappa, K. K., & Mahale, R. S. (2022). A Review on Cyclic Voltammetric Investigation of Toxic Heavy Metals. Frontiers in Voltammetry, 1–10.

Ramakrishna, D., Saravu, S., & Rajendrachari, S. (2022). A Palladium Complex Dispersed in Ionic Liquid as an Efficient Catalytic Combination for the Synthesis of Benzazoles. *Topics in Catalysis*, *65*, 1–12. doi:10.1007/s11244-021-01554-6

Ravikanth Reddy, C., Satya Prasad, K., & Srinivasarao, B. (2023). Microstructure and mechanical properties of Mg–Ni–Gd alloy synthesised by powder metallurgy. *Powder Metallurgy*, *66*(3), 270–277. Advance online publication. doi:10.1080/00325899.2022.2163553

Salleh, E. M., Ramakrishnan, S., & Hussain, Z. (2016). Synthesis of Biodegradable Mg-Zn Alloy by Mechanical Alloying: Effect of Milling Time. *Procedia Chemistry*, *19*, 525–530. doi:10.1016/j.proche.2016.03.048

Sarraf, M., Rezvani, E., & Saeid, G. (2022). A state - of - the - art review of the fabrication and characteristics of titanium and its alloys for biomedical applications. *Bio-Design and Manufacturing*, *5*(2), 371–395. doi:10.1007/s42242-021-00170-3 PMID:34721937

Seth, P. P., Parkash, O., & Kumar, D. (2020). Structure and mechanical behavior ofin situdeveloped Mg2Si phase in magnesium and aluminum alloys - a review. *RSC Advances*, *10*(61), 37327–37345. doi:10.1039/D0RA02744H PMID:35521263

Sharma, S. K., Saxena, K. K., Malik, V., Mohammed, K. A., Prakash, C., Buddhi, D., & Dixit, S. (2022). Significance of Alloying Elements on the Mechanical Characteristics of Mg-Based Materials for Biomedical Applications. *Crystals*, *12*(8), 1138. Advance online publication. doi:10.3390/cryst12081138

Shashanka, R. (2022). Carbon composite voltammetric sensors for food quality assessment. Electrochemical Sensors Based on Carbon Composite Materials: Fabrication, properties and applications, 1–13. doi:10.1088/978-0-7503-5127-0ch8

Sidambe, A. T. (2014). Biocompatibility of Advanced Manufactured Titanium Implants—A Review. *Materials (Basel)*, *7*(12), 8168–8188. doi:10.3390/ma7128168 PMID:28788296

Somashekharappa, K. K., & Halappa, R. B. Sr. (2023). Green Chemistry Applications in Electrochemical Sensors. In *Recent Developments in Green Electrochemical Sensors* (pp. 23–37). Design, Performance, and Applications.

Sun, P., Fang, Z. Z., Zhang, Y., & Xia, Y. (2017). Review of the Methods for Production of Spherical Ti and Ti Alloy Powder. *Journal of the Minerals Metals & Materials Society*, *69*(10), 1853–1860. doi:10.1007/s11837-017-2513-5

Tabie, V. M., Li, C., Saifu, W., Li, J., & Xu, X. (2020). Mechanical properties of near alpha titanium alloys for high-temperature applications - a review. *Aircraft Engineering and Aerospace Technology*, *92*(4), 521–540. doi:10.1108/AEAT-04-2019-0086

Tun, K. S., Kumar, A., & Gupta, M. (2019). Introducing a High Performance Mg-Based Multicomponent Alloy as an Alternative to Al-Alloys. *Frontiers in Materials*, *6*, 1–6. doi:10.3389/fmats.2019.00215

Vo, P., Goldbaum, D., & Wong, W. (2015). *Cold-spray processing of titanium and titanium alloys*. Elsevier Inc. doi:10.1016/B978-0-12-800054-0.00022-8

Weiss, I., & Semiatin, S. L. (1998). Thermomechanical processing of beta titanium alloys — An overview. *Materials Science and Engineering A*, *243*(1-2), 46–65. doi:10.1016/S0921-5093(97)00783-1

Weiss, I., & Semiatin, S. L. (1999). Thermomechanical processing of alpha titanium alloys — An overview. *Materials Science and Engineering A*, *263*(2), 243–256. doi:10.1016/S0921-5093(98)01155-1

Williams, J. C., & Boyer, R. R. (2020). Opportunities and Issues in the Application of Titanium Alloys for Aerospace Components. *Metals*, *10*(6), 705. doi:10.3390/met10060705

Yang, R., Hao, Y., & Li, S. (2014). Development and Application of Low-Modulus Biomedical Titanium Alloy Ti2448. Biomedical Engineering: Trends in Materials Science, 1–25.

Yolton, C. F., & Froes, F. H. S. (2015). *Conventional titanium powder production*. Elsevier Inc. doi:10.1016/B978-0-12-800054-0.00002-2

Zhang, L., & Chen, L. (2019). A Review on Biomedical Titanium Alloys : Recent Progress and Prospect. *Advanced Engineering Materials*, *21*(4), 1–29. doi:10.1002/adem.201801215

Zhou, H., Hu, L., Sun, Y., Zhang, H., Duan, C., & Yu, H. (2016). Synthesis of nanocrystalline AZ31 magnesium alloy with titanium addition by mechanical milling. *Materials Characterization*, *113*, 108–116. doi:10.1016/j.matchar.2016.01.014

Zhou, Y., Niinomi, M., Akahori, T., Nakai, M., & Fukui, H. (2007). Comparison of Various Properties between Titanium-Tantalum Alloy and Pure Titanium for Biomedical Applications. *Materials Transactions*, *48*(3), 380–384. doi:10.2320/matertrans.48.380

Zhou, Y. L., Niinomi, M., & Akahori, T. (2004). Effects of Ta content on Young's modulus and tensile properties of binary Ti – Ta alloys for biomedical applications. *Materials Science and Engineering A*, *371*(1-2), 283–290. doi:10.1016/j.msea.2003.12.011

Zhu, S-S., Kan, H-M., & Feng, X-J. (2016). *The Preparation Method of Aluminum-magnesium Alloys*. doi:10.2991/ame-16.2016.50

Chapter 6
Processing and Applications of Shape Memory Alloys

Sudipto Datta
https://orcid.org/0000-0003-2161-6878
Indian Institute of Science, Bangalore, India

Nipun Jain
https://orcid.org/0009-0008-8270-9449
Indian Institute of Science, Bengaluru, India

Yusif Olatunji Waidi
https://orcid.org/0000-0002-5728-1769
Indian Institute of Science, Bangalore, India

Ranjit Barua
https://orcid.org/0000-0003-2236-3876
Indian Institute of Engineering Science and Technology, India

ABSTRACT

SMA (shape memory alloys) materials are a kind of smart material that undergoes shape changes with respect to temperature variations. In this chapter, the authors focus on the microstructure, classifications, fabrication, and future applications of these materials in brief. They discuss the four main types of SMA processing techniques, which include mechanical, thermomechanical, powder metallurgy, and melting route. Finally, its applications in various fields like transportation, civil structures, robotics, aerospace, automation, and medical fields will be briefly discussed.

DOI: 10.4018/978-1-6684-9385-4.ch006

1. INTRODUCTION

Shape Memory Alloys (SMAs) are a special material that can regain shape when the temperature rises (Hu et al., 2005). SMAs are also widely referred to as memory alloy, memory metal, muscle wire, and smart metal. These one-of-a-kind materials continued to enhance the achievement in accordance with the demand for the engineering sector in general; form Because of qualities such as increased strength, higher strain recovery, lightweight, high stability, and so on, memory alloys are used in heavy engineering (Zhao et al., 2019). Ni-Ti is the most often used SMAs (Liu et al., 2014). Ag-Cd, In-Th, and Cu-based alloys are some further examples. The sophisticated material called form memory alloy exhibits a shape-memory effect and exceptional elasticity. These qualities distinguish them from other materials (Li et al., 2016). The shape memory effect is concerned with recovering a distorted material to its previous shape during heating; in other words, the material's capacity to restore its shape under thermal load. This feature allows them to restore their original shape after being distorted by heating to a crucial temperature (Velvaluri et al., 2021). During a mechanical loading and unloading cycle, super elasticity deals with non-linear recoverable stresses (Kim et al., 2021). Figure 1 shows the shape memory effect's microscopic phenomenology, and Figure 2 shows the pseudo elastic effect connected with microscopic phenomena.

Figure 1. Phenomenology of the microscopic shape memory effect

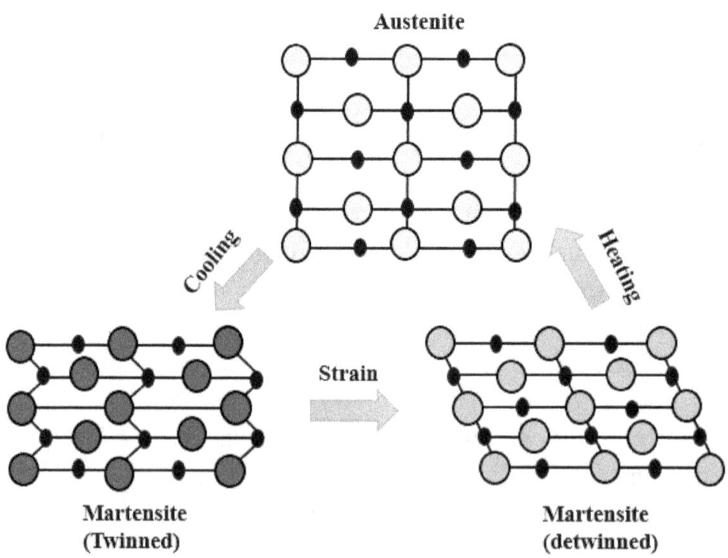

Figure 2. Pseudo elastic effect connected with microscopic phenomena

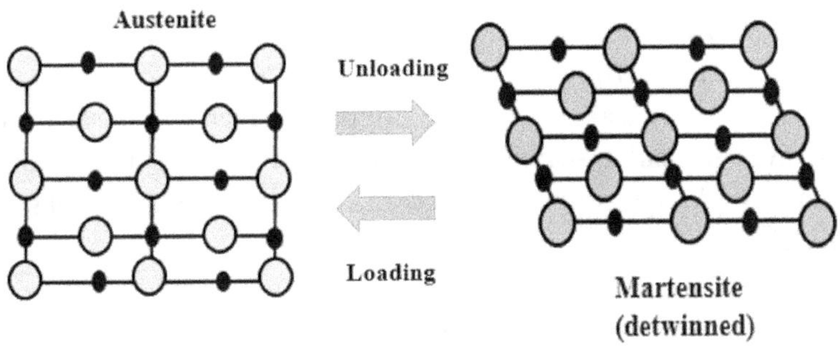

Shape memory effect-wise, SMAs are classified as one-way and two-way shape memory alloys. Previous demonstrations show that the metal may be bent while unconscious. This will keep the form until the heating level exceeds what is known as the changeover temperature. The materials in the latter exhibit one of their qualities while cold and another when heated (Li et al., 2016) (Kim et al., 2021). Due to the SMAs materials availability having improved characteristics and low manufacturing cost, the use has grown in recent years. Therefore, it has recently prompted an expansion in many technical domains, new economic prospects establishment and research disciplines (Zhang et al., 2017). Among the various uses are civil engineering, medical devices, mechanical engineering, aerospace, and military.

Difference between SMAs and conventional alloys

The martensite to austenite transition is solely affected by temperature and stress (Shashanka et al., 2014). It is not affected by time since diffusion is not present. Steel lacks shape memory because martensite may be produced by the fast-cooling carbon steel from austenite, an irreversible process. In addition, ordinary steel has a higher yield strength than SMAs; however, some are superior to aluminium or plastic. The amount of recoverable plastic strain that can be created via SMAs makes them stand out (Mohd Jani et al., 2014). Some SMAs can bear the highest recoverable strain of 8% without permanent deformation, which is much greater than the maximum strain in typical steels. Stainless steel has a greater density, thermal conductivity, heat capacity, melting temperature, and vaporisation temperature than Nitinol. Both materials have the same latent heat of melting and evaporation.

2. PROPERTIES OF SMA

As previously stated, Shape Memory Alloys have distinct characteristics prior to the Martensitic and Austenitic transition. Due to this change, the SMAs material's mechanical macroscopic behaviour may be divided into two groups, as shown below.

1) In the “shape memory effect (SME)” form, a shape memory alloy specimen suffers significant residual strain after loading and unloading. The alloy may entirely recover this remaining deformation after raising the temperature (Exarchos et al., 2018).
2) For the “pseudo elastic effect (PEE)” form at which SMAs samples display a massive substantial (seemingly plastic) deformation upon load application, which may subsequently be recovered via a hysteretic loop unloading (Exarchos et al., 2018).

Beginning with a zero-stress state, the alloy material is stressed. The behaviour is initially linear because of 100% elasticity in the austenite phase, with EA equal to Young’s modulus. Figures 1 and 2 show the microscopic phenomenology of the SME. It explains the impact of a material’s loading and unloading. Then, the phase change (from austenite to martensite) commences at the start stress of martensite (σM_s). The new phase has a linear elastic range, just like the original phase. However, the young’s modulus of the new phase is different, and the yielding limit is also different. The transition from the original phase to the new phase finishes when the alloy is 100% martensite, and this happens at the martensite finish stress (σM_f). The unloading process commences (usually at T (temperature) > σA_f). Until austenite begins to stress, the same linear martensite elastic unloading behaviour is followed, hence signifies the achievement of A_s. The applied load reduces below the stress level. The phase transitions prevail until the austenite finish stress reaches 100% austenite σ_{Af}. In a linear elastic pattern, further stress reduction returns to the beginning condition. During the process, a similar tendency to the preceding situation is seen. The zero-stress criterion is achieved in this situation due to alloy residual strain (res). The material regains shape and strain only when the heat T > A_f.

3. CLASSIFICATION OF SMA

Shape memory alloys (SMAs) have been classed in various ways in the past years. It is classed based on its qualities, such as form memory effect and superelasticity, and the material from which it is manufactured, such as Ni-Ti SMAs, Iron-based SMAs, and Copper-based SMAs. SMAs are categorised into three categories.

3.1. Alloys Caused by Temperature

Temperature-induced shape memory alloys have a form memory effect, and there are two types of shape memory alloys, namely one-way SMA and two-way SMA. One-way SMAs are materials which exhibit SME by reverting to their original shape after heating. In contrast, two-way form SMAs are materials which exhibit SME during both heating and cooling (Xue et al., 2016) (Ivošević et al., 2019).

After imposing a distorted shape and thermal activation, the initial configuration is obtained in one-way SMA. The material in two-way SMAs remembers two forms that can be restored at various temperatures regardless of the deformation. SMAs with lesser mechanical qualities produce a two-way shape memory effect after significant loading and unloading. (Frensemeier et al., 2015)

3.2. Alloys with Ferro Magnetic Shape Memory

Ferro Magnetic SMAs are SMAs which exhibit SME upon magnetic field subjection. Magnetic SMAs are another name for them (MSMAs) (Hînganu et al., 2018). Magnetoelastic deformation is the distortion caused by a magnetic field. Ferromagnetic SMAs include Ni2MnGa, Fe-Pd, and Fe3Pt. MSMAs are ferromagnetic substances that may generate forces and motion upon moderate magnetic field subjection. Manganese, Gallium, and Nickel are often utilised (Martins et al., 2016). Magnetic shape memory (MSM) technology is extremely appealing for constructing new actuators for application in robotics, pneumatics, mechatronics, and medical devices because of the substantial magnetically generated strain and rapid reaction times. MSMAs' magnetic characteristics vary based on deformation. This co-existing companion effect can aid in the design of speed, force sensors, mechanical energy harvesters, and displacement. The magnetic SME results from the specific geometry of the elementary tetragonal cells in the martensite phase. When the temperature is raised, and the alloy enters the austenite phase, the geometry of the elementary tetragonal cells changes to cubic, and the SME is lost. The martensite transition to austenite causes deformation and force (Mekki et al., 2011). As a result, MSMAs, like thermal shape memory alloys, may be activated thermally. The magnetic SME is a significant phenomenon in ferromagnetic shape memory alloys.

3.3. Magnetic Shape Memory Effect (MSME)

Magnetically induced reorientation (MIR) is responsible for the significant strain of MSMAs alloys, as seen in the image. Upon magnetic application externally, MSM alloys display magnetization macroscopically, such as other ferromagnetic substances, resulting from the elementary magnetizations alignment along the field heading. Unlike in normal ferromagnetic substances, however, the geometric rotation of the elementary cells that comprise the alloy rather than the magnetisation vectors in the cells (such as in magnetostriction) causes the alignment (Ragahavan et al., 2010) A similar effect happens once the magnetic alloy is exposed to force from an external source. The force operates as a magnetic field at the macroscopic level, favouring the elementary cell's rotation and accomplishing contraction or elongation depending on how it is used inside the source coordinate system. The figure illustrates the contraction or elongation, with elongation accomplished magnetically and contraction accomplished mechanically, respectively. The cells rotate due to the high mobility of the interior areas and the substantial magnetic anisotropy of MSM alloys. Simply put, the MSM framework comprises internal areas with varying orientations of the constituent cells (the figure depicts the areas in blue and green). The areas are known as twin variations. Applying external tension or magnetic field alters the dual limits of the variations, favouring either one or both. Once materials contract or extend completely, only one version forms it in a single variation condition (Chopade et al., 2013). If the MSM material is in the elongation or contraction single variant condition, the magnetization along a fixed path varies. Magnetic anisotropy is the energy difference needed to magnetise the single elongation variant state component and the single contraction variant state component. The anisotropy amount is proportional to the MSM alloy's maximal work-output and the available strain and force for application (Saha et al., 2018).

Figure 3. SMA orthodontic wire

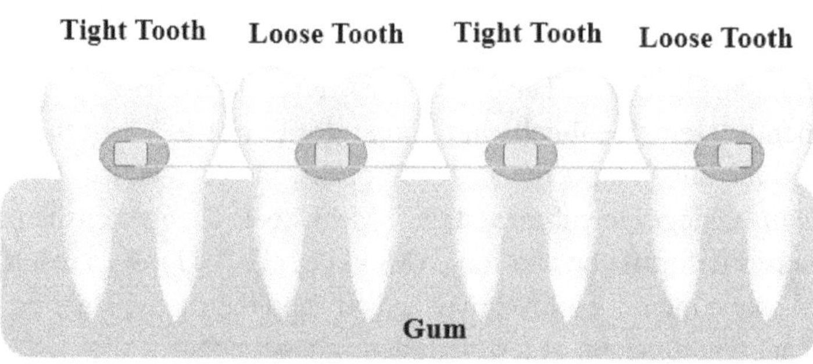

4. PROCESSING AND DEVELOPMENT OF SMAS

Numerous types of processing methods available for a particular material are available in this advanced, progressed, and competitive technological world. The processing routes vary in terms of manufacturing, uses, economics, efficiency, etc. In the same way, various processing methods are available to produce shape memory alloys. The problems with processing shape memory alloys include retaining the shape, mechanical property variation, microstructural alterations, and the production of intermetallic compounds. Chemical reactions and phase transitions are all impacted by temperature and mechanical processing. The processing includes melting routes, metal powder processing, and mechanical and thermo-mechanical testing. Some pre-processing steps are also employed before processing SMA's through various routes for accurate and efficient production. Pre-processing steps needed for shape memory include collecting the clean raw ingredients, fragmenting in needed proportion, and powder formation of the material.

4.1 Thermo-Mechanical Processing

It comprises a class of manufacturing techniques incorporating characteristics and physical shape change. Friction stir processing, hot forging, and hot rolling methods come under these. Friction stir works using solid-state processing. On optimizing the process conditions, it was viable to fabricate an aluminum-nitride composite that had strong bonding and homogenous dispersion (Laydi et al., 2015). Hot forging is another production technique that uses high temperature and pressure to compress the substance into its intended structure (Elahinia et al., 2012). Ni-Mn- Ga is one such SMA prepared using this technique. It is demonstrated that under the right conditions, a complicated alloy such as Ni-Mn-Ga can be created with exceptional memory effects and no significant flaws. "hot rolling" refers to the method that involves heating and compression with rollers. Crucial process parameters need to be taken care of, such as temperature, applied tension, lubricant under use, and some operations to manage memory effect and other mechanical characteristics (Kocich et al., 2013).

4.2 Mechanical Processing

In this instance, mechanical stress is used for the treatment, resulting in substance deformation and an alteration in its characteristics. It includes angular extrusion, cold forging, rolling, and drawing. In the angular extrusion technique, severe mechanical distortion results from elevated axial pressure. This method has key advantages, such as consistent microstructure, grain morphology, texture, and ease of operation. This technique has a significant impact on the microstructural alterations and transformation behavior of NiTi material. In cold forging, the substance is physically altered without being heated by using sufficient force. This process involves subjecting the material to compression under two rollers. It is proposed that a thin oxide layer helps absorb the force applied and shields the material beneath (Mehta et al., 2019). Also, annealing the NiTi material creates nanostructured microstructures with exceptional shape memory capabilities (Fernandes et al., 2013). Similarly, cold drawing involves pulling an alloy rod or wire through a die which causes the grains to get longer and their diameters to decrease. Applying lubricants or protective coatings is advised to improve finishing and processing.

4.3 Melting Route

The transaction of these SMA through the melting process is further classified into four routes: Vacuum Induction Melting, Selective Laser Melting, Electron Beam Melting, and Plasma Melting. Vacuum Induction Melting is one of the production processes used to prepare alloy composite comprising Nickel and Titanium. This methodology comprising graphite crucibles represents the current planning. The metal output has increased quickly because of this technology's success in achieving extremely tight tolerances, perfect heat regulation, and a sustainable approach (Lexcellent, 2013). Selective Laser Melting has been majorly used to fabricate compounds of Ni-Ti, Cu-Al-Ni-Mn, Fe–Mn–Al–Ni, and Al–Fe– V–Si. Creating porous alloys with improved shape memory characteristics is an effective fabrication method. It benefits from an f-theta lens scanning pattern that reduces distortion when scanning (Tu et al., 2019).

4.4 Powder Metallurgy Processing

This remains a traditional way of processing involving using metal powder to create a finished product without wasting material. Previously tested Ni-Ti, Cu–Al– Ni–Mn, and Fe-Mn-Si-Cr-Ni alloys are treated and researched using various powder metallurgical approaches such as sintering, isostatic pressing, and injection molding (Lexcellent, 2013). The primary disadvantage of the traditional sintering technique is that it does not improve density (Shashnka et al., 2017), while Hot isostatic pressing or pressure-enhanced sintering does that work by decreasing the porosity (Lexcellent, 2013). Sintering is manufacturing solid masses by compressing and shaping them under pressure or heat without completely melting them, such as Nitinol (Goryczka et al., 2013). In traditional sintering, the creation of porosity is governed by a small number of variables, including applied pressure, imbalance in diffusion, shrinkage, and capillary forces. Metal injection molding primarily utilizes plastic injection molding, which operates under multiple sequential steps such as feedstock fabrication, injection molding, debinding, and sintering. The production's excellent value, geometry accuracy, and low price are the key advantages of this process. It is used to fabricate NiTi materials, especially for medical devices (Kauffman et al., 1993)

5. THE ROLE OF POWDER METALLURGY IN PROCESSING OF SMAS

Powder metallurgy is crucial in the processing of shape memory alloys (SMAs). A unique family of materials known as shape memory alloys has the ability to "remember" and restore its original shape after deformation. This peculiar activity is attributed to a reversible solid-state phase transition between the two different crystal forms austenite and martensite. Due to the following advantages, powder metallurgy is widely used to create shape memory alloys:

5.1. *Homogeneity:* Uniform distribution of alloying elements for consistent shape memory properties.

5.2. *Composition Control:* Precise tuning of chemical composition for specific applications.

5.3. *Particle Size Control:* Fine-grained microstructure to enhance shape memory effect and mechanical properties.

5.4. *Mechanical Alloying:* Mixing powders to induce solid-state reactions and create a more uniform structure.

5.5. *Consolidation Techniques:* Conversion of powders into desired shapes with good mechanical properties.

5.6. *Near-Net Shaping:* Fabrication close to final shape, reducing machining needs.

5.7. *Complex Compositions:* Production of SMAs with intricate alloy combinations.

6. APPLICATION OF SMAS

These alloys have a breakthrough in a wide range of utilization worldwide. They impart multiple uses in microscale and nanoscale industries. Traditional characteristics like pseudoelasticity or spontaneous restoration are more important in aiding applications. SMA is used in several application areas for high densities, significant reversible deformations, elevated stress, and excellent biocompatibility. This tire can be deformed 30 times more than that of conventional alloy (Yamauchi et al., 2011). Deployment of this idea in the aerospace sector has included fixed-wing aircraft and spacecraft, and research in several of these fields is still ongoing. Research is being performed to investigate embedding techniques for beam components inside chevrons. When flying at low altitudes or slow speeds, the SMA beams bend the chevrons into the flow, boosting mixing and lowering noise (Mirzaeifar et al., 2011). These SMA beam parts will cool into high-altitude martensite, straightening the chevrons and enhancing the engine output. Figure 3 shows the SMA wire and many brackets are used in fixed orthodontic therapy.

Shape memory alloys have their potential in the following fields:

1. Medicine
2. Automotives
3. Aerospace
4. Robotics
5. Transportation
6. Civil structures

Medicines

Shape memory alloys are a boon to the medical industry as they have numerous applications in several fields. Biocompatibility is a crucial prerequisite for using an SMA or any additional substance in human tissue. The material to be used should not trigger any inflammation-based pathways. Another requirement is bio-functionality, which governs the substance's capability to perform desirable functions for the duration of its anticipated service (Peterson et al., 2017). These two conditions are essential for using SMAs in the medical sector. Cardiovascular treatment therapies also employ these alloys. Simon Filter was the initial device utilized in this process. Titanium-based shape memory alloy is often used in eyeglass frames as it returns to its original position after bending where room temperature is above the transition temperature (Wang et al., 2017) Shape-memory alloys are used as a fixation device for osteotomies in orthopedic surgery. The orthopedic staples and shape memory plates may attach to shattered or damaged bones to speed healing (Chlanda et al., 2018). They have also found their use in orthodontic applications as Nitinol braces are employed in teeth alignment while NiTi drill has a major impact on root canal surgery.

Automotives

Automobile manufacturers are also administering the use of shape memory alloys. It has excellent potential in various automobile duties because of its mechanical simplicity, better performance, attractive attributes, and flexibility (Hua et al., 2021). The first commercial SMA introduced in 1989 is the temperature-controlled pressure regulator valve in Mercedes-Benz automated transmission for smooth gear switching. After that, actuators were developed that substituted the heavy ones to open the vent, releasing air. This makes sealing easy and was first used in the 2014 Chevrolet Corvette (Motors, 2013). Alfmeier Prazision AG successfully created shape-based pneumatic valves in large quantities used in Daimler Mercedes Benz (Jani et al., 2014).

Aerospace

SMA's have significant use in aerospace industries. Using NiTi SMA, Variable Geometry Chevron was commenced by NASA, Texas A&M University, Goodrich Corporation, All Nippon Airways, and General Electric Aircraft Engines. Recently

NASA has also developed Spring Tire, made up of Ni-Ti SMA (Barbarino et al., 2014). Other uses are also being considered, such as power generation to produce current from dissipated energy that can improve system dynamics at different rates.

Robotics

One of the main fields of application of shape memory alloy is robotics. With this advanced technology, SMAs are a boon in this industry.

Transportation

The transportation facilities for SMAs include everything from impact absorption to sensors and actuation. (Loh et al., 2006) An efficient system for dissipating vibrations and impacts is the pseudo-elastic behavior hysteresis. This characteristic has been applied to armored vehicles for impact attenuation for military and industrial purposes. The gearbox of the Shinkansen bullet train has an analogous system in place, where the temperature is tracked, and an SMA spring opens a valve to change the oil (Mantovani., 2000). Other uses for trains involve steam traps in the heating systems and thermally activated switches for radiator fans in diesel engines.

Civil Structures

Shape memory alloys are also useful in civil engineering. Intelligent Reinforced Concrete (IRC) encompasses SMAs implanted in concrete building materials for its usage. This can sense and heal the cracks within it. They have a series of applications in structures like bridges and buildings. SMAs are also advantageous for building establishments in earthquake-prone areas. When reinforced with SMA, it will buckle under heavy earthquake loads but won't sustain severe irreversible deformation (Sharabash et al., 2009). In refinery and piping industries, shape memory alloys were used in piping coupling for oil pipelines. In commercial and consumer applications, they are used in water pipelines.

7. CONCLUSION

This book chapter covered the primary characteristics of shape memory alloys, like properties, history, classification, and microstructure. The various processing techniques of shape memory alloys include melting route, powder metallurgy processing, and mechanical and thermo-mechanical treatments were focused. Due to their precious properties like recoverable strain, corrosion resistance, good ductility, lightweight, etc., are utilized in aerospace, automotive, robotics, medical, transportation, and civil structures. The properties and uses of SMA ensure the continued use of these alloys. The future is bright, and its application will be incorporated into each field due to technological advancements. Many research works are currently underway, but their processing is quite difficult due to poor fatigue properties, difficult formation, and a high installation cost.

REFERENCES

Barbarino, S., Flores, E. S., Ajaj, R. M., Dayyani, I., & Friswell, M. I. (2014). A review on shape memory alloys with applications to morphing aircraft. *Smart Materials and Structures*, *23*(6), 063001. doi:10.1088/0964-1726/23/6/063001

Chlanda, A., Witkowska, J., Morgiel, J., Nowińska, K., Choińska, E., Swieszkowski, W., & Wierzchoń, T. (2018). Multi-scale characterization and biological evaluation of composite surface layers produced under glow discharge conditions on NiTi shape memory alloy for potential cardiological application. *Micron, 114*, 14–22. doi:10.1016/j.micron.2018.07.009

Chopade, J., Barjibhe, R., & Kumar, B. (2013). Critical review on shape memory alloy. *International Journal of Engineering*, *2*(2), 1–6.

Elahinia, M. H., Hashemi, M., Tabesh, M., & Bhaduri, S. B. (2012). Manufacturing and processing of NiTi implants: A review. *Progress in Materials Science*, *57*(5), 911–946. doi:10.1016/j.pmatsci.2011.11.001

Exarchos, D. A., Dalla, P. T., Tragazikis, I. K., Dassios, K. G., Zafeiropoulos, N. E., Karabela, M. M., De Crescenzo, C., Karatza, D., Musmarra, D., Chianese, S., & Matikas, T. E. (2018). Development and Characterization of High Performance Shape Memory Alloy Coatings for Structural Aerospace Applications. *Materials (Basel)*, *11*(5), 832. doi:10.3390/ma11050832 PMID:29783626

Fernandes, F. B., Mahesh, K. K., & dos Santos Paula, A. (2013). Thermomechanical treatments for Ni-Ti alloys. *Shape Memory Alloys-Processing, Characterization and Applications*, 3-26.

Frensemeier, M., Kaiser, J. S., Frick, C. P., Schneider, A. S., Arzt, E., Fertig, R. S. III, & Kroner, E. (2015). Temperature-Induced Switchable Adhesion using Nickel-Titanium-Polydimethylsiloxane Hybrid Surfaces. *Advanced Functional Materials*, *25*(20), 3013–3021. doi:10.1002/adfm.201500437 PMID:26120295

Goryczka, T. (2013). Ni25Ti50Cu25 shape memory alloy produced by nonconventional techniques. *Shape Memory Alloys: Processing, Characterization and Applications, Dr. Francisco Manuel Braz Fernandes*, 53-76.

Hînganu, D., Scutariu, M. M., & Hînganu, M. V. (2018). The existence of labial SMAS - Anatomical, imaging and histological study. *Annals of Anatomy = Anatomischer Anzeiger: Official Organ of the Anatomische Gesellschaft, 218*, 271–275. doi:10.1016/j.aanat.2018.04.009

Hua, P., Xia, M., Onuki, Y., & Sun, Q. (2021). Nanocomposite NiTi shape memory alloy with high strength and fatigue resistance. *Nature Nanotechnology*, *16*(4), 409–413. doi:10.1038/s41565-020-00837-5 PMID:33479541

Ivošević, Š., & Rudolf, R. (2019). Materials with Shape Memory Effect for Applications in Maritime. *Maritime Technical Journal*, *218*(3), 25–41. doi:10.2478/sjpna-2019-0016

Jani, J. M., Leary, M., & Subic, A. (2014). Shape memory alloys in automotive applications. *Applied Mechanics and Materials*, *663*, 248–253. doi:10.4028/www.scientific.net/AMM.663.248

Kauffman, G. B., & Mayo, I. (1993). The metal with a memory. *American Heritage of Invention & Technology*, *9*(2), 18–23.

Kim, D., Hong, K., Sim, J., Lee, J., & Lee, W. (2021). Shape Memory and Mechanical Properties of Cold Rolled and Annealed Fe-17Mn-5Si-5Cr-4Ni-1Ti-0.3C Alloy. *Materials (Basel)*, *14*(2), 255. doi:10.3390/ma14020255 PMID:33419156

Kocich, R., Szurman, I., & Kursa, M. (2013). The methods of preparation of Ti–Ni–X alloys and their forming. *Shape Memory Alloys-Processing, Characterization and Applications*, 28-35.

Laydi, M. R., & Lexcellent, C. (2015). Determination of phase transformation surfaces around crack tip in shape memory alloys. In *MATEC Web of Conferences* (Vol. 33, p. 02010). EDP Sciences. 10.1051/matecconf/20153302010

Lexcellent, C. (2013). *Shape-memory alloys handbook*. John Wiley & Sons. doi:10.1002/9781118577776

Li, H. F., Qiu, K. J., Zhou, F. Y., Li, L., & Zheng, Y. F. (2016). Design and development of novel antibacterial Ti-Ni-Cu shape memory alloys for biomedical application. *Scientific Reports*, *6*(1), 37475. doi:10.1038/srep37475 PMID:27897182

Liu, Q., Ding, D., & Ning, C. (2014). Anodic Fabrication of Ti-Ni-O Nanotube Arrays on Shape Memory Alloy. *Materials (Basel)*, *7*(4), 3262–3273. doi:10.3390/ma7043262 PMID:28788616

Loh, C. S., Yokoi, H., & Arai, T. (2006, January). New shape memory alloy actuator: design and application in the prosthetic hand. In *2005 IEEE Engineering in Medicine and Biology 27th Annual Conference* (pp. 6900-6903). IEEE.

Mantovani, D. (2000). Shape memory alloys: Properties and biomedical applications. *Journal of the Minerals Metals & Materials Society*, *52*(10), 36–44. doi:10.1007/s11837-000-0082-4

Martins, A. R., Cunha, J. F., Patrício, J., & Caravana, J. (2016). Familial superior mesenteric artery syndrome. *BMJ Case Reports, 2016*. Advance online publication. doi:10.1136/bcr-2016-214784 PMID:27126095

Mehta, K., & Gupta, K. (2019). *Fabrication and processing of shape memory alloys*. Springer International Publishing. doi:10.1007/978-3-319-99307-2

Mekki, O. B., & Auricchio, F. (2011). Performance evaluation of shape-memory-alloy superelastic behavior to control a stay cable in cable-stayed bridges. *International Journal of Non-linear Mechanics*, *46*(2), 470–477. doi:10.1016/j.ijnonlinmec.2010.12.002

Mirzaeifar, R., DesRoches, R., & Yavari, A. (2011). A combined analytical, numerical, and experimental study of shape-memory-alloy helical springs. *International Journal of Solids and Structures*, *48*(3-4), 611–624. doi:10.1016/j.ijsolstr.2010.10.026

Mohd Jani, J., Leary, M., Subic, A., & Gibson, M. A. (2014). A review of shape memory alloy research, applications and opportunities. *Materials & Design, 56*, 1078-1113. doi:10.1016/j.matdes.2013.11.084

Motors, G. (2013). *Chevrolet debuts lightweight "smart material" on corvette*. Corporate Newsroom. Available at: http://media. gm. com/media/us/en/gm/home. detail. html/content/Pages/news/us/en/2013/Feb/0212-corvette. html

Peterson, G. I., Dobrynin, A. V., & Becker, M. L. (2017). Biodegradable Shape Memory Polymers in Medicine. *Advanced healthcare materials, 6*(21). doi:10.1002/adhm.201700694

Raghavan, J., Bartkiewicz, T., Boyko, S., Kupriyanov, M., Rajapakse, N., & Yu, B. (2010). Damping, tensile, and impact properties of superelastic shape memory alloy (SMA) fiber-reinforced polymer composites. *Composites. Part B, Engineering*, *41*(3), 214–222. doi:10.1016/j.compositesb.2009.10.009

Ren, X., & Otsuka, K. (1997). Origin of rubber-like behaviour in metal alloys. *Nature*, *389*(6651), 579–582. doi:10.1038/39277

Saha, S., & Islam, A. (2018). Shape memory material-concepts, recent trends and future directions. In *Student Conference on Electronics, Informatics and Vision 2018* (pp. 1-6). Academic Press.

Sharabash, A. M., & Andrawes, B. O. (2009). Application of shape memory alloy dampers in the seismic control of cable-stayed bridges. *Engineering Structures*, *31*(2), 607–616. doi:10.1016/j.engstruct.2008.11.007

Shashanka, R., & Chaira, D. (2014). Phase transformation and microstructure study of nano-structured austenitic and ferritic stainless steel powders prepared by planetary milling. *Powder Technology*, *259*, 125–136. doi:10.1016/j.powtec.2014.03.061

Shashanka, R., & Chaira, D. (2017). Effect of Sintering Temperature and Atmosphere on Nonlubricated Sliding Wear of Nano-Yttria-Dispersed and Yttria-Free Duplex and Ferritic Stainless Steel Fabricated by Powder Metallurgy. *Tribology Transactions*, *60*(2), 324–336. doi:10.1080/10402004.2016.1168897

Tu, J., & Reeves, N. (2019). Feasibility study of microneedle fabrication from a thin nitinol wire using a CW single-mode fiber laser. *Open Engineering*, *9*(1), 167–177. doi:10.1515/eng-2019-0023

Velvaluri, P., Soor, A., Plucinsky, P., de Miranda, R. L., James, R. D., & Quandt, E. (2021). Origami-inspired thin-film shape memory alloy devices. *Scientific Reports*, *11*(1), 10988. doi:10.1038/s41598-021-90217-3 PMID:34040022

Wang, W., & Ahn, S. H. (2017). Shape Memory Alloy-Based Soft Gripper with Variable Stiffness for Compliant and Effective Grasping. *Soft Robotics*, *4*(4), 379–389. doi:10.1089/soro.2016.0081 PMID:29251571

Xue, D., Yuan, R., Zhou, Y., Xue, D., Lookman, T., Zhang, G., Ding, X., & Sun, J. (2016). Design of High Temperature Ti-Pd-Cr Shape Memory Alloys with Small Thermal Hysteresis. *Scientific Reports*, *6*(1), 28244. doi:10.1038/srep28244 PMID:27328764

Yamauchi, K., Ohkata, I., Tsuchiya, K., & Miyazaki, S. (Eds.). (2011). *Shape memory and superelastic alloys: Applications and technologies.* Elsevier. doi:10.1533/9780857092625

Zhang, Y., Wang, P., Xia, Y., Zhou, P., Xie, Y., Xu, S., & Zhang, C. (2017). Application of a Shape-Memory Alloy Concentrator in Displaced Patella Fractures: Technique and Long-Term Results. *The Journal of Knee Surgery*, *30*(2), 166–173. doi:10.1055/s-0036-1584187 PMID:27206070

Zhao, W., Liu, L., Zhang, F., Leng, J., & Liu, Y. (2019). Shape memory polymers and their composites in biomedical applications. *Materials Science and Engineering C*, *97*, 864–883. doi:10.1016/j.msec.2018.12.054 PMID:30678978

Zhu, G., & Liu, Z. (2005). Article. *Sheng wu yi xue gong cheng xue za zhi = Journal of biomedical engineering = Shengwu yixue gongchengxue zazhi*, *22*(5), 1082–1084.

Chapter 7
Properties and Applications of Bainitic Railway Steel Prepared by Mechanical Alloying

Rahul Samanta
Kazi Nazrul University, India

Gaurav Kumar Bansal
CSIR-National Metallurgical Laboratory, India

Swarup Kumar Ghosh
https://orcid.org/0000-0002-2365-4854
Indian Institute of Engineering Science and Technology, Shibpur, India

Gurudas Mandal
https://orcid.org/0000-0002-4140-2271
Kazi Nazrul University, India

ABSTRACT

Researchers face a significant challenge in designing low-carbon bainite rail steel, which offers better mechanical properties and corrosion resistance than high-carbon pearlitic steel tracks. High-carbon bainitic steel takes a long time to manufacture, so low-carbon bainite steel is preferred for heavy-haul railway tracks due to its shorter production time and cost-effectiveness. The steel is prepared through a specific heat treatment process, and although carbide formation is limited at lower temperatures, precipitation hardening is used to improve its properties. Mechanical alloying with the elements like Si, Ni, and Co is also considered to enhance further the mechanical and chemical properties of the bainitic rail steel. This chapter aims to find an economical solution to reduce accidents caused by the failure of pearlitic steel and improve safety in rail transportation.

DOI: 10.4018/978-1-6684-9385-4.ch007

1. INTRODUCTION

In modern times, the demand for lightweight and high-strength to low-weight ratio materials has increased significantly. Researchers are keen to develop mechanically alloyed bainitic rail steel for heavy haul railway tracks. Railway transportation is considered one of the most important modes worldwide (Zhang et al., 2019). At present, pearlitic steels with high carbon percentages are used widely as rail material (Caballero et al., 2009). In this modern era, high-carbon pearlitic rail steel has become a key component that significantly affects efficiency and, most importantly, railway transportation safety (Beynon & Perez-Unzueta, 1993; Gui et al., 2016; Satoh & Iwafuchi, 2005). In this scenario, the development of heavy-haul railways worldwide effectively increases traffic density and train speed (Hasan et al., 2021). Conventional pearlitic rail steel has faced more severe damage problems, such as rolling contact fatigue (RCF), wear, and catastrophic fracture (Gao et al., 2022). Pearlitic steel becomes brittle due to the presence of cementite (Fe_3C), which acts as a notch and tends to initiate cracks at low fracture toughness portions of the matrix. Meanwhile, bainitic steel, due to its non-lamellar structure of ferrite and carbide, reveals higher ductility and fracture toughness. Advanced bainitic steels consisting of a multiphase structure of bainitic ferrite can effectively retain both martensite and austenite, which shows a characteristic fatigue crack initiation behavior during high cycle fatigue/very high cycle fatigue (HCF/VHCF) regimes (Gao et al., 2022). For the fatigue crack initiation, inclusion and microstructure play vital roles. However, the non-inclusion-induced fatigue crack initiation is a chief failure mode for the advanced carbide-free bainitic (C.F.B.) steels. This article deals with an insight into the properties and applications of bainitic railway steel prepared by mechanical alloying.

2. RATIONALE

The failure of currently used pearlitic rail steels can significantly threaten the safety measures of train operations and is followed by an increase in railway maintenance costs. Researchers have found that bainitic steels have the exciting potential to become at the forefront of the development of the steel industry (Garcia-Mateo & Caballero, 2007; Takahashi & Bhadeshia, 1991). Moreover, compared to traditional pearlitic rail steels with high carbon percentages, low-alloyed bainitic rail steels effectively showed a much better combination of toughness, wear resistance, and strength (Adamczyk-Cieślak et al., 2021; Pereloma & Edmonds, 2012; Tasak et al., 2014). Therefore, a significant need has occurred to fill the gap or mitigate the above-discussed problem. Thus, developing a new generation of rail materials, such as low alloyed bainitic steel with high performance, is very much required.

The essential purpose of this article is to analyze the development of a lower bainite structure that consists of a dispersion of very fine carbide at the inner side of bainitic ferrite plates from chemical composition less modified or unmodified traditional pearlite steel under bainitic transformation to investigate its impact on compressive and fatigue properties (Liu et al., 2022). As discussed above, the development of low-carbon bainitic steel for application in heavy-haul railway tracks has already been started (Perez-Unzueta & Beynon, 1993; Tressia et al., 2020). However, all the research regarding this matter is in the early stages, and to perform an in-depth analysis of low-bainitic steel and its RCF behavior, the article has been prepared (Chen et al., 2019; Johnson & Becker, 1993; Lee & Polycarpou, 2005). The key issue with current or conventional pearlitic rail steel is severe damage problems, like fracture, RCF, and wear (Aglan et al., 2004; Zhou et al., 2017). This causes inefficiency or threat to the safety measures of several train operations. Moreover, it significantly increases the maintenance costs of railways (Hasan et al., 2020; Tenaglia et al., 2020; Zhang et al., 2020). However, the ease of availability of raw materials, easy manufacturing process and low manufacturing cost make it suitable for continuous use. According to the researchers, compared to conventional pearlitic rail steels with high carbon percentage, low-alloyed bainitic rail steels effectively showed a much better combination of strength, toughness, RCF, and wear resistance. However, the high manufacturing cost of bainitic steel makes it less preferable in the industry, although, in the long run, it will be beneficial as the life cycle cost is lower due to its lower maintenance compared to pearlitic steel.

3. INSIGHTS OF MECHANICAL ALLOYING AND PREPARATION OF BAINITIC STEEL BY MECHANICAL ALLOYING

3.1 Insights of Mechanical Alloying

About 50 years ago, mechanical alloying (MA) was invented. Initially, the method successfully manufactured superalloys that were strengthened due to oxide dispersion and showed an exceptional combination of strength and oxidation resistance at temperatures above 1000°C (Benjamin, 1970). Subsequently, alloy development produced many systems like aluminium-based alloys, ceramics, intermetallic compounds, and composite materials (Benjamin, 1990). Mechanical alloying deals with the severe deformation of mixtures of different powders to form atomic solutions without melting (Bhadeshia, 2000). Figure 1(a) shows the deformation characteristics of starting powders in MA. The metal powders (metals A and B), which are ductile, get flattened, while both the dispersoid particles and brittle intermetallic get fragmented into finer particles (Suryanarayana, 2001). In addition, Figure 1(b)

Figure 1. (a) Deformation characteristics of starting powders and (b) ball-powder-ball collision of powder mixture during mechanical alloying

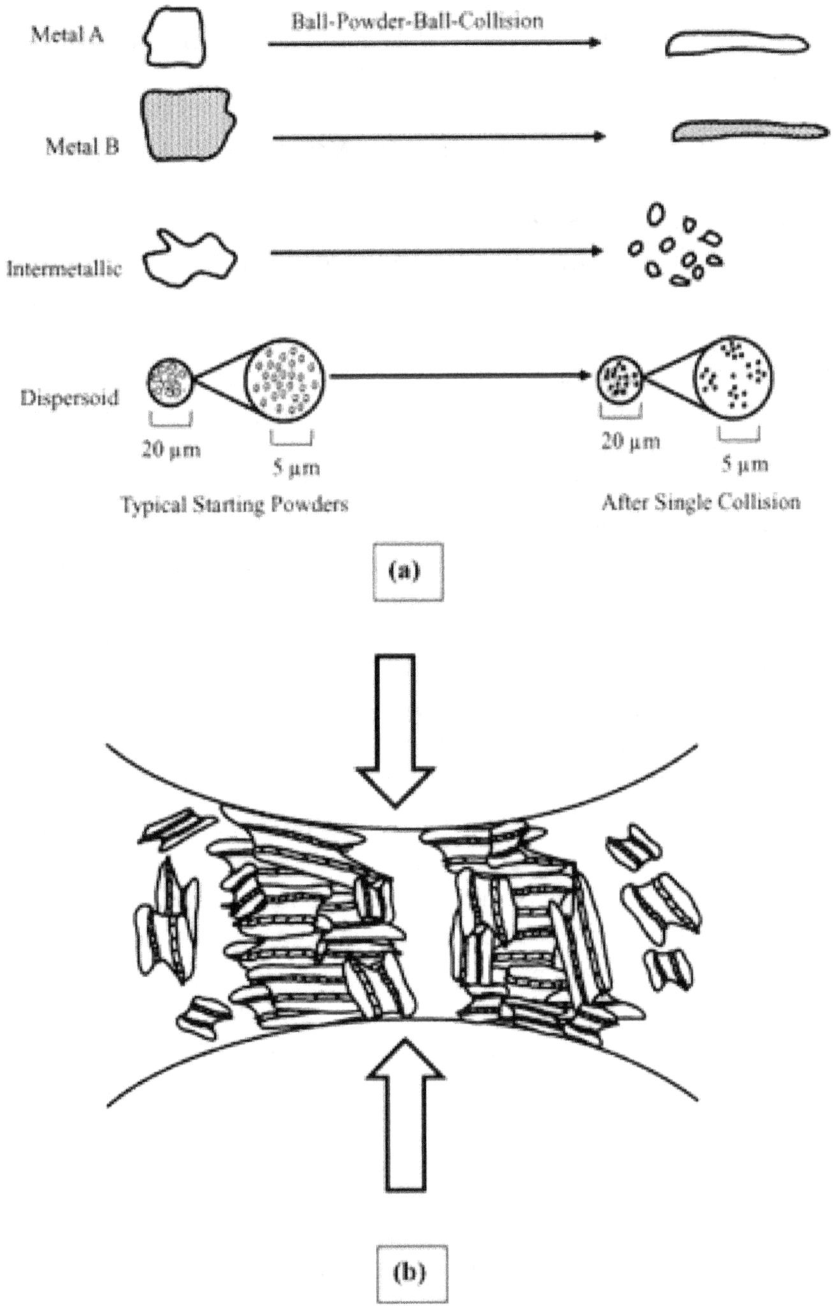

shows the process of MA by powder mixing through ball-powder-ball collision. Finally, in the same proportion as the initial constituent powders, homogenized and finer microstructures are obtained. The dispersion-strengthened alloy powders are consolidated using hot extrusion, cold pressing and sintering, hot rolling, and hot isostatic pressing, followed by recrystallization annealing to produce a solid with a fine grain structure.

3.2 Preparation of Bainitic Steel by Mechanical Alloying

In the modern day, mechanical alloying has become a very popular method, especially in steel manufacturing (Shashanka, 2016; Shashanka, 2017; Shashanka et al., 2016). Methods like hydraulic pressings, ultrasonic shot peening, and, most importantly, synthesis of very fine-grained steel using machining have been done to manufacture steel (Chaira, 2014; Shashanka & Chaira, 2015; Shashanka et al., 2015). However, mechanical alloying gets the upper hand due to its ability to produce bulk materials using simple equipment at room temperature in a solid-state condition (Chaira, 2015; Mahale & Shashanka, 2022; Nayak et al., 2016; Shashanka & Chaira, 2016; Shashanka & Chaira, 2017). On the other hand, the process also reduces the processing time by decreasing the oxidation of the constituents (Mahale, Rajendrachari, Vasanth et al, 2022; Rajendrachari et al., 2013; Shashanka, 2019). Mechanical alloying conducted in a planetary ball mill followed by spark plasma sintering to achieve almost full density and isothermal heat treatment (240°C, 4 h) were reported to process high strength (>1800 MPa) nano-structured (<100 nm) medium carbon bainitic steel (Fe–0.48C–1.92Si–2.62Ni–1.47Cr–0.24Mo, wt.%) (Lonardelli et al., 2012). Various nano-structured steels could be manufactured by mechanical alloying, and the structures and properties of such steels are comparable to or even better than those produced by traditional methods (Suryanarayana, 2022). To ensure the safety of travelers and to avoid fatal train accidents due to the failure of railway tracks, designing bainite rail steel has become a significant aspect. In the foreseeable future, the use of bainite rail steel in place of existing pearlitic steel can effectively decrease accidents due to catastrophic fracture, RCF, and wear of the heavy haul railway tracks (Adamczyk-Cieślak et al., 2021; Satoh & Iwafuchi, 2005; Tressia et al., 2020). It also increases the lifespan of railway tracks, and with improved performance, safety measures of railway tracks can also increase effectively (Gui et al., 2016; Lai et al., 1974; Long, Kang, Lv et al, 2014). These superior bainitic railway steels have a multiphase bainitic ferrite structure that retains martensite and austenite (Lee & Polycarpou, 2005). Carbide-free low-carbon bainitic rail steel showed much higher performance than existing pearlitic rail steel in toughness and strength, which motivated the researchers to focus on it more (Hasan et al., 2018). Microstructural evaluations showed that a suitable distribution in retained austenite

(RA), stability, and RA morphology with multiphase structure in the case of high carbon bainite steel, and thus, low carbon carbide-free bainite is much more suitable for this application (Pacyna, 2008).

Zhang et al. (2019) examined the properties of bainitic steel with different silicon content (ranging from 1.0% to 2.5%), focusing on its microstructure, hardness, and resistance to sliding wear. Various methods such as optical microscopy, scanning electron microscopy (SEM), X-ray diffraction (XRD), energy-dispersive spectroscopy (EDS), Rockwell hardness testing, microhardness testing, and M-200 wear testing were utilized to analyze the steel samples. The findings demonstrated that the original structure of the steel, in its as-cast state, consisted primarily of acicular lower bainite and retained austenite. As the silicon content increased, the bainitic lath became finer, the amount of retained austenite decreased, and the hardness showed an upward trend after normalizing the steel at 900°C, the microstructure predominantly comprised acicular lower bainite and film-like retained austenite. Increasing the silicon content further refined the bainitic needles, reduced the retained austenite content, and resulted in an approximate 20% increase in hardness. Additionally, the as-cast and normalized steel samples exhibited improved wear resistance as the silicon content increased. Moreover, the wear loss of the normalized steel was notably lower than that of the as-cast steel.

Hasan et al. (2021) have reported that the microstructure and mechanical properties of carbide-free bainitic steel are influenced by thermo-mechanical treatment. The steel, which contains 0.2% C-1.53% Mn-1.1% Si-1% Cr-0.3% Mo-0.5% N, is subjected to different variants of deformation and analyzed using impact tests, metallographic examination, and X-ray diffraction. The researchers examined the impact toughness of the steel after it underwent bainite treatment, focusing on the size of the initial austenite grains and the Vickers hardness. They find that the hot plastic deformation process has a positive effect on the impact toughness of the steel, indicating an improvement in its ability to withstand impact forces.

4. HEAT TREATMENT OF BAINITIC STEEL

Previous studies have shown that heat treatment can optimize the mechanical properties of bainitic rail steel, including resilience, ductility, and toughness. Specifically, a heat treatment process involving austempering at 900°C for 45 minutes, followed by heat treatment at 200°C for ten days and air cooling, has been selected to produce carbide-free bainitic rail steel (Aglan et al., 2004; Hajizad et al., 2019; Sawley & Kristan, 2003). This steel exhibits high performance and is well-suited for heavy-haul railway tracks (Magel et al., 2005; Satoh et al., 1999; Singh et al., 2001). The resulting bainitic rail steel has a complex structure comprising bainitic ferrite with both martensite and austenite phases (Su & Clayton, 1996). Compared to traditional pearlitic low-carbon steel, low-carbon bainitic rail steel demonstrates superior toughness and strength (Barbacki, 1995; Li & Milbourn, 2011). This improvement is attributed to factors such as the appropriate distribution and stability of retained austenite, as well as the refinement of grain size, which reduces the occurrence of fine cracks (Strey et al., 2021; Tunna et al., 2007). Retained austenite (RA) plays a crucial role in crack initiation and propagation during cyclic loading. Therefore, understanding and controlling the morphology, stability, and disbandment of RA is vital for enhancing the localized cyclic plastic deformation in bainitic rail steel with a multiphase structure (Aglan & Fateh, 2006; Ishida, 2013; Lin & Zhou, 2018; Ray, 2017). In the steel industry, particularly for heavy haul railway tracks, there is a demand for developing low-carbon, carbide-free bainitic rail steel to address issues related to low toughness, rail-induced fatigue cracking, wear resistance, and strength (Anis et al., 2018; Girsch & Heyder, 2006; Stock, 2012; Yang & Zhang, 2020). The advancements emphasize the importance of designing bainitic rail steel with these desirable characteristics and implementing appropriate heat treatment processes to ensure high strength and toughness for heavy-haul railway tracks.

5. MICROSTRUCTURE AND MECHANICAL PROPERTIES OF BAINITIC RAIL STEEL

Figures 2(a) and (b) show the transmission electron microscope (TEM) bright field (BF) images of thermomechanically controlled processed (TMCP) air-cooled (AC) steel followed by full austenitization (930°C, 30 min)-isothermal bainitic treatment (IBT) (440°C, 30 min)-AC sample, in two different locations of the same sample. The microstructure is reasonably homogeneous and consists of fine bainitic ferrite(B)

Figure 2. Transmission electron microscope (TEM) bright field image (BF) of (a) and (b) thermomechanically controlled processed (TMCP) - air-cooled (AC) steel + full austenitization (930°C, 30 min)-IBT-AC sample

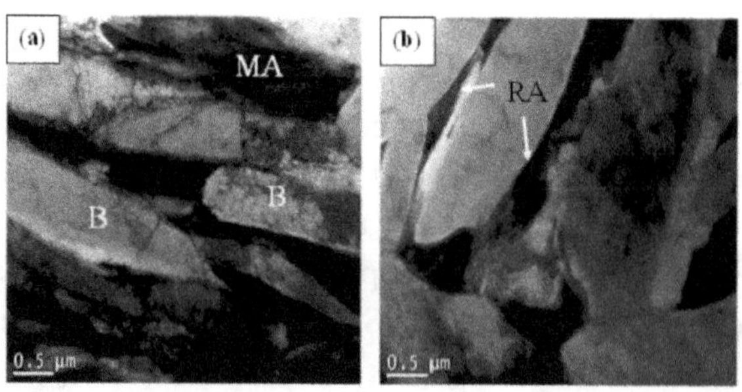

laths, martensite-austenite (MA) constituent/island, and interlath filmy retained austenite (RA) (<0.5 μm) where the carbon concentration is expected to increase from the base composition.

The existing pearlitic railway steel exhibits hardness (310-340 HV), yield strength (700-1100 MPa), tensile strength (1000-1400 MPa), and ductility (10-12%) (Masoumi et al., 2021; Sahay Satyam et al., 2009). On the other hand, the effective processing/heat treatment of the bainitic steels can produce adequate mechanical properties, and some of those are listed in Table 1. The hardness and tensile results are commonly used to compare mechanical properties. The attractive combination of hardness (400-550 HV), yield strength (800-1200 MPa), tensile strength (1300-1700 MPa), and ductility (9-17%) are attributed to multiphase microstructure comprising bainite, martensite, martensite-austenite constituent along with filmy high carbon retained austenite between the bainite/martensite laths (Gurudas et al., 2017; Hasan et al., 2018; Li et al., 2019; Miao et al., 2022; Tressia et al., 2020). To achieve high-strain hardening and uniform elongation, the role of RA and its stress or strain-induced martensitic transformation are noteworthy (Samek et al., 2006).

Table 1. Mechanical properties of bainitic steels after various heat treatments

Ref.	Chemical Composition(wt.%)	Heat-treatment	Hardness (HV)	YS (MPa)	UTS (MPa)	TEL (%)
[15]	0.75C-1.03Mn-0.22Si-0.2Cr-0.05Ni	Austenitization (900°C/300s) + Quenching (300°C/600s) +Tempering (400°C/300s) and air-cooling	415±12	1060±10	1700±15	9.5±0.5
[46]	0.23C-1.5Mn-1.45Si-1.2Cr-0.35Mo-0.85Ni	Soaking (1250°C/2h) + finish hot rolling (900°C) and air-cooling	419±7	819±15	1586±20	16±3
[66]	0.3C-1.4Si-1.8Mn-1.3Cr-0.3Mo	Austenitization (900°C/20min) + Quenching (320°C/60min) and air-cooling	472	1180	1480	17.3
[67]	0.2C-2.2Mn-1Si-1Cr-0.3Mo-0.5Ni	Hot rolling and natural air-cooling	580	1208±12	1335±8	14.5±0.5

6. APPLICATIONS OF BAINITIC STEEL IN THE RAILWAY INDUSTRY

The use of current pearlitic rail steels poses a significant risk to train safety and results in higher railway maintenance costs. On the other hand, low-alloyed bainitic rail steels offer improved toughness, wear resistance, and strength compared to high-carbon pearlitic steels (Ekberg, 1996; Moon et al., 2015). Therefore, there is a pressing need to develop high-performance, low-alloyed bainitic steel to address this problem (Ghodrati et al., 2018; Kang et al., 2017; Pointner, 2008). The aim is to create a lower bainite structure by incorporating fine carbides within bainitic ferrite plates without altering the chemical composition of standard pearlitic steel. The effects of this structure on compressive and fatigue properties have been examined (Gao et al., 2019; Liu, Pan, Liu et al, 2020; Onal et al., 2012). The development of low-carbon bainitic steel for heavy-haul railway tracks has already begun, offering exceptional strength and elasticity compared to existing pearlitic rail steel (Li et al., 2006; Long et al., 2020; Yu et al., 2021; Zheng et al., 2014). As heavy-haul railways expand worldwide, realizing the relationship between mechanical properties and microstructure under continuous cooling and isothermal conditions for carbide-free, low-carbon bainitic steel is crucial. Various experiments, including metallographic and X-ray diffraction analyses and impact tests, have been conducted to investigate this issue (Gui et al., 2016; Kumar & Singh, 2021; Long, Zhang, Kang et al, 2014; Zheng et al., 2016). The findings suggest that carbide-free, low-alloyed bainitic rail steels outperform high-carbon pearlitic rail steels in terms of strength, toughness, and wear resistance (Green et al., 2007; Klaasen et al., 2007; Lee & Polycarpou, 2005; Shur et al., 2005). However, the research in this field is still in the early stages, and a comprehensive analysis of low-bainitic steel and its behavior under rolling contact fatigue (RCF) conditions is warranted (Garnham & Davis, 2008; Huang et al., 2018; Zapata et al., 2011).

The worldwide railway transport development is based on increasing speed and growth of railway axel load. The fundamental motto of introducing bainitic steel in the railway industry is generating safety in this transportation process, and bainitic steel, with high wear resistance and strength, can fulfill the demand quickly. The design of rail accessories needs to be suitable for resisting plastic deformation, RCF, wear, thermal, and bending stress during the rail resurfacing and welding process (Rezende et al., 2020). The high strength and wear-resistant bainitic steels are considered suitable for heavy load rails, railway wheels, railway crossings, points, and level crossing at low material and production costs (Bhadeshia, 2001; Chang, 2005; Hlavatý et al., 2009; Kapito et al., 2012).

7. RECENT DEVELOPMENT

Low-carbon bainite rail steel development has emerged as a promising solution for heavy haul railway tracks, offering superior mechanical properties and corrosion resistance compared to the existing high-carbon pearlitic steel tracks. However, designing this type of steel presents a significant challenge for researchers (Mahale, Shamanth, Hemanth et al, 2022; Rajendrachari, 2018; Shashanka & Kumara Swamy, 2020). Although possessing desirable properties, high-carbon bainite steel suffers from lengthy manufacturing times, making it less practical for industrial applications (Adimule et al., 2022; Basavarajappa et al., 2023). Consequently, low-carbon bainite steel has been favored for heavy-haul railway tracks due to its shorter preparation time and cost-effectiveness. The production of low-carbon bainite steel involves a specific heat treatment schedule that enables its rapid formation (Rezende et al., 2021). Austempering, conducted at temperatures ranging from 900-950°C, promotes bainite formation, although carbide formation is limited at these lower temperatures (Zhang et al., 2021). To compensate for this, precipitation hardening techniques are employed to enhance the mechanical and chemical properties of the steel. Additionally, microalloying, which involves the addition of alloying elements like Si, Ni, and Co, has been found to further improve the properties of bainitic rail steel. The combination of heat treatment and microalloying allows for the achievement of superior mechanical and chemical characteristics in the design of bainite rail steel. This research contributes to understanding an economically viable approach for reducing fatal accidents caused by the failure of pearlitic steel in railway tracks (Aglan & Fateh, 2007; Bhadeshia, 1997; Qiu et al., 2019). By addressing these issues and developing low-carbon bainitic steel, the study aims to provide a solution that enhances safety and performance in rail transportation systems.

Recent studies showed that initial grain refinement tends to increase the strength of the steel. The Vickers hardness results also indicated that the hardness of bainitic steel increased significantly compared to pearlitic steel (Liu, Li, Zhang et al, 2020). Hence, the bainitic microstructure prevented crack nucleation and propagation to greater depths. Moreover, hot plastic deformation positively impacts the toughness of the steel that was treated for bainite (Kostryzhev et al., 2018). The formation of small grains was attributed to the advancement of fine cracks (Shashanka et al., 2015). It is also observed that the role of RA, especially in crack initiation and propagation, can be considered a major cause during cyclic loading. The development of low-carbon bainitic steel, suitable for RCF in terms of RA morphology, stability, and distribution, can determine the improvement of localized cyclic plastic deformation in bainitic steel with multiphase structure (Zhu et al., 2020). To counter low toughness, RCF, wear resistance, and strength-related issues in currently used high carbon pearlitic steel, a development in the steel industry, especially in the making of heavy haul railway tracks, is very much required. The study highlighted the importance of low alloyed bainitic steel to get high performance in toughness, RCF, wear resistance, and strength.

8. CONCLUSION

As the railway has become an essential way to travel worldwide, safety measures have become a primary concern for researchers. For that, safety issues concerning railway accessories and, most importantly, rail tracks have gained the most attention. Nowadays, high carbon pearlitic rail steel is used for manufacturing railway tracks, which causes damage problems like rolling contact fatigue (RCF), catastrophic fracture, and wear. Eventually, all those factors effectively raise safety issues of railway transportation and degrade the proficiency and longevity of heavy haul rail tracks. However, to counter low wear resistance, RCF, low strength, and low toughness of pearlitic steel development in heavy haul rail tracks become essential. The advancement of bainitic rail steels has been started to ensure longevity and proficiency and enhance safety measures. To fill the gap, a design of low-carbon carbide-free bainitic steel, especially for applying rail tracks with better chemical and mechanical properties, is required. The necessity of designing low carbon carbide-free bainitic rail steel and its preparation by mechanical alloying and heat treatment schedule has been discussed to secure high performance in heavy haul rail tracks. Although the industrial heat treatment process is expensive, the prolonged service of bainitic steel requires reduced maintenance compared to the existing pearlitic steel. Thus, this study supports the idea that the bainitic microstructure can be a better substitute than the pearlitic microstructure for railway applications.

9. FUTURE PROSPECTS

In this study, attempts have been made to analyze the development and design of high-strength carbide-free bainitic steel with low-carbon and low alloy for heavy-haul rail track applications. The study has started with an analytical approach for understanding the impacts of various alloying elements on the kinetic and thermodynamic aspects of bainite transformation and its effect on the resultant microstructure and mechanical properties. Based on those analyses, two or more steel compositions can be designed for further investigation. The low-carbon bainitic alloys (0.1-0.3wt.%C) are most suitable for this application. Moreover, high entropy bainitic steel alloys can also be used for further development of the steel industry as they pose ultra-high strength levels with high ductility (Rajendrachari, 2022).

The advancement of the properties of the powder metallurgy (PM) product and its microstructures mainly relies on alloy composition and manufacturing processes. To obtain the desired phase transformation, morphology, and microstructure, alloy design plays a significant role. Inclinations to research high entropy alloys are increasing day-to-day as their unique properties can mitigate considerable challenges and produce a better scope of applications.

REFERENCES

Adamczyk-Cieślak, B., Koralnik, M., Kuziak, R., Majchrowicz, K., & Mizera, J. (2021). Studies of bainitic steel for rail applications based on carbide-free, low-alloy steel. *Metallurgical and Materials Transactions. A, Physical Metallurgy and Materials Science*, *52*(12), 5429–5442. doi:10.1007/s11661-021-06480-6

Adimule, V., Koutsonas, S. S., Kaloop, M., & Shashanka, R. (2022). *Materials and Production Technologies in Machinery*. Trans Tech Publications. doi:10.4028/b-sG2xM9

Aglan, H., & Fateh, M. (2007). Fracture and fatigue crack growth analysis of rail steels. *Journal of Mechanics of Materials and Structures*, *2*(2), 335–346. doi:10.2140/jomms.2007.2.335

Aglan, H. A., & Fateh, M. (2006). Fatigue damage tolerance of bainitic and pearlitic rail steels. *International Journal of Damage Mechanics*, *15*(4), 393–410. doi:10.1177/1056789506060775

Aglan, H. A., Liu, Z. Y., Hassan, M. F., & Fateh, M. (2004). Mechanical and fracture behavior of bainitic rail steel. *Journal of Materials Processing Technology*, *151*(1-3), 268–274. doi:10.1016/j.jmatprotec.2004.04.073

Anis, M. A., Srivastava, J. P., Duhan, N. R., & Sarkar, P. K. (2018, June). Rolling contact fatigue and wear in rail steels: An overview. *IOP Conference Series. Materials Science and Engineering*, *377*(1), 012098. doi:10.1088/1757-899X/377/1/012098

Barbacki, A. (1995). The role of bainite in shaping mechanical properties of steels. *Journal of Materials Processing Technology*, *53*(1-2), 57–63. doi:10.1016/0924-0136(95)01961-D

Basavarajappa, P. N., Hegde, M. M. R., Rajendrachari, S., & Surendranathan, A. O. (2023). Investigation of structural and Mechanical properties of Nanostructured TiMgSr Alloy for Biomedical applications, *Biointerface Research in Applied Chemistry*, *13*(2), 1-12.

Benjamin, J. S. (1970). Dispersion Strengthened Superalloys by Mechanical Alloying. *Metallurgical Transactions, 1*(10), 2943–2951. doi:10.1007/BF03037835

Benjamin, J. S. (1990). Mechanical Alloying-A Perspective. *Metal Powder Report, 45*(2), 122–127. doi:10.1016/S0026-0657(10)80124-9

Beynon, J. H., & Perez-Unzueta, A. J. (1993). Microstructure and wear resistance of pearlitic rail steels. *Wear, 162/164*, 173–182. doi:10.1016/0043-1648(93)90498-B

Bhadeshia, H. K. D. H. (1997). Martensite and bainite in steels: transformation mechanism & mechanical properties. *J. Phys. IV France, 7*, C5-367-C5-376.

Bhadeshia, H. K. D. H. (2000). Mechanically Alloyed Metals. *Materials Science and Technology, 16*(11-12), 1404–1411. doi:10.1179/026708300101507361

Bhadeshia, H. K. D. H. (2001). *Bainite in steels* (2nd ed.). University Press.

Caballero, F. G., Santofimia, M. J., García-Mateo, C., Chao, J., & De Andres, C. G. (2009). Theoretical design and advanced microstructure in super high strength steels. *Materials & Design, 30*(6), 2077–2083. doi:10.1016/j.matdes.2008.08.042

Chaira, D. (2014). Phase transformation and microstructure study of nano-structured austenitic and ferritic stainless steel powders prepared by planetary milling. *Powder Technology, 259*, 125–136. doi:10.1016/j.powtec.2014.03.061

Chaira, D. (2015). Development of nano-structured duplex and ferritic stainless steels by pulverisette planetary milling followed by pressurelesssintering. *Materials Characterization, 99*, 220–229. doi:10.1016/j.matchar.2014.11.030

Chang, L. C. (2005). The rolling/sliding wear performance of high silicon carbide-free bainitic steels. *Wear, 258*(5-6), 730–743. doi:10.1016/j.wear.2004.09.064

Chen, G., Xu, G., Zurob, H. S., Hu, H., & Wan, X. (2019). Effect of strain rate on the bainitic transformation in Fe-C-Mn-Si medium-carbon bainitic steels. *Metallurgical and Materials Transactions. A, Physical Metallurgy and Materials Science, 50*(2), 573–580. doi:10.1007/s11661-018-5051-z

Ekberg, A. (1996, July). Rolling contact fatigue of railway wheels–computer modeling and in-field data. In *Proceedings of 2nd mini conf. Contact mechanics and wear of railway systems* (pp. 154-163). Academic Press.

Gao, B., Tan, Z., Liu, Z., Gao, G., Zhang, M., Zhang, G., & Bai, B. (2019). Influence of non-uniform microstructure on rolling contact fatigue behavior of high-speed wheel steels. *Engineering Failure Analysis, 100*, 485–491. doi:10.1016/j.engfailanal.2019.03.002

Gao, G., Liu, R., Fan, Y., Qian, G., Gui, X., Misra, R. D. K., & Bai, B. (2022). Mechanism of subsurface microstructural fatigue crack initiation during high and very-high cycle fatigue of advanced bainitic steels. *Journal of Materials Science and Technology, 108*, 142–157. doi:10.1016/j.jmst.2021.08.060

Garcia-Mateo, C., & Caballero, F. G. (2007). Design of carbide-free low-temperature ultra-high strength bainitic steels. *International Journal of Materials Research, 98*(2), 137–143. doi:10.3139/146.101440

Garnham, J. E., & Davis, C. L. (2008). The role of deformed rail microstructure on rolling contact fatigue initiation. *Wear, 265*(9-10), 1363–1372. doi:10.1016/j.wear.2008.02.042

Ghodrati, M., Ahmadian, M., & Mirzaeifar, R. (2018). Modeling of rolling contact fatigue in rails at the microstructural level. *Wear, 406*, 205–217. doi:10.1016/j.wear.2018.04.016

Girsch, G., & Heyder, R. (2006, June). Advanced pearlitic and bainitic high-strength rails promise to improve rolling contact fatigue resistance. In *7th World Congress on Railway Research (WCRR2006)*, Montreal, Canada.

Green, M. R., Rainforth, W. M., Frolish, M. F., & Beynon, J. H. (2007). The effect of microstructure and composition on the rolling contact fatigue behaviour of cast bainitic steels. *Wear, 263*(1-6), 756–765. doi:10.1016/j.wear.2007.01.070

Gui, X., Wang, K., Gao, G., Misra, R. D. K., Tan, Z., & Bai, B. (2016). Rolling contact fatigue of bainitic rail steels: The significance of microstructure. *Materials Science and Engineering A, 657*, 82–85. doi:10.1016/j.msea.2016.01.052

Gurudas, M., Kumar, T.N., & Ghosh. (2017). Enhancement of Mechanical Properties in Bainitic Steel Processed from Different Austenitization Temperatures. *Steel Research International, 87*, 1700259 (1-9). doi:10.1002/srin.201700259.

Hajizad, O., Kumar, A., Li, Z., Petrov, R. H., Sietsma, J., & Dollevoet, R. (2019). Influence of microstructure on mechanical properties of bainitic steels in railway applications. *Metals, 9*(7), 1-19.

Hasan, S., Chakrabarti, D., & Singh, S. B. (2021). Thermomechanical Treatment of Carbide-Free Bainitic Steel. *Metal Science and Heat Treatment, 63*(7), 354–362. doi:10.1007/s11041-021-00695-5

Hasan, S. M., Chakrabarti, D., & Singh, S. B. (2018). Dry rolling/sliding wear behaviour of pearlitic rail and newly developed carbide-free bainitic rail steels. *Wear, 408*, 151–159. doi:10.1016/j.wear.2018.05.006

Hasan, S. M., Ghosh, M., Chakrabarti, D., & Singh, S. B. (2020). Development of continuously cooled low-carbon, low-alloy, high strength carbide-free bainitic rail steels. *Materials Science and Engineering A*, *771*, 138590. doi:10.1016/j.msea.2019.138590

Hlavatý, I., Sigmund, M., Krejčí, L., & Mohyla, P. (2009). The Bainitic Steels for Rails Applications. *Materials Engineering*, *16*(4), 44–50.

Huang, Y. B., Shi, L. B., Zhao, X. J., Cai, Z. B., Liu, Q. Y., & Wang, W. J. (2018). On the formation and damage mechanism of rolling contact fatigue surface cracks of wheel/rail under the dry condition. *Wear*, *400*, 62–73. doi:10.1016/j.wear.2017.12.020

Ishida, M. (2013). Rolling contact fatigue (RCF) defects of rails in Japanese railways and its mitigation strategies. *Electronic Journal of Structural Engineering*, *13*(1), 67–74. doi:10.56748/ejse.131621

Johnson, D. R., & Becker, W. T. (1993). Toughness of tempered upper and lower bainitic microstructures in a 4150 steel. *Journal of Materials Engineering and Performance*, *2*(2), 255–263. doi:10.1007/BF02660294

Kang, J., Zhang, F. C., Yang, X. W., Lv, B., & Wu, K. M. (2017). Effect of tempering on the microstructure and mechanical properties of a medium carbon bainitic steel. *Materials Science and Engineering A*, *686*, 150–159. doi:10.1016/j.msea.2017.01.044

Kapito, A., Stumpf, W., & Papo, M. J. (2012). On the development of bainitic alloys for railway wheel applications. *Journal of the Southern African Institute of Mining and Metallurgy*, *7A*, 539–544.

Klaasen, H., Kollo, L., & Kübarsepp, J. (2007). Mechanical properties and wear performance of compression sintered TiC based cermets. *Powder Metallurgy*, *50*(2), 132–136. doi:10.1179/174329007X153305

Kostryzhev, A., Singh, N., Chen, L., Killmore, C., & Pereloma, E. (2018). Comparative effect of Mo and Cr on microstructure and mechanical properties in NbV-microalloyed bainitic steels. *Metals*, *8*(2), 1-19.

Kumar, A., & Singh, A. (2021). Mechanical properties of nanostructured bainitic steels. *Materialia*, *15*, 101034. doi:10.1016/j.mtla.2021.101034

Lai, G. Y., Wood, W. E., Clark, R. A., Zackay, V. F., & Parker, E. R. (1974). The effect of austenitizing temperature on the microstructure and mechanical properties of as-quenched 4340 steel. *Metallurgical and Materials Transactions. B, Process Metallurgy and Materials Processing Science*, *5*, 1663–1670.

Lee, K. M., & Polycarpou, A. A. (2005). Wear of conventional pearlitic and improved bainitic rail steels. *Wear*, *259*(1-6), 391–399. doi:10.1016/j.wear.2005.02.058

Li, Q., Huang, X., & Huang, W. (2019). Fatigue property and microstructure deformation behavior of multiphase microstructure in a medium-carbon bainite steel under rolling contact condition. *International Journal of Fatigue*, *125*, 381–393. doi:10.1016/j.ijfatigue.2019.04.019

Li, Y., Kang, G., Wang, C., Dou, P., & Wang, J. (2006). Vertical short-crack behavior and its application in rolling contact fatigue. *International Journal of Fatigue*, *28*(7), 804–811. doi:10.1016/j.ijfatigue.2005.06.047

Li, Y., & Milbourn, D. (2011). Vanadium in bainitic steels: A review of recent developments. *Advanced Steels: The Recent Scenario in Steel Science and Technology*, 303-308.

Lin, Y., & Zhou, Q. (2018, February). The effect of V Ni microalloying on microstructure and properties of bainite steel. In *2017 3rd International Forum on Energy, Environment Science and Materials (IFEESM 2017)* (pp. 1656-1659). Atlantis Press. 10.2991/ifeesm-17.2018.302

Liu, C. P., Pan, J. Z., Liu, P. T., Ren, R. M., & Zhao, X. J. (2020). Influence of original microstructure on rolling contact fatigue property of D2 wheel steel. *Wear*, *456*, 203380. doi:10.1016/j.wear.2020.203380

Liu, J., Li, Y., Zhang, Y., Hu, Y., Shi, L., Ding, H., Wang, W., Liu, F., Zhou, S., & Shi, T. (2020). Dry rolling/sliding wear of bainitic rail steels under different contact stresses and slip ratios. *Materials (Basel)*, *13*(4678), 1–15. doi:10.3390/ma13204678 PMID:33092302

Liu, M., Fan, Y., Gui, X., Hu, J., Wang, X., & Gao, G. (2022). Relationship between Microstructure and Properties of 1380 MPa Grade Bainitic Rail Steel Treated by Online Bainite-Based Quenching and Partitioning Concept. *Metals, 12*(2), 1-15.

Lonardelli, I., Bortolotti, M., van Beek, W., Girardini, L., Zadra, M., & Bhadeshia, H. K. D. H. (2012). Powder metallurgical nanostructured medium carbon bainitic steel: Kinetics, structure, and in situ thermal stability studies. *Materials Science and Engineering A*, *555*, 139–147. doi:10.1016/j.msea.2012.06.043

Long, X. Y., Branco, R., Zhang, F. C., Berto, F., & Martins, R. F. (2020). Influence of Mn addition on cyclic deformation behaviour of bainitic rail steels. *International Journal of Fatigue*, *132*, 105362. doi:10.1016/j.ijfatigue.2019.105362

Long, X. Y., Kang, J., Lv, B., & Zhang, F. C. (2014). Carbide-free bainite in medium carbon steel. *Materials & Design*, *64*, 237–245. doi:10.1016/j.matdes.2014.07.055

Long, X. Y., Zhang, F. C., Kang, J., Lv, B., & Shi, X. B. (2014). Low-temperature bainite in low-carbon steel. *Materials Science and Engineering A*, *594*, 344–351. doi:10.1016/j.msea.2013.11.089

Magel, E., Sroba, P., Sawley, K., & Kalousek, J. (2005). *Control of rolling contact fatigue of rails.* Center for Surface Transportation Technology, National Research Council Canada.

Mahale, R. S., Rajendrachari, S., Vasanth, S., Krishna, H., Kapanigowda, N. S., Chikkegowda, S. P., & Patil, A. (2022). Technology and challenges in additive manufacturing of duplex stainless steels. *Biointerface Research in Applied Chemistry*, *12*(1), 1110–1119.

Mahale, R. S., Shamanth, V., Hemanth, K., Nithin, S. K., Sharath, P. C., Shashanka, R., Patil, A., & Shetty, D. (2022). Processes and applications of metal additive manufacturing. *Materials Today: Proceedings*, *54*, 228–233. doi:10.1016/j.matpr.2021.08.298

Mahale, R. S., & Shashanka, R. (2022, March). Mechanical testing of spark plasma sintered materials: A review. In AIP Conference Proceedings (Vol. 2469, No. 1). AIP Publishing. doi:10.1063/5.0080183

Masoumi, M., Tressia, G., Centeno, D. M. A., & Goldenstein, H. (2021). Improving the mechanical properties and wear resistance of a commercial pearlitic rail steel using a two-step heat treatment. *Metallurgical and Materials Transactions. A, Physical Metallurgy and Materials Science*, *52*(11), 4888–4906. doi:10.1007/s11661-021-06432-0

Miao, Fan, Gui, Hu, Wang, & Gao. (2022). Relationship between Microstructure and Properties of 1380 MPa Grade Bainitic Rail Steel Treated by Online Bainite-Based Quenching and Partitioning Concept. *Metals*, *12*(330), 1–15.

Moon, A. P., Sangal, S., Layek, S., Giribaskar, S., & Mondal, K. (2015). Corrosion behavior of high-strength bainitic rail steels. *Metallurgical and Materials Transactions. A, Physical Metallurgy and Materials Science*, *46*(4), 1500–1518. doi:10.1007/s11661-014-2732-0

Nayak, A. K., Shashanka, R., & Chaira, D. (2016). Effect of NanosizeYittria and Tungsten Addition to Duplex Stainless Steel During High Energy Planetary Milling. *IOP Conf. Series. Materials Science and Engineering*, *115*, 012008.

Onal, O., Canadinc, D., Sehitoglu, H., Verzal, K., & Jiang, Y. (2012). Investigation of rolling contact crack initiation in bainitic and pearlitic rail steels. *Fatigue & Fracture of Engineering Materials & Structures*, *35*(11), 985–997. doi:10.1111/j.1460-2695.2012.01686.x

Pacyna, J. (2008). The microstructure and properties of the new bainitic rail steels. *Journal of Achievements in Materials and Manufacturing Engineering*, *28*(1), 19–22.

Pereloma, E., & Edmonds, D. V. (Eds.). (2012). *Phase transformations in steels: Diffusionless transformations, high strength steels, modelling and advanced analytical techniques*. Elsevier.

Perez-Unzueta, A. J., & Beynon, J. H. (1993). Microstructure and wear resistance of pearlitic rail steels. *Wear*, *162*, 173–182. doi:10.1016/0043-1648(93)90498-B

Pointner, P. (2008). High strength rail steels-The importance of material properties in contact mechanics problems. *Wear*, *265*(9-10), 1373–1379. doi:10.1016/j.wear.2008.03.015

Qiu, J., Zhang, M., Tan, Z., Gao, G., & Bai, B. (2019). Research on the microstructures and mechanical properties of bainite/martensite rail treated by the controlled-cooling process. *Materials (Basel)*, *12*(3061), 1–9. doi:10.3390/ma12193061 PMID:31547092

Rajendrachari, S. (2018). Effect of sintering temperature on the pitting corrosion of ball milled duplex stainless steel by using linear sweep voltammetry. *Analytical and Bioanalytical Electrochemistry*, *10*(3), 349–361.

Rajendrachari, S. (2022). An overview of high-entropy alloys prepared by mechanical alloying followed by the characterization of their microstructure and various properties. *Alloys*, *1*(2), 116–132. doi:10.3390/alloys1020008

Rajendrachari, S., Swamy, B. K., Reddy, S., & Chaira, D. (2013). Synthesis of silver nanoparticles and their applications. *Analytical and Bioanalytical Electrochemistry*, *5*(4), 455–466.

Ray, A. (2017). Niobium microalloying in rail steels. *Materials Science and Technology*, *33*(14), 1584–1600. doi:10.1080/02670836.2017.1309111

Rezende, A. B., Fonseca, S. T., Fernandes, F. M., Miranda, R. S., Grijalba, F. A. F., Farina, P. F. S., & Mei, P. R. (2020). Wear behavior of bainitic and pearlitic microstructures from microalloyed railway wheel steel. *Wear*, *456*, 203377. doi:10.1016/j.wear.2020.203377

Rezende, A. B., Fonseca, S. T., Miranda, R. S., Fernandes, F. M., Grijalba, F. A. F., Farina, P. F. S., & Mei, P. R. (2021). Effect of niobium and molybdenum addition on the wear resistance and the rolling contact fatigue of railway wheels. *Wear, 466*, 203571. doi:10.1016/j.wear.2020.203571

Satyam, S., Goutam, M., & Totten G. (2009). Overview of Pearlitic Rail Steel: Accelerated Cooling, Quenching, Microstructure, and Mechanical Properties. *Journal of ASTM International, 6*(7), 1–26.

Samek, L., de Moor, E., Penning, J., & De Cooman, B. C. (2006). Influence of Alloying Elements on the Kinetics of Strain-Induced Martensitic Nucleation in Low-Alloy, Multiphase High-Strength Steels. *Metallurgical and Materials Transactions. A, Physical Metallurgy and Materials Science, 37A*(1), 109–124. doi:10.1007/s11661-006-0157-0

Satoh, Y., & Iwafuchi, K. (2005). Effect of Rail/Wheel Rolling Contact on Microstructure in Surface Layer of Rail. *Quarterly Report of RTRI, 46*(3), 178–183. doi:10.2219/rtriqr.46.178

Satoh, Y., Tatsumi, M., Kasiwaya, K., Ueda, M., & Yokoyama, H. (1999). Development of anti-darkspot bainitic steel rail. *Quarterly Report of RTRI, 40*(2), 86–91. doi:10.2219/rtriqr.40.86

Sawley, K., & Kristan, J. (2003). Development of bainitic rail steels with potential resistance to rolling contact fatigue. *Fatigue & Fracture of Engineering Materials & Structures, 26*(10), 1019–1029. doi:10.1046/j.1460-2695.2003.00671.x

Shashanka, R. (2016). *Fabrication of nano-structured duplex and ferritic stainless steel by planetary milling followed by consolidation* [Doctoral dissertation]. NIT, Rourkela, India.

Shashanka, R. (2017). Synthesis of nano-structured stainless steel powder by mechanical alloying-an overview. *International Journal of Scientific and Engineering Research, 8*, 588–594.

Shashanka, R. (2019). Non-lubricated dry sliding wear behavior of spark plasma sintered nano-structured stainless steel. *Journal of Materials & Environmental Sciences, 10*(8), 767–777.

Shashanka, R., & Chaira, D. (2015). Optimization of milling parameters for the synthesis of nano-structured duplex and ferritic stainless steel powders by high energy planetary milling. *Powder Technology, 278*, 35–45. doi:10.1016/j.powtec.2015.03.007

Shashanka, R., & Chaira, D. (2016). Effects of nano-Y_2O_3 and sintering parameters on the fabrication of PM duplex and ferritic stainless steels. *Acta Metallurgica Sinica, 29*(1), 58–71. doi:10.1007/s40195-015-0362-1

Shashanka, R., & Chaira, D. (2017). Effect of sintering temperature and atmosphere on nonlubricated sliding wear of nano-yttria-dispersed and yttria-free duplex and ferritic stainless steel fabricated by powder metallurgy. *Tribology Transactions, 60*(2), 324–336. doi:10.1080/10402004.2016.1168897

Shashanka, R., Chaira, D., & Kumara Swamy, B. K. (2016). Fabrication of yttria dispersed duplex stainless steel electrode to determine dopamine, ascorbic and uric acid electrochemically by using cyclic voltammetry. *International Journal of Scientific and Engineering Research, 7*(3), 1275–1285.

Shashanka, R., Chaira, D., & Swamy, B. K. (2015). Electrochemical investigation of duplex stainless steel at carbon paste electrode and its application to the detection of dopamine, ascorbic and uric acid. *International Journal of Scientific and Engineering Research, 6*(9), 1863–1871.

Shashanka, R., & Kumara Swamy, B. E. (2020). Simultaneous electro-generation and electro-deposition of copper oxide nanoparticles on glassy carbon electrode and its sensor application. *SN Applied Sciences, 2*(5), 1–10. doi:10.1007/s42452-020-2785-1

Shur, E. A., Bychkova, N. Y., & Trushevsky, S. M. (2005). Physical metallurgy aspects of rolling contact fatigue of rail steels. *Wear, 258*(7-8), 1165–1171. doi:10.1016/j.wear.2004.03.027

Singh, U. P., Roy, B., Jha, S., & Bhattacharyya, S. K. (2001). Microstructure and mechanical properties of as rolled high strength bainitic rail steels. *Materials Science and Technology, 17*(1), 33–38. doi:10.1179/026708301101509098

Stock, R. F. (2012). *Influencing rolling contact fatigue and wear by different rail grades and contact conditions* [Doctoral dissertation]. University of Leoben, Austria.

Strey, N. F., Rezende, A. B., da Silva Miranda, R., da Fonseca, S. T., Mei, P. R., & Scandian, C. (2021). Comparison of rolling contact fatigue damage between railway wheels and twin-disc test specimens. *Tribology International, 160*, 107037. doi:10.1016/j.triboint.2021.107037

Su, X., & Clayton, P. (1996). Surface-initiated rolling contact fatigue of pearlitic and low carbon bainitic steels. *Wear, 197*(1-2), 137–144. doi:10.1016/0043-1648(95)06888-0

Suryanarayana, C. (2001). Mechanical Alloying and Milling. *Progress in Materials Science*, *46*(1-2), 1–184. doi:10.1016/S0079-6425(99)00010-9

Suryanarayana, C., Al Joubori Ahmed A.& Wang Zhi. (2022). Nanostructured Materials and nanocomposites by Mechanical Alloying: An Overview. *Metals and Materials International*, *28*, 41–53. doi:10.1007/s12540-021-00998-5

Takahashi, M., & Bhadeshia, H. K. D. H. (1991). A model for the microstructure of some advanced bainitic steels. *Materials Transactions. JIM*, *32*(8), 689–696.

Tasak, E., Ziewiec, A., & Tuz, L. (2014). Resistance Welding of High-Manganese Cast Steel with Unalloyed Steel Rails. *Archives of Foundry Engineering*, *14*, 115–120.

Tenaglia, N. E., Massone, J. M., Boeri, R. E., & Speer, J. G. (2020). Effect of microsegregation on carbide-free bainitic transformation in a high-silicon cast steel. *Materials Science and Technology*, *36*(6), 690–698. doi:10.1080/02670836.2020.1732076

Tressia, G., Alves, L. H., Sinatora, A., Goldenstein, H., & Masoumi, M. (2020). Effect of bainitic transformation on the microstructure and wear resistance of pearlitic rail steel. *Industrial Lubrication and Tribology*, *72*(9), 1095–1102. doi:10.1108/ILT-07-2019-0282

Tunna, J., Sinclair, J., & Perez, J. (2007). A review of wheel wear and rolling contact fatigue. *Proceedings of the Institution of Mechanical Engineers. Part F, Journal of Rail and Rapid Transit*, *221*(2), 271–289. doi:10.1243/0954409JRRT72

Yang, Z., & Zhang, F. (2020). Nanostructured bainitic bearing steel. *ISIJ International*, *60*(1), 18–30. doi:10.2355/isijinternational.ISIJINT-2018-693

Yu, X., Wu, H., Zhang, Y., Yuan, R., Gu, Y., & Feng, Y. (2021). Influence of niobium microalloying on the thermal stability of medium-carbon carbide-free bainitic steel. *Journal of Materials Research and Technology*, *12*, 487–499. doi:10.1016/j.jmrt.2021.03.033

Zapata, D., Jaramillo, J., & Toro, A. (2011). Rolling contact and adhesive wear of bainitic and pearlitic steels in low load regime. *Wear*, *271*(1-2), 393–399. doi:10.1016/j.wear.2010.10.009

Zhang, C., Fu, H., Lin, J., & Lei, Y. (2019). The effect of silicon on microstructure and wear resistance in bainitic steel. *Transactions of the Indian Institute of Metals*, *72*(5), 1231–1244. doi:10.1007/s12666-019-01611-5

Zhang, C., Li, S., Fu, H., & Lin, Y. (2020). Microstructure evolution and wear resistance of high silicon bainitic steel after austempering. *Journal of Materials Research and Technology*, *9*(3), 4826–4839. doi:10.1016/j.jmrt.2020.03.003

Zhang, R., Zheng, C., Chen, C., Lv, B., Gao, G., Yang, Z., Yang, Y., & Zhang, F. (2021). Study on fatigue wear competition mechanism and microstructure evolution on the surface of a bainitic steel rail. *Wear*, *482-483*, 203978. doi:10.1016/j.wear.2021.203978

Zheng, C., Dan, R., Zhang, F., Lv, B., Yan, Z., Shan, J., & Long, X. (2014). Effects of retained austenite and hydrogen on the rolling contact fatigue behaviours of carbide-free bainitic steel. *Materials Science and Engineering A*, *594*, 364–371. doi:10.1016/j.msea.2013.11.072

Zheng, C., Lv, B., Zhang, F., Yang, Z., Kang, J., She, L., & Wang, T. (2016). A novel microstructure of carbide-free bainitic medium carbon steel observed during rolling contact fatigue. *Scripta Materialia*, *114*, 13–16. doi:10.1016/j.scriptamat.2015.11.023

Zhou, M., Xu, G., Tian, J., Hu, H., & Yuan, Q. (2017). Bainitic transformation and properties of low carbon carbide-free bainitic steels with Cr addition. *Metals*, *7*(7), 1-13.

Zhu, J., Tan, Z., Tian, Y., Gao, B., Zhang, M., Wang, J., & Weng, Y. (2020). Effect of tempering temperature on microstructure and mechanical properties of bainitic railway wheel steel with thermal damage resistance by alloy design. *Metals*, *10*(9), 1-13.

Chapter 8
Sintering of Mechanically Alloyed Powders

B. N. Sharath
https://orcid.org/0000-0001-7521-1759
Malnad College of Engineering, India

R. Raghavendra Rao
Malnad College of Engineering, India

K. S. Madhu
Malnad College of Engineering, India

S. Pradeep
Malnad College of Engineering, India

ABSTRACT

The process of sintering mechanically alloyed powders entails subjecting a powder mixture of diverse metals or alloys, which have undergone mechanical milling to elevated temperatures to produce a compacted mass. This procedure is employed to fabricate substances possessing distinctive mechanical, physical, and chemical characteristics. The process of mechanical alloying entails the milling of two or more metals or alloys together in a high-energy ball mill. The procedure yields a powder amalgamation that exhibits a uniform dispersion of its constituent elements, thereby engendering a microstructural morphology characterized by fine-grained features.

DOI: 10.4018/978-1-6684-9385-4.ch008

1. INTRODUCTION

Materials scientists have always focused their scientific inquiries on finding ways to enhance the characteristics and performance of materials. Notable enhancements in mechanical, chemical, and physical characteristics have been attained via chemical alterations and traditional techniques of thermal, mechanical, and thermomechanical techniques. The pursuit of materials that possess greater strength, stiffness, and reduced weight compared to conventional materials has resulted in the creation and advancement of sophisticated materials. The endeavours have been further stimulated by the high-technology sectors. Advanced materials can be defined as materials that prioritise the methodical synthesis and regulation of their structure, with the aim of producing a precisely customised range of properties that are suitable for challenging applications. Currently, it is widely acknowledged that advanced materials can be more effectively regulated in terms of their structure and composition through non-equilibrium (sometimes referred to as "far from equilibrium") processing methods (Suryanarayana, 1996). Sintering is a method of manufacturing whereby metal or ceramic powders are sintered together under high temperatures to create materials and components with a specified density. As a result, sintering falls within the category of synthesis/processing, one of the four foundational sub-fields of materials science and engineering (Figure 1). Sintering's usefulness as a processing method

Figure 1. The fundamental constituents of materials science and engineering (Kang, 2004)

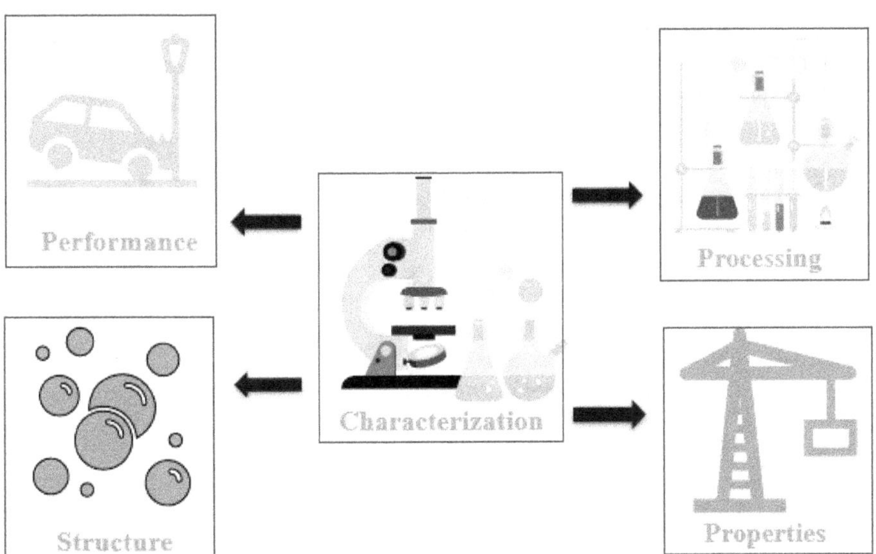

for materials is growing as material synthesis and processing have taken on more significance in recent years. The process of sintering is considered to be one of the earliest forms of human technology, dating back to prehistoric times when pottery was first fired. Sintering made it feasible to manufacture tools out of sponge iron, which was another benefit of the process. Despite this, basic and scientific research on sintering didn't begin until well after the 1940s.There have been significant leaps in in scientific understanding of the sintering process since that time (Fang, 2010; German, 1996; Kang, 2004).

The utilisation of sintering in contemporary times is deemed to be highly significant and advantageous, particularly in the production of diverse sintered parts such as powder-metallurgical parts and bulk ceramic components. Figure 2. illustrates the overall process that is used to fabricate sintered parts. When producing such components, in contrast to the manufacturing of parts using conventional processing methods, several processing stages and factors need to be taken into consideration. For instance, the stage of shaping can be accomplished by the use of basic die compaction, isostatic pressing, slip casting, injection moulding, and many more methods.in accordance with the desired form and characteristics of the final product. Not only the sintering circumstances, but also the characteristics of the sintered material might be rather different from one another depending on the shaping procedures that are applied. There are many different procedures and processing factors involved in the sintering stage, which can lead to variances in the microstructure and characteristics of the sintered product. The purpose of sintering, in general, is to generate sintered components with a repeatable and, if at all feasible,

Figure 2. Manufacturing procedure for sintered components in general (Rahaman, 2017)

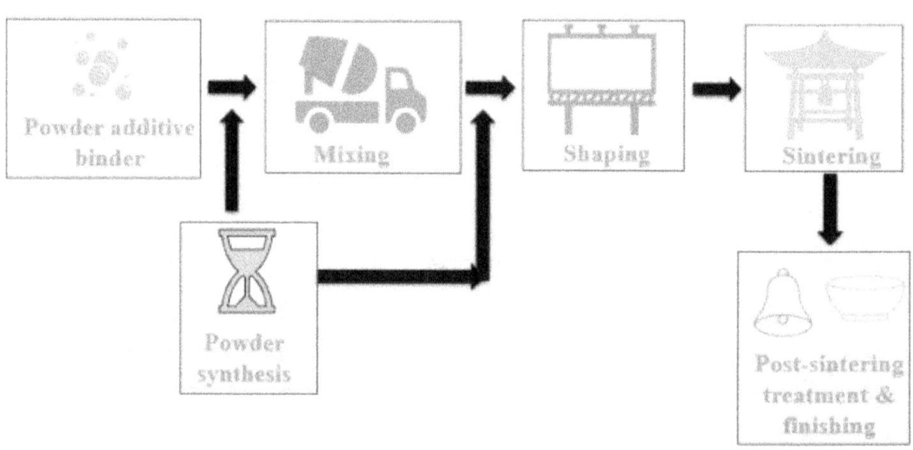

specified microstructure through the control of sintering factors. Sintering is a thermal treatment method employed to consolidate and shape solid substances by subjecting them to a temperature that is lower than their melting point. Sintering is a process whereby the constituent particles of a material are fused together, resulting in the formation of a solid object that exhibits enhanced mechanical properties, including heightened strength and durability. The process of sintering is extensively employed in the manufacturing of ceramics, metals, and various other materials. The manufacturing process is of significant importance in the production of a diverse range of products, encompassing electronics components, automotive parts, and medical devices. In contemporary times, the significance of sintering as a material processing technology has escalated due to the critical role that material synthesis and processing play in the advancement of materials (Cross et al., 2015; German et al., 2009; Rahaman, 2017).

Various sintering techniques exist, such as solid-state sintering, liquid-phase sintering, and gas-phase sintering. The selection of a sintering process is contingent upon the characteristics of the material undergoing sintering and the intended properties of the resultant product.

1.1 Classification of Sintering Processes

In their most fundamental form, sintering operations may be broken down into two categories: solid state sintering and liquid phase sintering. Solid state the process of sintering takes place when the powder compact is heated to the sintering temperature and undergoes complete densification in a solid form. In contrast, the occurrence of liquid phase sintering is contingent upon the presence of a liquid phase within the powder compact. Apart from the conventional solid state and liquid phase sintering, alternative sintering techniques such as transient liquid phase sintering and viscous flow sintering can also be employed. The phenomenon of viscous flow sintering takes place when the proportion of liquid volume reaches a certain threshold, enabling the compact to attain complete densification through the viscous flow of a mixture of grain and liquid, without any alteration in the shape of the grain during the process of densification. The process of transient liquid phase sintering involves the amalgamation of solid-state sintering and liquid phase sintering techniques. The sintering process involves the formation of a liquid phase within the compact during the initial stages of sintering. However, as the sintering process progresses, the liquid phase dissipates, and the densification process is completed through solid-state mechanisms.

Material variables and process variables are the primary factors that play a role in determining the sinterability and sintered microstructure of a powder compact. These fundamental factors may be split into two groups (Figure 3).

Figure 3. Sintering variables (German, 2014)

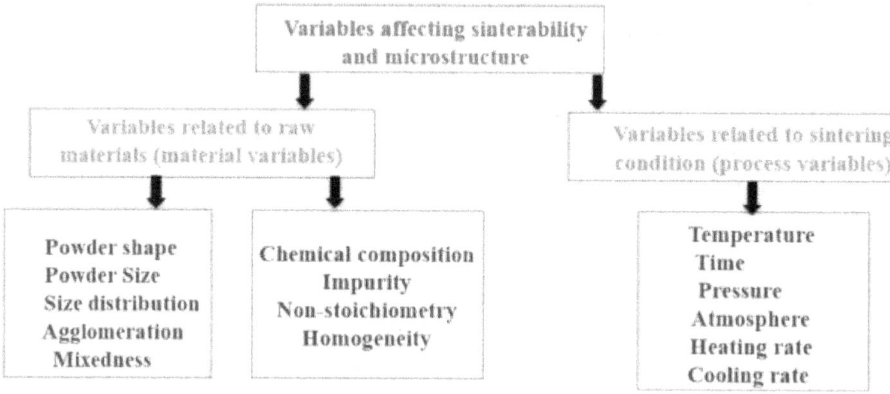

The variables that are associated with the raw materials are significant. These variables have an impact on the compressibility and sinterability of powders. Possessing a powder composition that is homogeneous is especially crucial for multi-powder compacts. In order to enhance homogeneity, various methods have been explored and implemented, including mechanical milling as well as chemical techniques such as sol-gel and coprecipitation processes. Numerous prior investigations on sintering have explored the impact of sintering temperature and duration on the sinterability of powder compacts. However, it is evident that in practical processing, the impacts of sintering atmosphere and pressure are considerably intricate and significant. There has been extensive research and development focused on unconventional methods for controlling these variables (Braginsky et al., 2005; Ertug, 2013; German, 2014; Robert & Haasen, n.d.).

1.2 Solid-State Sintering

Solid-state sintering is the term used to describe the procedure of consolidating and compacting particles through the use of heat, while maintaining a temperature below the material's melting point.

The process of sintering is distinguished by a decrease in the available surface area of the compact, frequently succeeded by a rise in its density. However, the compact may reduce its surface area and enhance its cohesiveness without observable density change. In addition to these structural modifications, the compact's electrical and thermal conductivities will improve. The increased density achieved during sintering is often what is sought after in practical applications. The driving force behind any physical process is the reduction of the system's internal energy. The process of sintering involves a decrease in energy, which can be attributed to the reduction in the overall surface energy of the powder particles. This reduction occurs as a result of the replacement of surfaces with grain boundaries and the process of coarsening. In general, the diminution of internal energy can be expressed as the disparity between the complete surface energy of the powder compact and the residual surface energy, along with the entire grain boundary energy of the sintered body. Sintering is a thermal treatment method that involves subjecting a substance in either a powdered or compacted state to a temperature that is lower than its melting point. This process induces atomic diffusion, resulting in the formation of a solid material that exhibits a more uniform and denser structure. This method is utilised extensively throughout the production of a variety of materials, including ceramics, metals, and composites. Solid-state sintering is a process that takes place exclusively in the solid state, without the involvement of a liquid phase in the material. The procedure commonly encompasses the subsequent stages (Figure 4). Densification is a process whereby the particles of a material are subjected to heat, leading to the formation of inter-particle bonds, and ultimately resulting in a more compact and dense structure. The precise bonding mechanism is contingent upon the specific material undergoing sintering and the process parameters. In the context of material processing, the cooling process is a crucial step that follows sintering. The material is subjected to a controlled cooling process to gradually reduce its temperature to ambient levels. This is done to mitigate the risk of undesirable outcomes such as

Figure 4. Solid-state sintering steps (Bordia et al., 2017; Castro & Van Benthem, 2012)

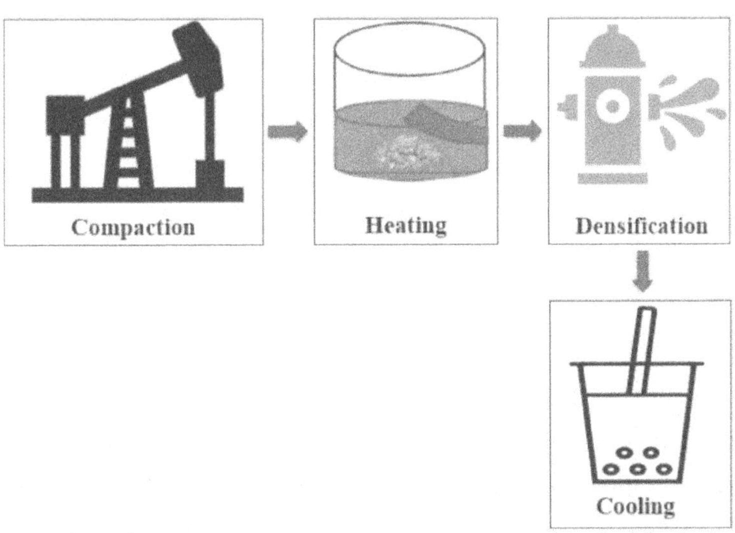

cracking or warping that may arise due to rapid cooling. The manipulation of the sintering process can result in the attainment of desired characteristics, including but not limited to, porosity, strength, and conductivity, in the ultimate product. The compaction process can be subject to modification through the introduction of dopants or other substances into the powder prior to its compaction, thereby altering the characteristics of the ultimate output (Bordia et al., 2017; Castro & Van Benthem, 2012).

Various solid-state sintering techniques are frequently employed, contingent upon the materials undergoing sintering and the intended characteristics of the ultimate product. Several categories can be identified, including:

Figure 5. Types of solid-state sintering

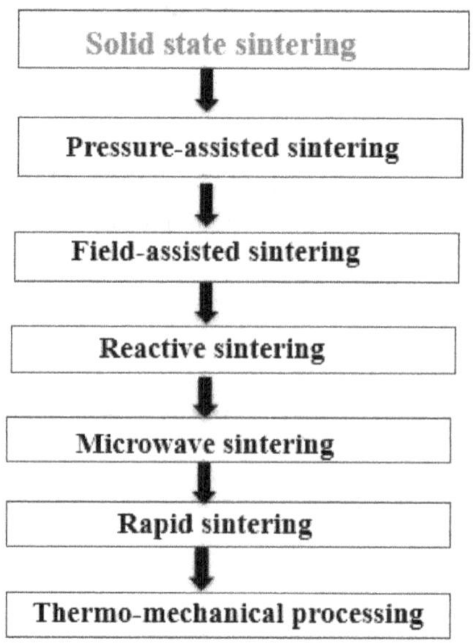

1.2.1 Pressure-Assisted Sintering

The process of pressure-assisted sintering is a solid-state sintering technique that entails the application of pressure during sintering with the aim of enhancing densification and minimising porosity. The aforementioned method is frequently employed for the compaction of substances that pose challenges in terms of densification through traditional sintering processes, including but not limited to ceramics and opt for metals. The utilisation of pressure-assisted sintering methodologies has the potential to yield materials possessing elevated density, uniform microstructure, and superior mechanical characteristics in comparison to those produced through traditional sintering techniques. Nevertheless, these methodologies may incur higher costs and necessitate specialised equipment and proficiency (Diantonio & Ewsuk, 2010; German, 2014; Pomeroy, 2021). There exist various pressure-assisted sintering methodologies, such as hot pressing, spark plasma sintering, and hot isostatic pressing. Various techniques possess their own set of benefits and drawbacks and are appropriate for diverse materials and applications.

Hot Pressing

The process of hot pressing entails the application of both heat and pressure to the material in order to facilitate densification. The substance is inserted into a mould and subjected to thermal energy at a level below its point of liquefaction, yet sufficiently elevated to encourage the process of molecular diffusion. Densification is promoted in the material through the application of pressure during the heating process. The procedure may be conducted under vacuum or in an inert environment to avert oxidation or contamination.

Spark Plasma Sintering

The Spark Plasma Sintering (SPS) method, also referred to as pulsed electric current sintering, utilises high current pulses to facilitate densification of the material. The substance is inserted into a mould and subjected to a temperature that is lower than its melting point, yet sufficiently elevated to facilitate the process of diffusion. Subsequently, the material is subjected to high current pulses in order to facilitate densification. The procedure may be conducted under vacuum conditions or in the presence of an inert atmosphere to avert oxidation or contamination.

Hot Isostatic Pressing

The process of Hot Isostatic Pressing (HIP) entails the application of heat and pressure to a given material through the utilisation of an inert gas or liquid medium. The substance is introduced into a vessel and subjected to thermal energy, maintaining a temperature that is below its melting threshold but sufficiently elevated to facilitate the process of diffusion. The process of densification is facilitated by the application of pressure to the material through the utilisation of an inert gas or a liquid medium. The procedure may be executed under vacuum or in an inert atmosphere to avert oxidation or contamination.

1.2.2 Field-Assisted Sintering

Field-assisted sintering, which is also referred to as flash sintering or electric field-assisted sintering, is a solid-state sintering method that entails the application of an electric field during sintering. This technique is aimed at reducing the sintering time and temperature, thereby leading to a more energy-efficient process. The aforementioned method is deemed appropriate for substances that pose challenges in terms of densification via traditional sintering processes, including but not limited to ceramics, metals, and composites. In the process of field-assisted sintering, the material is subjected to an electric field by placing it between two electrodes. The phenomenon of heat generation within a material due to the presence of an electric field is known to facilitate the processes of diffusion and densification. The procedure may be conducted under vacuum conditions or in the presence of an inert atmosphere to inhibit oxidation or contamination. Field-assisted sintering offers a notable benefit in terms of reducing both the sintering time and temperature in comparison to traditional sintering techniques, thereby leading to a more energy-efficient process. The utilisation of this particular technique has the potential to generate materials with distinct microstructures and properties that cannot be attained through traditional sintering methods. Nevertheless, the utilisation of this method necessitates specific equipment and proficiency and may not be appropriate for all substances. Direct current (DC) sintering pulsed current (PC) sintering, and microwave-assisted sintering are some of the many forms that field-assisted sintering may take. DC sintering refers to the use of direct current, while PC sintering and microwave-assisted sintering use pulsed current. Each method offers a unique set of benefits and drawbacks, and it may be utilised with a diverse range of materials and in a variety of contexts.

1.2.3 Reactive Sintering

During reactive sintering, powders are combined with reactive components before being sintered to produce chemical reactions that give the final material its distinctive characteristics. Materials including ceramics, composites, and intermetallic are often synthesised using this process since they are challenging to create using more traditional approaches. Reactive sintering involves subjecting the material to a temperature that is lower than its melting point, but sufficiently high to facilitate diffusion. The powder mixture is supplemented with reactive elements that undergo a chemical reaction with the base material, resulting in the formation of novel compounds. The exothermic nature of the chemical reactions results in the production of thermal energy, thereby facilitating the process of diffusion and densification. The procedure may be conducted under vacuum or in an inert environment to inhibit oxidation or pollution. There are a few distinct approaches to reactive sintering, the most common of which being combustion synthesis and self-propagating high-temperature synthesis (SHS).In the process known as combustion synthesis, the reactive materials are set ablaze to begin the chemical processes, which results in a procedure that is both rapid and effective. Self-sustaining reactions in solid-state materials exhibit a high-energy and self-propagating process, wherein the reaction propagates throughout the material. This phenomenon is commonly observed in SHS. The utilisation of reactive sintering methodologies can yield materials possessing distinctive microstructural features and properties that are unattainable through traditional sintering techniques. Nevertheless, the methodology necessitates meticulous management of the chemical reactions and sintering parameters to guarantee homogeneity and consistency. Reactive sintering methodologies are typically associated with higher costs and necessitate specialised equipment and expertise.

1.2.4 Microwave Sintering

The utilisation of microwave energy to heat and sinter materials is known as microwave sintering. The sintering process involves the utilisation of a microwave generator to produce electromagnetic waves that are assimilated by the material undergoing sintering, thereby inducing a rapid increase in temperature and subsequent sintering. Microwave sintering presents various benefits when compared to conventional sintering techniques. Primarily, it facilitates accelerated heating rates, leading to reduced sintering durations and enhanced energy efficacy. Furthermore, it facilitates targeted heating, thereby enabling meticulous regulation of temperature and homogeneous heating of the substance, thereby enhancing the calibre and characteristics of the sintered output. Furthermore, microwave sintering has the potential to decrease the reliance on sintering additives, such as binders or lubricants, due to its ability to expedite the heating process and facilitate densification without the requirement of external pressure. In addition to its widespread application in the electronic component and device manufacturing industries, microwave sintering is also widely utilised in the manufacture of ceramics, composites, and metals. Fortunately, the technology is still in its infancy and pricey, so it may need further work before it can be extensively used in industry.

1.2.5 Rapid Sintering

The process of rapid sintering endeavours to minimise the duration and energy expenditure necessary for the creation of sintered products that are both fully dense and of superior quality. The objective is to attain the intended degree of compaction within a reduced timeframe, typically through the implementation of elevated temperatures and/or pressures, or the incorporation of supplementary substances or alternate heating modalities. Various methods can be employed for expediting sintering, such as microwave sintering, spark plasma sintering (SPS), and hot isostatic pressing (HIP). The techniques commonly employed for this purpose entail subjecting the materials to elevated temperatures, pressures, or electric fields, facilitating their rapid densification. A notable benefit of rapid sintering is the decreased duration of the processing period, resulting in potentially diminished manufacturing expenses and heightened efficiency. Furthermore, it can lead to enhanced material characteristics, such as heightened strength and resilience, alongside improved uniformity and microstructural features. Notwithstanding its advantages, rapid sintering may entail certain drawbacks such as heightened energy consumption, augmented equipment and maintenance expenses, and the possibility of amplified material defects or non-uniformities. As is the case with any sintering procedure, it is crucial to meticulously weigh the advantages and disadvantages and opt for the suitable technique depending on the particular material and application prerequisites.

1.2.6 Thermo-Mechanical Sintering

The production of superior sintered products can be achieved through the implementation of thermo-mechanical sintering, a technique that involves the application of both thermal energy and mechanical force. The accepted procedure entails utilizing a hydraulic press to exert a substantial amount of force on the substance while simultaneously subjecting it to elevated temperatures for sintering. The process of thermo-mechanical sintering necessitates meticulous control of both pressure and temperature to attain consistent densification of the material and eradication of any pre-existing pores or defects. The simultaneous application of thermal energy and mechanical force facilitates expeditious compaction of the substance, culminating in a sintered output characterised by elevated levels of durability and rigidity. Thermo-mechanical sintering is a widely employed technique in the manufacturing of ceramics, metals, and composites. It has been observed to be especially efficacious for materials that pose challenges in sintering through conventional means. The utilisation of this material is frequently employed in the fabrication of components intended for high-performance applications, including but not limited to cutting tools, bearings, and engine components. Thermo-mechanical sintering confers the benefit of enabling the production of sintered goods characterised by elevated degrees of homogeneity and regularity, notwithstanding intricate geometries. Furthermore, it may aid in diminishing the extent of secondary processing needed, such as grinding or machining, thereby resulting in decreased manufacturing expenses and lead times. Notwithstanding, the process of thermo-mechanical sintering can be intricate and costly, necessitating specialised equipment and expertise for its efficient implementation. Achieving optimal process parameters for a given material and application may necessitate a significant amount of trial and error (Fang & Wang, 2013; Manogar et al., 2022; Pomeroy, 2021; Sullivan et al., 2012; Xu, 2019; Yang et al., 2020).

1.2.7 Liquid Phase Sintering

The liquid phase sintering technique involves the utilisation of a liquid phase to enhance the atomic diffusion and densification of a material during the sintering process. The technique is frequently employed for the consolidation of substances that pose challenges in achieving densification through solid-state sintering in isolation. Examples of such materials include ceramics, specific types of metals, and composites. The process of liquid phase sintering involves the incorporation of a minor quantity of a secondary material, which is usually a metal or glass possessing a lower melting point than the primary material, into the powder mixture prior to compaction. Upon heating, the secondary material undergoes a phase transition into a molten state, which subsequently wets the surface of the primary material particles. This phenomenon results in the formation of a liquid phase that facilitates atomic diffusion and bonding between the two materials. The presence of a liquid phase can serve as a lubricant, facilitating inter-particle sliding and promoting densification. The procedure generally encompasses the subsequent stages (Figure 6).The homogeneity of the powder mixture consisting of the primary and secondary materials is attained by means of a comprehensive blending process, ensuring a uniform dispersion throughout the mixture. Subsequently, the amalgamation is compressed into the intended configuration, commonly accomplished through the application of force into a cast. The densified substance is subjected to thermal

Figure 6. Liquid phase sintering steps (Yoon et al., 2005)

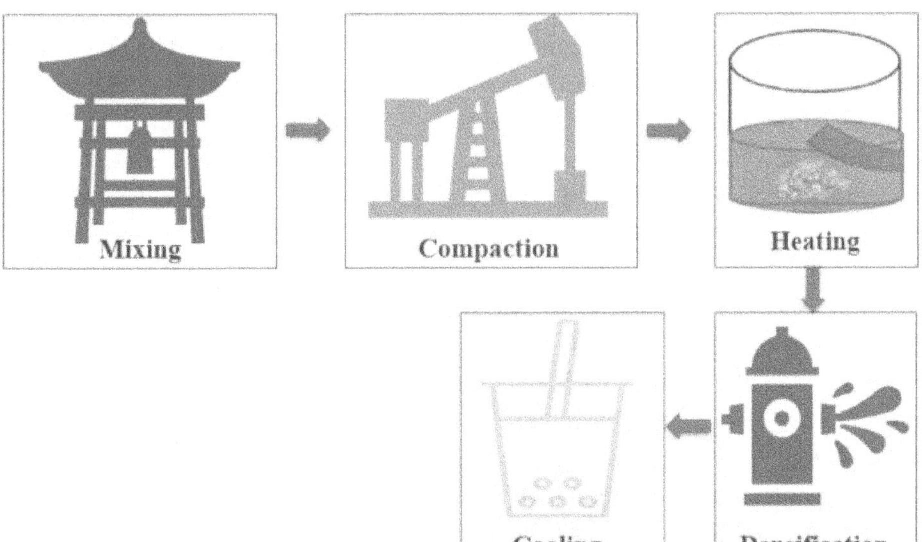

energy at a level surpassing the liquefaction threshold of the secondary constituent yet remaining beneath that of the primary constituent. Within this temperature range, the secondary material is able to undergo a phase transition into a liquid state, while the primary material remains in a solid state. As the temperature rises, the transition from the solid to the liquid phase contributes to the acceleration of atomic diffusion and bonding, which ultimately leads to a denser consistency in the material. After that, the material is gradually cooled until it reaches room temperature in order to avoid breaking or warping. The process of liquid phase sintering is a viable method for customising the characteristics of the product. This can be achieved by regulating the dimensions and arrangement of pores or incorporating alloying components into the primary substance. Nevertheless, the procedure may entail greater intricacy and cost compared to solid-state sintering, owing to the requirement of an extra material and regulation of the melting and cooling phases (de Oro Calderon et al., 2022; Dobrzański et al., 2017; Kang, 2010; Yoon et al., 2005).

There are primarily two distinct types of liquid phase sintering processes.

1.2.8 Infiltration Liquid Phase Sintering

Infiltration liquid phase sintering is a variant of liquid phase sintering that involves the introduction of the liquid phase into the material subsequent to its compaction and formation into a green body. Subsequent to infiltration, the liquid permeates the material via capillary action, thereby occupying the interstices and facilitating compaction. In the process of infiltration liquid phase sintering, the green body is heated to a temperature that is higher than the melting point of the liquid phase but lower than the melting point of the substance that is being sintered. Subsequently, the fluid phase permeates the substance, occupying the interstices and cavities amidst the constituent particles. The presence of the liquid phase facilitates the process of grain growth and particle rearrangement, which ultimately results in enhanced densification and strength. The process of infiltration liquid phase sintering is frequently employed in the production of wear-resistant ceramics and cermets that exhibit high strength. This method is particularly useful in cases where sintering may prove challenging in the absence of a liquid phase. The utilisation of this procedure can also enhance the microstructural and mechanical characteristics of sintered materials, while simultaneously mitigating porosity and enhancing dimensional precision. An inherent drawback associated with infiltration liquid phase sintering pertains to the intricate management of the quantity and dispersion of the liquid phase, thereby resulting in non-homogeneous densification and imperfections in the ultimate output. Furthermore, the implementation of the process can be a laborious and costly endeavour, necessitating the utilisation of specialised equipment and proficiency to execute it efficiently (De Oro Calderón et al., 2016; Ibrahim et al., 2014; Oro et al., 2012).

1.2.9 Powder Liquid Phase Sintering

Powder liquid phase sintering is a variant of liquid phase sintering, whereby the liquid phase is incorporated into the initial powder blend prior to its compaction into a green body. The phenomenon of wetting by the liquid phase facilitates the rearrangement of powder particles and promotes their bonding during the process of sintering. In the process of powder liquid phase sintering, a supplementary substance with a lower melting point than the primary material is incorporated into the initial powder blend. Subsequently, the amalgamation is compressed into a verdant structure and subjected to a temperature surpassing the liquefaction point of the fluid phase, yet lower than the liquefaction points of the substance undergoing sintering. Subsequent to this, the liquid phase exhibits wetting behaviour towards the powder particles, thereby facilitating the process of particle rearrangement and inter-particle bonding. Powder liquid phase sintering is a technique that is often utilised for the processing of metals and metal-ceramic composites because to its high thermal and electrical conductivity. This technique is also utilised for the processing of other materials that have a high sinterability. The aforementioned procedure can also be employed to enhance the microstructural and mechanical characteristics of sintered substances, while simultaneously mitigating porosity and augmenting dimensional precision. A drawback associated with powder liquid phase sintering pertains to the potential reduction in the strength and toughness of the end product, which can occur if the quantity and arrangement of the liquid phase are not effectively regulated. Furthermore, the implementation of the process can be a laborious and costly endeavour, necessitating the utilisation of specialised equipment and proficiency to execute it efficiently (Corker et al., 2000).

1.3 Mechanical Alloying

The process of mechanical alloying (MA) is a powder-based method that facilitates the creation of uniform substances by utilising a combination of elemental powders. This approach was initially implemented in the beginning of 1966. The methodology employed was a culmination of extensive exploration aimed at generating a nickel-based superalloy suitable for employment in gas turbine contexts, which was anticipated to amalgamate the elevated-temperature potency of oxide dispersion with the moderate-temperature potency of gamma-prime precipitate. The alloy was rendered with the necessary corrosion and oxidation resistance through the incorporation of appropriate alloying additions (Benjamin & Arzt, 1989). Furthermore, it has been acknowledged that the mechanical activation of powder mixtures can prompt chemical reactions, known as mechanochemical reactions, to occur at ambient temperature or significantly lower temperatures than those typically necessary for the production of pure metals, nanocomposites, and various materials with commercial applications (McCormick, 1995). The utilisation of mechanical alloying presents numerous benefits in contrast to traditional powder metallurgy methods. These advantages include the capacity to generate powders that are fine-grained and homogeneous, the adaptability to produce a diverse array of alloys and composites, and the possibility of cost-effective processing. The methodology has been extensively employed across diverse sectors such as aerospace, automotive, electronics, and biomedical engineering. One limitation associated with mechanical alloying pertains to its potential to incur significant costs and time investments, necessitating the use of specialised equipment and technical proficiency to execute proficiently. Moreover, the high-energy milling procedure may result in the incorporation of impurities or flaws in the ultimate product, thereby adversely impacting its characteristics (Atazadeh et al., 2016; Soni, 2000; Suryanarayana, 2011; Suryanarayana & Al-Aqeeli, 2013). The Figure7 displays the characteristics of mechanical alloying.

Figure 7. Attributes of mechanical alloying (Suryanarayana, 2001)

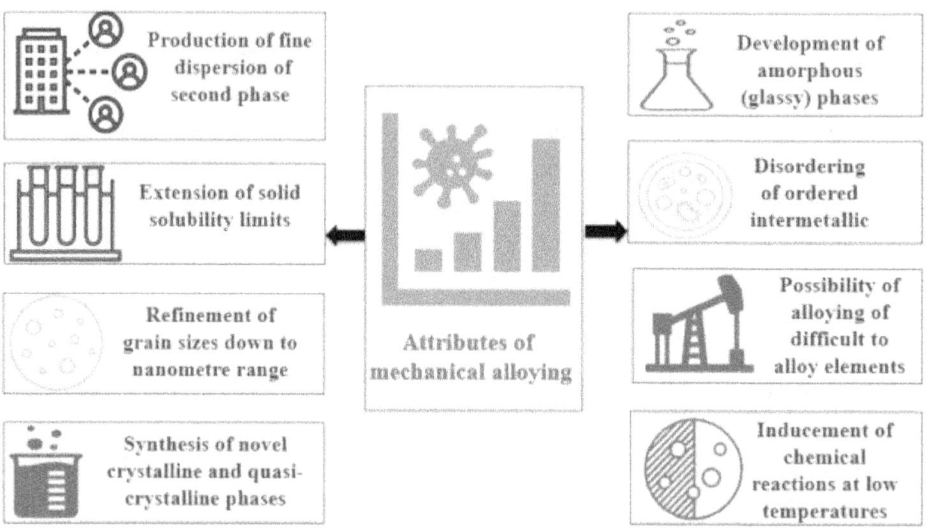

The process of mechanical alloying has the capability to generate powders that are uniform in composition and possess a fine-grained structure, exhibiting particle dimensions that span from a few nanometres to microns. The process of reducing the size of particles can result in improved characteristics, including heightened strength, hardness, and ductility. The process of mechanical alloying guarantees a comprehensive blending and homogenization of powders, resulting in a consistent dispersion of constituent elements within the material. Ensuring consistent properties and performance is a crucial aspect of this endeavour. The process of mechanical alloying has the capability to generate metastable phases which exhibit instability when subjected to equilibrium conditions. The aforementioned stages may manifest distinctive characteristics that are absent in traditional alloys. The process of mechanical alloying has the capability to generate a diverse array of alloys and composites, some of which may be challenging or unfeasible to fabricate using alternative techniques. The process of mechanical alloying can be executed utilising equipment that is relatively low in cost, thereby rendering it a cost-efficient approach for the production of sophisticated materials. Mechanical alloying, which employs high-energy milling, has the potential to decrease processing durations relative to traditional powder metallurgy methods, thereby resulting in expedited production and reduced expenses.

1.4 Process of Mechanical Alloying

The initial step in the process of MA involves the precise blending of powders and subsequent introduction of the resulting mixture into the mill, where it is combined with a grinding medium, typically comprised of steel balls. Subsequently, the mixture undergoes a milling process for a specific duration until it attains a state of equilibrium, wherein the elemental composition of each powder particle corresponds to the proportion of elements present in the initial powder mixture. Subsequently, the powder that has undergone milling is subjected to consolidation into a shape of larger dimensions, followed by a heat treatment process aimed at achieving the intended microstructural and property characteristics. The crucial elements of the MA procedure consist of the input materials, the milling equipment, and the operational parameters (Figure 8).

1.5 Temperature Rises During Mechanical Alloying

Crystal defects are created when powders undergo significant mechanical deformation, and the relative abundance of cold welding and fracturing procedures among powder particles is thought to influence the structural modifications that result. The final characteristics of the powder product can be influenced by the temperature incurred during the milling process, which is contingent upon the kinetic energy of the milling balls. This is an additional significant parameter to consider. If the temperature that is created is high, the increased diffusivity that is

Figure 8. Process flowchart of double mechanical alloying (Suryanarayana, 2001)

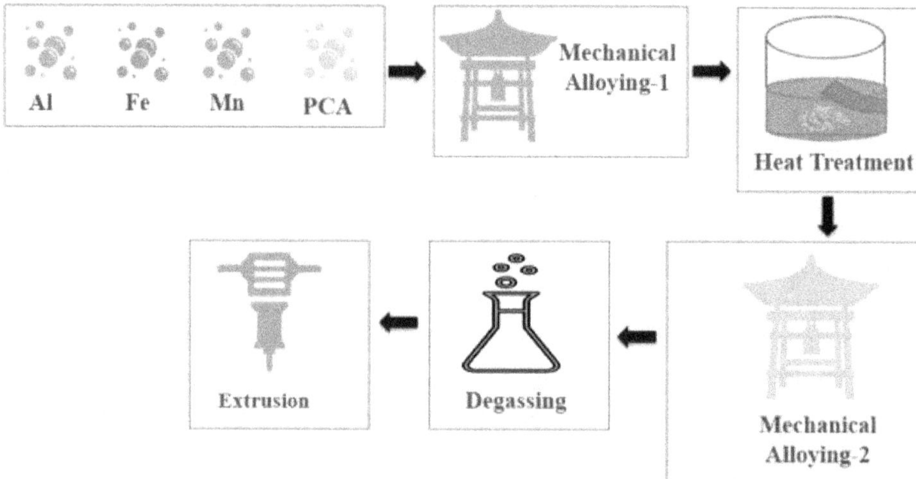

linked with it will lead to processes that result in recovery (and recrystallization). If such a scenario were to occur, a phase that is stable, such as an intermetallic compound, would be generated. Conversely, in the event that the temperature is decreased, the process of defect recovery is hindered, resulting in the formation of an amorphous or nanocrystalline phase. Two distinct factors can contribute to elevated temperatures of powders during the milling process. Initially, as previously stated, the cause of this phenomenon can be attributed to the kinetic energy possessed by the grinding medium. Furthermore, it is plausible that exothermic phenomena taking place throughout the milling procedure produce thermal energy. In practical applications, the measurement of temperature in the powder or milling container can be attributed to a potential combination of both aforementioned factors (Suryanarayana, 2001; Suryanarayana et al., 2001). The temperature increases observed during mechanical alloying are depicted in Figure 9.

2. PERFORMANCE ANALYSIS OF SINTERING OF MECHANICALLY ALLOYED POWDERS

The process of sintering mechanically alloyed powders has the potential to yield materials with improved properties, owing to the attainment of fine particle size and homogeneity through the mechanical alloying procedure. Several methods can be employed to evaluate the performance of sintered materials.

Figure 9. Temperature rises during mechanical alloying (Suryanarayana, 2001)

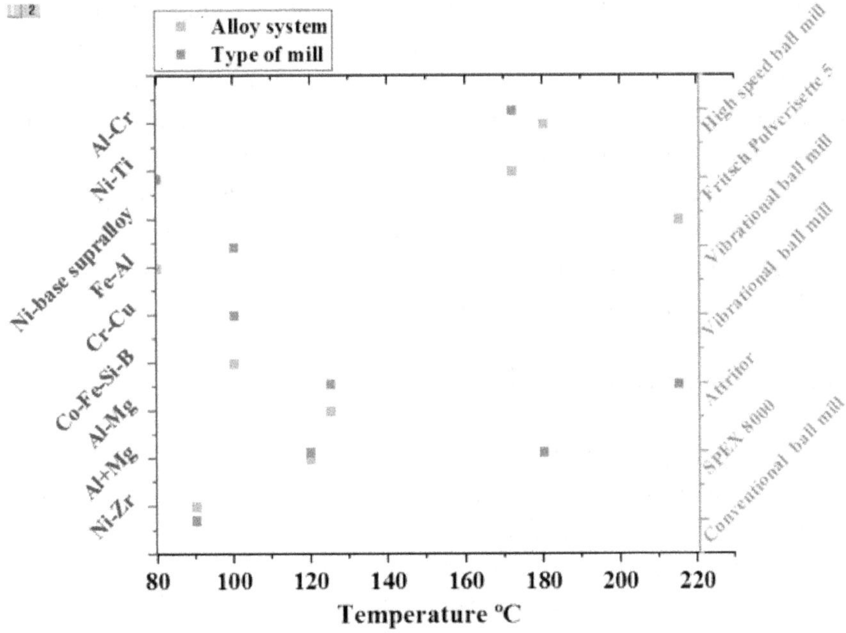

2.1 Density Measurement

The evaluation of density is an essential factor in the assessment of the efficacy of sintered materials that are derived from powders that have undergone mechanical alloying. The quantification of the packing efficiency of a sintered material through its density is a crucial parameter in assessing its characteristics and appropriateness for particular uses. There are a variety of methods available for determining the density of sintered materials. Some examples of these methods are Archimedes' method, gas psychometry, and the mercury intrusion porosimeter method. The method developed by Archimedes includes submerging the sintered material in a liquid and determining how much the weight of the liquid changes as a result. It is possible to determine the density of the material that has been sintered by combining the weight of the material that has been sintered with the volume of the liquid that has been displaced. The process of gas psychometry entails the determination of the volume of gas that is displaced by the sintered material within a restricted area. The determination of the density of the sintered material can be achieved by utilising the mass of the sintered material and the volume of the gas that has been displaced. Mercury intrusion is a technique commonly used in materials science and engineering to determine the pore size distribution and porosity of a material. The process of porosimetry entails the quantification of the amount of mercury that is displaced by the sintered material within a restricted area. The determination of the density of the sintered material can be accomplished by utilising the mass of the sintered material and the volume of the mercury that has been displaced. It is essential to be aware that the density of a sintered material may be influenced by a number of different aspects, including the sintering temperature, duration, and pressure, in addition to the particle size and composition of the mechanically alloyed powders. Hence, it is imperative to meticulously regulate the sintering process parameters to achieve materials possessing the intended density and characteristics (Awotunde et al., 2019; Sharath, Venkatesh, Afzal, Aslfattahi et al, 2021).

2.2 Microstructure Analysis

Microstructure analysis pertains to the scrutiny and delineation of the configuration and attributes of substances at a minuscule scale. The subject matter pertains to the examination and analysis of the organisation, structure, and characteristics of the fundamental components, stages, and imperfections present in each substance.

The examination of microstructure is a usual practice that involves the utilization of diverse methodologies such as SEM, XRD etc.

Optical Microscopy: The methodology entails utilising the electromagnetic radiation within the visible spectrum to analyse the microarchitecture of a substance. The technique furnishes data pertaining to the surface characteristics, grain boundaries, and general morphology of the sample.

The Scanning Electron Microscope (SEM) utilises an electron beam that is concentrated to produce images of a material's surface with exceptional resolution. The method furnishes comprehensive data regarding the topographical and compositional aspects of the specimen, enabling the scrutiny of microstructural characteristics at a superior level of magnification compared to optical microscopy.

Transmission electron microscopy (TEM) is a technique that utilises the transmission of a beam of electrons through a thin specimen to generate an image. The aforementioned offers comprehensive insights into the internal microstructural features, encompassing the crystal structure, grain boundaries, and dislocations. The Transmission Electron Microscope (TEM) is a valuable tool for investigating materials and interfaces at the nanoscale level.

X-ray Diffraction (XRD) is a technique employed to ascertain the crystallographic arrangement and phase composition of a given substance. Through the examination of the diffraction patterns generated by the interaction between X-rays and the specimen, scholars are able to ascertain the crystal configuration and quantify the lattice parameters.

Electron Backscatter Diffraction (EBSD) is a technique that integrates electron microscopy and diffraction methodologies to furnish insights into the crystallographic orientation and grain boundaries of a given material. Polycrystalline materials are frequently subjected to analysis of their texture and microstructure through the use of this technique.

Atomic Force Microscopy (AFM) is a technique that employs a diminutive probe to conduct a surface scan of a material and determine its topography with exceptional precision. The technique has the capability to furnish data pertaining to surface roughness, grain boundaries, and features at the nanoscale level.

Energy-dispersive X-ray spectroscopy (EDS/EDX) is a technique that is commonly employed in tandem with scanning electron microscopy (SEM) or transmission electron microscopy (TEM) to investigate the elemental composition of a given material. The technique furnishes both qualitative and quantitative data pertaining to the chemical composition and spatial distribution of elements within the microstructural features.

The utilisation of various methodologies enables scientists and engineers to scrutinise and comprehend the microstructural attributes of materials. This is a critical aspect in enhancing their properties and efficacy in diverse applications, including metals, ceramics, polymers, semiconductors, and composites (Ballupete Nagaraju et al., 2022; Chikkegouda et al., 2022; Sharath, Venkatesh, Afzal, Baig et al, 2021).

2.3 Mechanical Testing

The assessment of mechanical properties and behaviour of sintered mechanically alloyed powders under varying loading conditions is of paramount importance, necessitating the conduction of mechanical testing. Several mechanical tests can be conducted on sintered mechanically alloyed powders, including those commonly used in the field.

The mechanical property of hardness is considered to be fundamental as it offers valuable insights into a material's ability to withstand indentation or penetration. The hardness of the sintered material can be evaluated through the utilisation of methods such as Vickers or Rockwell hardness testing. The process of hardness testing is utilised to evaluate the mechanical properties of a material, including its strength, wear resistance, and overall mechanical behaviour.

The process of conducting tensile testing is aimed at ascertaining the tensile strength, yield strength, and elongation of a given material. In the realm of materials testing, it is customary to fabricate test specimens in the shape of dog-bones and subject them to a regulated tensile load until the point of fracture. The process of tensile testing is utilised to assess the ductility, deformation characteristics, and ultimate strength of a given material.

The process of compression testing entails the application of a compressive force onto a specimen, with the aim of determining its compressive strength, yield strength, and deformation properties. Since sintered materials are frequently used in applications that involve compression, this is an important consideration. Compression testing may reveal how well a material holds together under pressure and how much it can withstand deformation.

The process of flexural testing, which is also referred to as bending testing, is utilised to assess the capacity of a material to withstand bending or flexure. The experimental procedure entails subjecting a specimen to either a three-point or four-point bending load and subsequently assessing the corresponding deflection and stress distribution. Flexural testing enables the evaluation of a material's stiffness, fracture characteristics, and capacity to withstand bending.

The evaluation of a material's ability to withstand sudden or dynamic loading is conducted through impact testing. The Charpy or Izod test is the prevalent form of impact testing, which involves the use of a pendulum to strike a notched sample, leading to its fracture. The process of impact testing is utilised to assess the toughness, energy absorption capacity, and ability to withstand sudden loading of a given material (Madhu et al., 2021; Madhu et al., 2022).

The process of fatigue testing is utilised to assess the capacity of a material to withstand cyclic stress or repeated loading. The experimental procedure entails subjecting a specimen to repeated loading cycles and quantifying either the count of cycles until the point of failure or the correlation between stress and the count of cycles leading to failure. The process of fatigue testing enables the acquisition of valuable insights pertaining to the durability and capacity of materials to endure cyclic loading conditions (Murakami & Ritchie, 2012).

In order to ensure that the findings of the mechanical testing are precise and dependable, it is imperative that the specimens be meticulously prepared. It is possible for the specific testing procedures and standards to change in accordance with the material and the attributes that are of interest. In addition, the methodologies for analysing the microstructure of the sintered mechanically alloyed powders may be connected with the findings of the mechanical testing in order to acquire a thorough knowledge of the structure-property connection of these powders, as was described before.

3. CHARACTERIZATION OF SINTERING OF MECHANICALLY ALLOYED POWDERS

The alloys that were sintered from powders that underwent mechanical alloying exhibited tungsten particles of small size, measuring approximately 3 μm in diameter, and a relative density exceeding 99%. The alloys in dispute exhibited a reduced proportion of matrix phase and a notable proportion of tungsten/tungsten interfaces, resulting in elevated yield strengths of approximately 1100 MPa. However, the alloys' elongation and impact energy were suboptimal (Walunj et al., 2020). The microhardness and compressive strength of Ni-TiC composites were observed to be greater than those of pure nickel. This can be attributed to the uniform dispersion of TiC particles throughout the nickel matrix, refinement of the matrix grain, and the effective interfacial bonding between the nickel and TiC reinforcement. Mechanical alloying (MA) was used to create the nickel (Ni) matrix composites, and then spark plasma sintering (SPS) was used to strengthen the composites (Abderrazak et al., 2011). Consolidated carbide was observed to have varied physical and mechanical qualities depending on the initial particle size powders used in the process. The Vickers hardness was observed to have an enhancement of approximately 2700 Hv, corresponding to a maximum density of 95.1% of the bulk material (Wang et al., 2010). Mechanical alloying was studied to see how it affected the microstructure and mechanical characteristics of the hot-pressed alloys. Furthermore, the Vickers hardness and flexural strength of the bulk alloys exhibited a gradual increase as a result of the augmentation of the volume fraction of niobium silicides. The fracture toughness values fell within the range of 10.98-13.23 MPa $m^{1/2}$, indicating a superior performance compared to that of cast materials. The observed phenomenon was ascribed to the beneficial impact of the ductile phase and the toughening effect of fine-grained microstructure (Yadav et al., 2021). To achieve the desired qualities, the mechanical behaviour of the binder phase in a ceramic-metal (cermet) composite must be optimised, in particular, hardiness and resilience to shock. In contrast to a WC-Co composite commonly used in commercial applications, a WC-10 wt.% HEA composite displays high values of both hardness and fracture toughness (Hrapkowicz et al., 2022). An enhanced amalgamation of mechanical characteristics was achieved subsequent to the sintering of powder subsequent to a duration of 20 hours of milling. The observed increase in hardness was found to be directly proportional to the duration of milling, up to a maximum of 70 hours. This phenomenon can be attributed to the combined effects of grain refinement and the emergence of secondary phases (Gökçe & Öveçoğlu, 2023). The results indicate that an increase in both the duration of mechanical alloying and the content of TiB2 led to an increase in the microhardness of the milled and sintered materials (Soares et al., 2021). The specimens fabricated through the utilisation of ball-milled powder demonstrate elevated Vickers microhardness and transverse fracture strength values in comparison to those produced using the raw powder. This outcome is attributed to the more refined microstructure (Saheb et al., 2016). Grain growth was observed as a result of sintering of the synthesised powders. The presence of Al_2O_3 resulted in growth inhibition. Samples hardened after being sintered for 20 minutes, although densification suffered (Eymann et al., 2012). The present study focuses on the consolidation of nanocrystalline Cu-Nb-ZrO_2 powder through the process of spark plasma sintering. The powder was produced through mechanical alloying. It was found that the micro hardness of the bulk specimens dropped from 409 HV0.025 (in their as-milled form) to 272 HV after the milling process (Xiong et al., 2009). The density of Nb/Nb5Si_3 composites improves as the Si content rises since this correlates with a larger maximal shrinkage displacement of the specimen (Ryu et al., 2000).

Sintered mechanically alloyed powders' mechanical properties are affected by a number of variables, including but not limited to their alloy composition, powder characteristics, milling circumstances, sintering parameters, and post-processing treatments. Powders that have undergone mechanical alloying frequently demonstrate enhanced strength and hardness in comparison to powders that have been processed conventionally. Sintered materials benefit from the high-energy milling process because the microstructure is improved, and the alloying components are distributed uniformly. As a result of their fine microstructure and homogeneous distribution of reinforcing phases, powders that have been sintered via a mechanical alloying process have demonstrated increased wear resistance. Materials that contain hard phases or reinforcements exhibit increased resistance to abrasive wear, rendering them appropriate for use in applications that necessitate wear-resistant components. The grain size of the sintered material, as well as its porosity, has a considerable impact on the mechanical characteristics of the material. Fine-grained structures, which may be obtained by mechanical alloying and the application of the appropriate sintering conditions, can contribute to improvements in mechanical performance. There are a number of variables that can make it difficult to achieve complete densification in sintered mechanically alloyed powders. Some of these difficulties include particle agglomeration, non-uniform microstructure, and residual porosity (Chaira, 2014; Chaira, 2015; Gupta et al., 2015; Nayak et al., 2016; Rajendrachari, 2018; Rajendrachari et al., 2019; Shashanka & Chaira, 2015).

4. CHALLENGES IN SINTERING OF MECHANICALLY ALLOYED POWDERS

Powders that have been mechanically alloyed are frequently susceptible to oxidation because of the increased surface area of the powder. During the sintering process, being subjected to high temperatures can result in oxidation, which can then lead to the production of oxides on the surface of the powder particles. Oxidation can have a negative impact on the sintering process, which can result in inadequate densification and subpar mechanical characteristics. Milling at high energies is required for mechanical alloying, and this process can result in the agglomeration of powder particles. Agglomerates refer to clusters of particles that are relatively larger in size and pose challenges in achieving uniform sintering. Insufficient treatment of agglomeration may lead to inadequate densification and suboptimal mechanical characteristics in the sintered product. The process of mechanical alloying frequently facilitates the achievement of a uniform and finely dispersed distribution of alloying constituents within the powder blend. But during sintering, phenomena including diffusion, particle rearrangement, and grain development can lead to the alloying elements being distributed unevenly. The sintered material's mechanical and physical qualities may suffer as a result of its microstructure being unevenly formed. At higher temperatures, grain development is a possible outcome of sintering. The microstructure, strength, and hardness of the sintered material might all suffer if the grains grow too large. To achieve the necessary characteristics, it is essential to regulate grain growth. The process of sintering has the potential to generate residual stresses within the material. These stresses may be attributed to non-uniform shrinkage or thermal expansion mismatch between distinct phases or grains. The mechanical properties and dimensional stability of the sintered product may be compromised by the presence of residual stresses. Sintering can cause cracks or deformation if the residual stresses are not minimised and managed. There is always the risk of contamination during mechanical alloying, whether from the milling medium or the surrounding air. The presence of contaminants may have adverse impacts on both the sintering behaviour and ultimate properties of the material. The implementation of appropriate protocols for handling, milling, and sintering in regulated environments can effectively mitigate the occurrence of contamination-related concerns. To tackle these challenges, it is imperative to implement meticulous process optimisation and control measures. The quality of sintered products can be enhanced by employing various techniques such as regulating the sintering atmosphere, utilising protective coatings or atmosphere, optimising sintering parameters such as temperature, time, and heating rate, and implementing post-processing treatments like hot isostatic pressing. These methods can effectively address the challenges associated with sintering and enhance the overall quality of the final product ().

5. APPLICATION OF SINTERING OF MECHANICALLY ALLOYED POWDERS

Materials that have been sintered or mechanically alloyed find use in a wide variety of applications across a variety of sectors due to the particular qualities and benefits that they provide. Here are some significant applications:

Automotive Industry: Applications in the automobile industry make use of materials that have been mechanically alloyed and sintered. Some examples of these applications are engine components, gearbox parts, and braking systems. Because of their great strength as well as their resistance to wear and their dimensional stability, these materials are appropriate for use in circumstances of rigorous operation.

Aerospace Industry: Materials that have been mechanically alloyed and sintered are advantageous for use in aerospace applications because of their low weight, high strength-to-weight ratio, and superior resistance to thermal and corrosive damage. Their application in turbine blades, structural components, and heat exchangers has led to greater efficiency and performance while using less fuel.

Cutting Tools: Materials that have undergone mechanical alloying and sintering processes, such as cemented carbides, are extensively utilised in the manufacturing of cutting tools. The aforementioned materials demonstrate notable attributes of elevated hardness, wear resistance, and thermal stability, rendering them apt for deployment in machining procedures, including but not limited to drilling, milling, and turning.

Biomedical Applications: Materials that have undergone mechanical alloying and sintering processes find application in the biomedical domain, specifically in orthopaedic implantation, dental prosthetics, and surgical instrumentation. These materials exhibit biocompatible properties, resistance to corrosion, and can be customised to suit particular applications.

Electrical and Electronics: Materials that have been mechanically alloyed and sintered have been observed to have utility in the field of electrical and electronics engineering. Since they are conductive, magnetic, and thermally stable, they are put to use in a variety of electrical contact, connector, and magnetic applications.

Wear and Corrosion Resistance: Materials that have undergone mechanical alloying and sintering processes are favoured in situations where it is imperative to have high levels of wear and corrosion resistance. Instances of such applications comprise of bearings, seals, and valves in industrial machinery, wherein these substances furnish enhanced efficacy and prolonged durability.

Additive manufacturing: Additive manufacturing, often known as 3D printing, frequently makes use of mechanically alloyed and sintered materials. Powders that are fine enough and homogeneous enough for additive manufacturing processes may be made with the use of mechanical alloying, allowing for more design freedom and individualised component creation.

The instances serve as a limited representation of the multifaceted uses of mechanically alloyed and sintered materials. The selection of a particular material and process is contingent upon the desired properties and application requirements, underscoring the adaptability and significance of these materials in diverse industrial sectors (Bas et al., 2003; Fujiki, 2001; Ivasishin et al., 2000; Mukesh et al., 2021; Pradeep, Venkatesha, Parameshwara et al, 2022; Shashanka & Chaira, n.d.-a; Shashanka & Chaira, n.d.-b; Srinivas et al., 2022; Storch et al., 2003; Zhao et al., 2005).

6. CONCLUSION

The process of sintering mechanically alloyed powders poses various obstacles that require resolution to attain the intended material characteristics. The challenges that are encountered in this context encompass a range of issues such as oxidation, particle agglomeration, non-uniform microstructure, grain growth, residual stresses, and contamination. To surmount these obstacles, it is imperative to implement meticulous process optimisation and control. In order to reduce the effects of oxidation, it is possible to employ protective atmospheres or coatings during the sintering process, which serve to impede the formation of oxides on the surface of the powder particles. The reduction of particle agglomeration can be achieved through the utilisation of appropriate milling methodologies and the optimisation of powder properties. The issue of non-uniform microstructure can be effectively dealt with by implementing controlled sintering parameters and heat treatments. The regulation of grain growth is of utmost importance in preserving the intended microstructural characteristics and associated properties. Optimisation of the sintering temperature, duration, and heating rate can lead to the attainment of this objective. The mitigation of residual stresses can be achieved through meticulous design of the sintering process, taking into account variables such as thermal expansion mismatch and shrinkage. The mitigation of contamination concerns can be achieved through the implementation of measures that ensure the cleanliness of milling environments and the utilisation of suitable protective measures. In order to attain sintered products of superior quality possessing the desired properties, it is crucial to comprehend and tackle the obstacles encountered during the sintering of mechanically alloyed powders. Further investigation and refinement of the sintering process for mechanically alloyed powders is imperative in order to enhance its efficiency and efficacy.

REFERENCES

Abderrazak, H., Schoenstein, F., Abdellaoui, M., & Jouini, N. (2011, March 1). Spark plasma sintering consolidation of nanostructured TiC prepared by mechanical alloying. *International Journal of Refractory & Hard Metals*, *29*(2), 170–176. doi:10.1016/j.ijrmhm.2010.10.003

Adimule, V., Keri, R., Rajendrachari, S., Kendrekar, P., & Kulkarni, C. V. (2023, February). Highly photoluminescence and wide band gap insulating metal hybrid nanoparticles array of samarium-dopedSrO: CoO: synthesis, characterizations and sensor characteristics. *Journal of Materials Science Materials in Electronics*, *34*(5), 442. doi:10.1007/s10854-023-09899-2

Atazadeh, N., Heydari, M. S., Baharvandi, H. R., & Ehsani, N. (2016, December 1). Reviewing the effects of different additives on the synthesis of the Ti3SiC2 MAX phase by mechanical alloying technique. *International Journal of Refractory & Hard Metals, 61*, 67–78. doi:10.1016/j.ijrmhm.2016.08.003

Awotunde, M. A., Adegbenjo, A. O., Obadele, B. A., Okoro, M., Shongwe, B. M., & Olubambi, P. A. (2019, April 1). Influence of sintering methods on the mechanical properties of aluminium nanocomposites reinforced with carbonaceous compounds: A review. *Journal of Materials Research and Technology, 8*(2), 2432–2449. doi:10.1016/j.jmrt.2019.01.026

Ballupete Nagaraju, S., Kodigarahalli Somashekara, M., Puttegowda, M., Manjulaiah, H., Kini, C. R., & Channarayapattana Venkataramaiah, V. (2022, December 15). Effect of B4C/Gr on Hardness and Wear Behavior of Al2618 Based Hybrid Composites through Taguchi and Artificial Neural Network Analysis. *Catalysts, 12*(12), 1654. doi:10.3390/catal12121654

Bas, J. A., Calero, J. A., & Dougan, M. J. (2003, January 1). Sintered soft magnetic materials. Properties and applications. *Journal of Magnetism and Magnetic Materials, 254*, 391–398. doi:10.1016/S0304-8853(02)00934-4

Benjamin, J. S., & Arzt, E. (1989). *New materials by mechanical alloying techniques*. DGM Informationgesellschaft.

Bordia, R. K., Kang, S. J., & Olevsky, E. A. (2017, June). Current understanding and future research directions at the onset of the next century of sintering science and technology. *Journal of the American Ceramic Society, 100*(6), 2314–2352. doi:10.1111/jace.14919

Braginsky, M., Tikare, V., & Olevsky, E. (2005, January 1). Numerical simulation of solid state sintering. *International Journal of Solids and Structures, 42*(2), 621–636. doi:10.1016/j.ijsolstr.2004.06.022

Castro, R., & Van Benthem, K. (Eds.). (2012 Sep 13). *Sintering: mechanisms of convention nanodensification and field assisted processes*. Springer Science & Business Media.

Chaira, D. (2014, June 1). Phase transformation and microstructure study of nano-structured austenitic and ferritic stainless steel powders prepared by planetary milling. *Powder Technology, 259*, 125–136. doi:10.1016/j.powtec.2014.03.061

Chaira, D. (2015, January 1). Development of nano-structured duplex and ferritic stainless steels by pulverisette planetary milling followed by pressureless sintering. *Materials Characterization, 99*, 220–229. doi:10.1016/j.matchar.2014.11.030

Chikkegouda, S. P., Gurudath, B., Sharath, B. N., Karthik, S., & Mahale, R. S. (2022). Mechanical and Tribological Characteristics of Aluminium 2618 Matrix Composite Reinforced with Boron Carbide. *Biointerface Research in Applied Chemistry, 12*(4).

Corker, D. L., Whatmore, R. W., Ringgaard, E., & Wolny, W. W. (2000, November 1). Liquid-phase sintering of PZT ceramics. *Journal of the European Ceramic Society, 20*(12), 2039–2045. doi:10.1016/S0955-2219(00)00089-3

Cross, J. O., Opila, R. L., Boyd, I. W., & Kaufmann, E. N. (2015, December). Materials characterization and the evolution of materials. *MRS Bulletin, 40*(12), 1019–1034. doi:10.1557/mrs.2015.271

De Oro Calderón, R., Bernardo, E., Campos, M., Gierl-Mayer, C., Danninger, H., & Torralba, J. M. (2016, January 1). Tailoring master alloys for liquid phase sintering: Effect of introducing oxidation-sensitive elements. *Powder Metallurgy, 59*(1), 31–40. doi:10.1080/00325899.2016.1148897

de Oro Calderon, R., Gierl-Mayer, C., & Danninger, H. (2022). Fundamentals of sintering: liquid phase sintering. *Encyclopedia of Materials: Metals and Alloys, 3*, 481-92.

Diantonio, C.B., & Ewsuk, K.G. (2010). Master sintering curve and its application in sintering of electronic ceramics. In *Sintering of Advanced Materials* (pp. 130-161). Woodhead Publishing.

Dobrzański, L.A., Dobrzańska-Danikiewicz, A.D., Achtelik-Franczak, A., Dobrzański, L.B., Hajduczek, E., & Matula, G. (2017). Fabrication technologies of the sintered materials including materials for medical and dental application. In *Powder Metallurgy–Fundamentals and Case Studies*. InTech.

Ertug, B. (Ed.). (2013 Feb 6). *Sintering applications*. BoD–Books on Demand. doi:10.5772/56064

Eymann, K., Riedl, T., Bram, A., Ruhnow, M., Boucher, R., Kirchner, A., & Kieback, B. (2012, September 15). Consolidation of mechanically alloyed nanocrystalline Cu–Nb–ZrO2 powder by spark plasma sintering. *Journal of Alloys and Compounds, 535*, 62–69. doi:10.1016/j.jallcom.2012.04.102

Fang, Z. Z. (Ed.). (2010 Sep 27). *Sintering of advanced materials*. Elsevier. doi:10.1533/9781845699949

Fang, Z.Z., & Wang, H. (2013). Sintering of ultrafine and nanosized ceramic and metallic particles. In *Ceramic Nanocomposites* (pp. 431-473). Woodhead Publishing.

Fujiki, A. (2001, January 15). Present state and future prospects of powder metallurgy parts for automotive applications. *Materials Chemistry and Physics, 67*(1-3), 298–306. doi:10.1016/S0254-0584(00)00455-7

German, R. (2014). *Sintering: From empirical observations to scientific principles.* Butterworth-Heinemann.

German, R.M. (1996). *Sintering theory and practice.* Academic Press.

German, R. M., Suri, P., & Park, S. J. (2009, January). Liquid phase sintering. *Journal of Materials Science, 44*(1), 1–39. doi:10.1007/s10853-008-3008-0

Gökçe, H., & Öveçoğlu, M.L. (2023). Microstructure-property evolution of mechanically alloyed Al-20 wt% Si matrix powders and sintered composites reinforced withTiB2 particulates. *Engineering Science and Technology, an International Journal, 39.*

Gupta, S., Shashanka, R., & Chaira, D. (2015). Synthesis of nano-structured duplex and ferritic stainless steel powders by planetary milling: An experimental and simulation study. In *IOP conference series: Materials science and engineering* (Vol. 75, No. 1, p. 012033). IOP Publishing.

Hrapkowicz, B., Lesz, S., Karolus, M., Garbiec, D., Wiśniewski, J., Rubach, R., Gołombek, K., Kremzer, M., & Popis, J. (2022, February 22). Microstructure and Mechanical Properties of Spark Plasma Sintered Mg-Zn-Ca-Pr Alloy. *Metals, 12*(3), 375. doi:10.3390/met12030375

Ibrahim, H., Aziz, A., & Rahmat, A. (2014, March 1). Enhanced liquid-phase sintering of W–Cu composites by liquid infiltration. *International Journal of Refractory & Hard Metals, 43*, 222–226. doi:10.1016/j.ijrmhm.2013.12.004

Ivasishin, O. M., Anokhin, V. M., Demidik, A. N., & Savvakin, D. G. (2000). *Cost-effective blended elemental powder metallurgy of titanium alloys for transportation application. In Key Engineering Materials* (Vol. 188). Trans Tech Publications Ltd.

Kang, S.J. (2004). *Sintering: Densification, grain growth and microstructure.* Elsevier.

Kang, S.L. (2010). Liquid phase sintering. In *Sintering of advanced materials* (pp. 110-129). Woodhead Publishing.

Kumar, A., Singh, A., & Suhane, A. (2022, March 1). Mechanically alloyed high entropy alloys: Existing challenges and opportunities. *Journal of Materials Research and Technology, 17*, 2431–2456. doi:10.1016/j.jmrt.2022.01.141

Madhu, K. S., Sharath, B. N., Venkatesh, C. V., & Pradeep, D. G. (2021). Evaluation of mechanical properties of ceramic reinforced aluminium-7029 hybrid composite. In *IOP Conference Series: Materials Science and Engineering* (Vol. 1189, No. 1, p. 012019). IOP Publishing. 10.1088/1757-899X/1189/1/012019

Madhu, K. S., Venkatesh, C. V., Sharath, B. N., & Karthik, S. (2022, November 2). Characterization and Evaluation of Mechanical Properties of Al-Zn Based Hybrid Metal Matrix Composites. *Applied Science and Engineering Progress.*, *16*(1), 5804. doi:10.14416/j.asep.2022.03.008

Mahale, R. S., & Shashanka, R. (2022). Mechanical testing of spark plasma sintered materials: A review. In *AIP Conference Proceedings* (Vol. 2469, No. 1). AIP Publishing. 10.1063/5.0080183

Manogar, B., Yang, F., & Bolzoni, L. (2022, September 15). Effect of Nb Addition on the Phase Stability, Microstructure, and Mechanical Properties of Powder Metallurgy Ti-5Fe-xNb Alloys. *Metals*, *12*(9), 1528. doi:10.3390/met12091528

McCormick, P. G. (1995). Application of mechanical alloying to chemical refining (overview). *Materials Transactions, JIM, 36*(2), 161-9.

Mogale, N. F., & Matizamhuka, W. R. (2020, August 11). Spark plasma sintering of titanium aluminides: A progress review on processing, structure-property relations, alloy development and challenges. *Metals*, *10*(8), 1080. doi:10.3390/met10081080

Mukesh, Y. B., Naik, P. K., & Raghavendra Rao, R. (2021). Impact of heat treatment on the mechanical performance of hot extruded Al6061-BN reinforced metal matrix composites. *Bulletin of the Polish Academy of Sciences. Technical Sciences*, *69*(3), 1–6.

Murakami, Y., & Ritchie, R.O. (2012). Effects of hydrogen on fatigue-crack propagation in steels. In *Gaseous Hydrogen Embrittlement of Materials in Energy Technologies* (pp. 379-417). Woodhead Publishing.

Nayak, A. K., Shashanka, R., & Chaira, D. (2016). Effect of nanosizeyittria and tungsten addition to duplex stainless steel during high energy planetary milling. In *IOP Conference Series: Materials Science and Engineering* (Vol. 115, No. 1, p. 012008). IOP Publishing.

Oro, R., Campos, M., & Torralba, J. M. (2012, July 1). Study of high temperature wetting and infiltration for optimising liquid phase sintering in low alloy steels. *Powder Metallurgy*, *55*(3), 180–190. doi:10.1179/1743290111Y.0000000007

Pavitra, V., Praveen, B. M., Nagaraju, G., & Shashanka, R. (2022, January 8). Energy storage, Photocatalytic and Electrochemical nitrite sensing of ultrasound-assisted stable Ta2O5 nanoparticles. *Topics in Catalysis*, 1–4. doi:10.1007/s11244-021-01553-7

Pomeroy, M. (Ed.). (2021 May 24). *Encyclopedia of materials: technical ceramics and glasses*. Elsevier.

Pradeep, N. B., Hegde, M. R., Rajendrachari, S., & Surendranathan, A. O. (2022, August 1). Investigation of microstructure and mechanical properties of microwave consolidated TiMgSr alloy prepared by high energy ball milling. *Powder Technology, 408*, 117715. doi:10.1016/j.powtec.2022.117715

Pradeep, N. B., Venkatesha, B. K., Parameshwara, S., Raghavendra Rao, R., & Raviprakash, M. (2022). Synthesis of nanostructured ternary Ti based alloy for bio-medical applications. *Materials Today: Proceedings, 54*(2), 372-377. doi:10.1016/j.matpr.2021.09.429

Rahaman, M.N. (2017). *Ceramic processing and sintering*. CRC Press.

Rajendrachari, S. (2018, January 1). Effect of sintering temperature on the pitting corrosion of ball milled duplex stainless steel by using linear sweep voltammetry. *Analytical and Bioanalytical Electrochemistry., 10*(3), 349–361.

Rajendrachari, S. (2022, June 21). An overview of high-entropy alloys prepared by mechanical alloying followed by the characterization of their microstructure and various properties. *Alloys., 1*(2), 116–132. doi:10.3390/alloys1020008

Rajendrachari, S., Adimule, V. M., Jayaprakash, G. K., & Pandith, A. (2023, May 17). Electrochemical oxidation of methylene blue dye in wastewater using mechanically alloyed high entropy alloy modified carbon paste electrode using cyclic voltammetry. *Materials Research Express, 10*(5), 054003. doi:10.1088/2053-1591/acd3d7

Rajendrachari, S., Kamacı, Y., Taş, R., Ceylan, Y., Bülbül, A. S., Uzun, O., & Karaoğlanlı, A. C. (2019, December 1). Antimicrobial investigation of CuO and ZnO nanoparticles prepared by a rapid combustion method. *Physical Chemistry Research., 7*(4), 799–812.

Ramakrishnan, P. (2013). Automotive applications of powder metallurgy. In*Advances in powder metallurgy* (pp. 493-519). Woodhead Publishing.

Robert, W., & Haasen, P. (n.d.). *Physical Metallurgy* (4th ed.). Academic Press.

Ryu, H. J., Hong, S. H., & Baek, W. H. (2000, October 31). Microstructure and mechanical properties of mechanically alloyed and solid-state sintered tungsten heavy alloys. *Materials Science and Engineering A, 291*(1-2), 91–96. doi:10.1016/S0921-5093(00)00968-0

Saheb, N., Khan, M. S., & Hakeem, A. S. (2016, January 1). Effect of processing on mechanically alloyed and spark plasma sintered Al-Al2O3 nanocomposites. *Journal of Nanomaterials, 16*(1), 377.

Sharath, B. N., Venkatesh, C. V., Afzal, A., Aslfattahi, N., Aabid, A., Baig, M., & Saleh, B. (2021, May 28). Multi ceramic particles inclusion in the aluminium matrix and wear characterization through experimental and response surface-artificial neural networks. *Materials (Basel), 14*(11), 2895. doi:10.3390/ma14112895 PMID:34071305

Sharath, B. N., Venkatesh, C. V., Afzal, A., Baig, M. A., & Kumar, A. P. (2021, January 1). Study on effect of ceramics on dry sliding wear behaviour of Al-Cu-Mg based metal matrix composite at different temperature. *Materials Today: Proceedings, 46*, 8723–8733. doi:10.1016/j.matpr.2021.04.034

Shashanka, R. (2017). Synthesis of nano-structured stainless steel powder by mechanical alloying-an overview. *International Journal of Scientific and Engineering Research, 8*, 588–594.

Shashanka, R., & Chaira, D. (2015, July 1). Optimization of milling parameters for the synthesis of nano-structured duplex and ferritic stainless steel powders by high energy planetary milling. *Powder Technology, 278*, 35–45. doi:10.1016/j.powtec.2015.03.007

Shashanka, R., & Chaira, D. (2016, January). Effects of nano-Y 2 O 3 and sintering parameters on the fabrication of PM duplex and ferritic stainless steels. *Acta Metallurgica Sinica, 29*(1), 58–71. doi:10.1007/s40195-015-0362-1

Shashanka, R., & Chaira, D. (2017a, March 4). Effect of sintering temperature and atmosphere on nonlubricated sliding wear of nano-yttria-dispersed and yttria-free duplex and ferritic stainless steel fabricated by powder metallurgy. *Tribology Transactions, 60*(2), 324–336. doi:10.1080/10402004.2016.1168897

Shashanka, R., & Chaira, D. (2017b). *Ball Milled Nano-Structured Stainless Steel Powders: Fabrication and Characterization*. Educreation Publishing.

Shashanka, R., & Chaira, D. (n.d.-a). Optimization of Milling Parameters. *Ball Milled Nano-Structured Stainless Steel Powders: Fabrication and Characterization, 32*.

Shashanka, R., & Chaira, D. (n.d.-b). Fabrication of NANO-Y2O3 Dispersed Stainless Steel By Conventional And Spark Plasma Sintering. *Ball Milled Nano-Structured Stainless Steel Powders: Fabrication and Characterization, 54.*

Shashanka, R., Chaira, D., & Chakravarty, D. (2016). Fabrication of nano-yttria dispersed duplex and ferritic stainless steels by planetary milling followed by spark plasma sintering and non-lubricated sliding wear behaviour study. *J Mater Sci Eng B.*, *6*, 111–125.

Shashanka, R., Uzun, O., & Chaira, D. (2020). Synthesis of nano-structured duplex and ferritic stainless steel powders by dry milling and its comparison with wet milling. *Archives of Metallurgy and Materials*, *65*(1).

Soares, E., Bouchonneau, N., Alves, E., Alves, K., Araújo Filho, O., Mesguich, D., Chevallier, G., Laurent, C., & Estournès, C. (2021, January 16). Microstructure and mechanical properties of AA7075 aluminum alloy fabricated by spark plasma sintering (SPS). *Materials (Basel)*, *14*(2), 430. doi:10.3390/ma14020430 PMID:33467156

Soni, P. R. (2000). *Mechanical alloying: fundamentals and applications.* Cambridge Int Science Publishing.

Srinivas, A., Pavan, D., Venkatesha, B. K., Rao, R. R., & Mohith, L. (2022). Study on mechanical properties of AZ91 magnesium alloy. *Materials Today: Proceedings*, *54*, 291–294. doi:10.1016/j.matpr.2021.09.171

Storch, S., Nellessen, D., Schaefer, G., & Reiter, R. (2003, October 1). Selective laser sintering: Qualifying analysis of metal based powder systems for automotive applications. *Rapid Prototyping Journal*, *9*(4), 240–251. doi:10.1108/13552540310489622

Sullivan, K. T., Piekiel, N. W., Wu, C., Chowdhury, S., Kelly, S. T., Hufnagel, T. C., Fezzaa, K., & Zachariah, M. R. (2012, January 1). Reactive sintering: An important component in the combustion of nanocomposite thermites. *Combustion and Flame*, *159*(1), 2–15. doi:10.1016/j.combustflame.2011.07.015

Suryanarayana, C. (1996, August). Recent advances in the synthesis of alloy phases by mechanical alloying/milling. *Metals and Materials*, *2*(4), 195–209. doi:10.1007/BF03026094

Suryanarayana, C. (2001, January 1). Mechanical alloying and milling. *Progress in Materials Science*, *46*(1-2), 1–84. doi:10.1016/S0079-6425(99)00010-9

Suryanarayana, C. (2011, June 1). Synthesis of nanocomposites by mechanical alloying. *Journal of Alloys and Compounds, 509*, S229–S234. doi:10.1016/j.jallcom.2010.09.063

Suryanarayana, C., & Al-Aqeeli, N. (2013, May 1). Mechanically alloyed nanocomposites. *Progress in Materials Science, 58*(4), 383–502. doi:10.1016/j.pmatsci.2012.10.001

Suryanarayana, C., Ivanov, E., & Boldyrev, V. V. (2001, May 31). The science and technology of mechanical alloying. *Materials Science and Engineering A, 304*, 151–158. doi:10.1016/S0921-5093(00)01465-9

Teber, A., Schoenstein, F., Têtard, F., Abdellaoui, M., & Jouini, N. (2012, January 1). Effect of SPS process sintering on the microstructure and mechanical properties of nanocrystalline TiC for tools application. *International Journal of Refractory & Hard Metals, 30*(1), 64–70. doi:10.1016/j.ijrmhm.2011.06.013

Walunj, G., Bearden, A., Patil, A., Larimian, T., Christudasjustus, J., Gupta, R. K., & Borkar, T. (2020, November 23). Mechanical and tribological behavior of mechanically alloyed Ni-TiC composites processed via spark plasma sintering. *Materials (Basel), 13*(22), 5306. doi:10.3390/ma13225306 PMID:33238641

Wang, X. L., Wang, G. F., & Zhang, K. F. (2010, May 25). Effect of mechanical alloying on microstructure and mechanical properties of hot-pressed Nb–16Si alloys. *Materials Science and Engineering A, 527*(13-14), 3253–3258. doi:10.1016/j.msea.2010.02.031

Xiong, B., Long, W., Chen, Z., Xia, C., Wan, H., & Yan, Y. (2009, March 5). Effects of element proportions on microstructures of Nb/Nb5Si3 in situ composites by spark plasma sintering. *Journal of Alloys and Compounds, 471*(1-2), 404–407. doi:10.1016/j.jallcom.2008.03.095

Xu, Z. (2019). Study on microwave sintering process and surface texture characteristics of ceramic materials. In *IOP Conference Series: Materials Science and Engineering* (Vol. 677, No. 2, p. 022078). IOP Publishing. 10.1088/1757-899X/677/2/022078

Yadav, S., Zhang, Q., Behera, A., Haridas, R. S., Agrawal, P., Gong, J., & Mishra, R. S. (2021, October 5). Role of binder phase on the microstructure and mechanical properties of a mechanically alloyed and spark plasma sintered WC-FCC HEA composites. *Journal of Alloys and Compounds, 877*, 160265. doi:10.1016/j.jallcom.2021.160265

Yang, L., Dai, Q., Liu, L., Shao, D., Luo, K., Jamil, S., Liu, H., Luo, Z., Chang, B., & Wang, X. (2020, June 1). Rapid sintering method for highly conductive Li7La3Zr2O12 ceramic electrolyte. *Ceramics International*, *46*(8), 10917–10924. doi:10.1016/j.ceramint.2020.01.106

Yoon, B. K., Lee, B. A., & Kang, S. J. (2005, October 1). Growth behavior of rounded (Ti, W) C and faceted WC grains in a Co matrix during liquid phase sintering. *Acta Materialia*, *53*(17), 4677–4685. doi:10.1016/j.actamat.2005.06.021

Zhao, N., Nash, P., & Yang, X. (2005, December 30). The effect of mechanical alloying on SiC distribution and the properties of 6061 aluminum composite. *Journal of Materials Processing Technology*, *170*(3), 586–592. doi:10.1016/j.jmatprotec.2005.06.037

Chapter 9
Trends, Applications, and Factors of Wet Sieving as a Particle Size Analysis Technique:
Insights on Mineral Processing and Mechanical Alloying

Khuthadzo Edna Mudzanani
https://orcid.org/0000-0003-3660-7376
Mintek, South Africa

Terence Thabiso Phadi
https://orcid.org/0000-0002-9751-0853
Mintek, South Africa

Deogratius Titus Maiga
Mintek, South Africa

ABSTRACT

In mineral processing, particle size is an important factor to consider since it affects plant design and performance, which later determines the plant's economic return. A certain particle size above and below the mine's optimal size range can adversely influence the recovery of valuable minerals. Particle size analysis is crucial in mechanical alloying controlling alloy composition and homogeneity of the mixture, affecting milling kinetics, reducing contamination, optimizing the milling process, and predicting properties. This chapter includes the understanding of particle size distribution, which helps researchers to tailor the alloying process for specific applications. The trend has been the wet-sieving analysis as a particle size analysis tool for the past 25 years. It provides an overview of different industries where the particle size analysis is used and associated end-users. The chapter also reports on the variables and principles governing factors that affect wet sieving analysis and the challenges that can give misleading and erroneous particle size distribution results.

DOI: 10.4018/978-1-6684-9385-4.ch009

1. INTRODUCTION

Particle size analysis frequently involves the use of sieves, especially for granular materials. Particle size analysis has become more accurate and effective thanks to several trends and developments in sieving methods and technology during the past 25 years. The key trends of sieving that remain fundamental and widely used in particle size analysis techniques; are advancements in automation, sieve design, standardization, instrumentation, software integration, and environmental considerations (Sanchez-Suarez et al., 2022). These have contributed to improving the accuracy, efficiency, and sustainability of sieving over the past 25 years.

Particle size analysis, including sieving, plays a vital role in various industries, such as pharmaceuticals, construction, and mining. In recent years, there has been a growing emphasis on standardization and adherence to regulations for particle size analysis. Organizations like the International Organization for Standardization (ISO) and the American Society for Testing and Materials (ASTM) have developed guidelines and standards to ensure consistency and comparability of sieving results across different laboratories and industries (American Standards & Test Methods, 2017). Data analysis and software integration on sieving technique has recently become more interesting. With the rise of data analysis and automation, software solutions for sieving have evolved. Dedicated software programs can now integrate with sieving equipment, enabling real-time monitoring, data capture, and analysis (Sanchez-Suarez et al., 2022). These software solutions offer features like data visualization, automated calculations, and customizable reporting, making the interpretation and sharing of sieve analysis results more efficient. Additionally, in recent years, there has been an increased focus on sustainability and reducing environmental impact. This trend has influenced particle size analysis techniques, including sieving. Efforts have been made to minimize waste generated during the sieving process, optimize the energy consumption of automated sieve shakers, and explore environmentally friendly sieve materials (Bertula, 2022).

The study conducted by Liu, (2009) demonstrated that parameters such as material type, milling process, moisture content, tapping, sieving method, sieving time and their interactions might all exert an impact on sieving efficiency and performance. The most important parameter was tapping, followed by sieving time, sieving method, milling process, material type and lastly moisture content. It was reported that the reverse sieving approach showed an improved PSD curve as opposed to the conventional stacked sieve method. The improvement was reported in terms of increased sieving rates, minimized sieve loss and increased final fraction mass under similar conditions of investigation. The advantage of bigger particles was linked to the observed difference in sieving efficiency, performance, and loss between the two sieving methods. It is known that the presence of coarse particles during sieving can decrease the sieve blinding effect resulting from the sub-sieve-sized particles.

According to Ulusoy, (2023), the properties of particulate materials cannot be represented solely by single particle attributes, but also by bulk properties and their influences. To make things easier, particle properties can be split into primary properties, such as material composition and secondary properties. These are related to systems of individual particle collections and how they interact with their surroundings, such as gas (air) or water. Secondary properties such as particle settling rate, compaction, powder rehydration rate, filter cake resistance, flowability, sintering and particle systems are controlled by primary particle properties such as shape and density, as well as fluid viscosity and density.

1.1. Important Elements of Particle Size Analysis

1.1.1. Description of Particle Diameter and Shape

Particles in several industries are generally irregular and do not have the ideal spherical shape which influences the prediction of particle size and PSD of any particulate material. Recently, particle shape has been deemed to be as important as particle size and thus combined to characterize raw materials (Osborne, 1972; Bhrany and Brown, 1962; Lee Black, 1996). This led to the sensitivity of a sizing technique to particle shape as an important factor in determining its applicability. As a result, statistics based on a single size value are insufficient to reflect the true performance of the particulate material. Particle size without shape analysis can result in significant under-prediction of average particle size and PSD inaccuracies for non-spherical particles (Ulusoy, 2023).

A variety of phrases have specific meanings about the description of particle form and size. Some of the more frequent particle shapes are spherical, irregular (no symmetry), and acicular (needle-shaped), whereas angular is sharp-edged, while crystalline has a geometrical outline (Lee Black,1996). Numerous particle diameters can be used to describe particle size. These developed because of the wide range of scenarios faced when using particle-size measurement equipment. Different size distribution parameters are crucial for different processes. Most particle diameter descriptions require comparing some aspect of the particle to a particle with an equivalent spherical diameter. If the particle in question is spherical, all diameters become equal (Wills & Finch, 2016, Rajendran & Murty, 2023).

Because of the great diversity in particle shape, terminologies like volume and surface diameter are used. The diameter of a sphere with the same volume as the actual particle is known as the volume diameter, while the diameter of a sphere with the same surface area as the actual particle is known as the surface diameter (Kursun, 2009). The most popular particle size analyzer measures an equivalent spherical diameter and assumes that this is a genuine particle measurement. The relationship between the actual particle form and the diameter measurement, on the other hand, is frequently multifaceted and not explicitly defined as it is for non-optical methods (Stresty & Venkateswar, 1980).

The particle size analysis can then be defined in different ways, but too often spherical components are used. However, particles are three-dimensional structures, and only a few of them are particularly spherical in shape. It has generally been accepted that the diameter of particles can be used to describe them in a two-dimensional spherical phase (Wills & Finch, 2016). Since particles do not exactly have a spherical shape, it is necessary to transform them into a sphere-equivalent shape using various methods, as shown in Figure 1. It is important to keep in mind that different particle size measuring approaches choose different equivalent concepts. Based on the diameter that symbolizes behaviour identical to a class of particles under assessment, the equivalent sphere principle is used (Lee Black, 1996). It has been demonstrated that measuring particle size using different techniques may yield different results for the same set of non-spherical particles. As indicated in Figure 1, some of the most common equivalent

Figure 1. Particle size measuring approaches choose different equivalent concepts
Source: Adapted from Singh & Singh (2022)

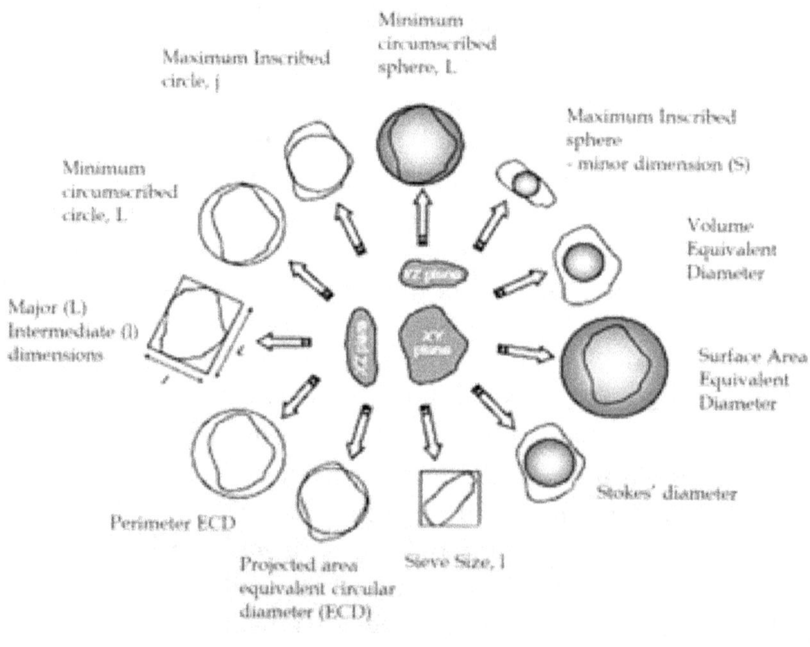

diameters for spheres are surface area equivalent, sieve size diameter, volume equivalent diameter, Stokes diameter, etc.

Several researchers have attempted to develop methods for determining the reference feature and calculating the circularity error in particles. To estimate the reference feature (e.g., circle or sphere), several geometry measurement tools are readily available. The Minimum Circumscribed Circle (MCC) and Minimum Circumscribed Sphere (MCS), as well as the Maximum Inscribed Circle (MIC) and Maximum Inscribed Sphere (MIS), are shown in Figure 1. Fanwua et al. (2011) defined MIC as the approach that fits the largest possible circle inside the profile. Liu (2016) defined MCC as the smallest circle that surrounds the section profile. Similarly, the MIS is the largest sphere that can be inscribed inside the section profile, while the MCS is the smallest sphere with the edge of the profile. The difference between the two methods is that MCC uses the outer circle as the reference circle, whilst MICI uses the inner circle. The out-of-roundness value, which is the maximum/minimum distance between the profile and the inscribed or circumscribed circle or sphere, can be used to identify the circle or sphere. MCC is appropriate for evaluating the roundness deviation of a cylindrical surface, while MIC is ideal for analyzing the roundness deviation of a bore surface.

Another method for assessing particle size characteristics is to use the mean diameter. The Sauter mean, the surface mean, the arithmetic mean, and the volume mean are the most common (Lee Black, 1996). Each mean diameter is defined as the most useful or practical metric to use for the phenomenon under investigation. The arithmetic mean is obtained from the traditional mathematical computation of the mean particle size. The diameter of a particle with the average area of the particles in the size distribution is described as the area mean, and the volume mean is calculated similarly to the area mean. The Sauter mean is the diameter of a particle that has the same mean surface area to volume ratio as the distribution's sizes (Singh & Singh, 2022).

1.1.2. The Functions of Particle Size Distribution

PSD Functions are mathematical expressions usually in the form of a curve, that establish a link between some aspect of the particles in the flow, such as volume or concentration, and particle diameter (Avramović et al., 2018; Singh & Singh, 2022). These distribution functions are a basic mathematical tool for describing particles when a comprehensive description of the experimental distribution is not required. They provide for succinct reporting of detailed particle information. The application of size distribution functions also allows for the extrapolation of particle size information beyond the measured range of values. This facilitates the calculation of crucial information such as mean particle diameters and variables that determine the PSD shape, hence providing insight into the procedures connected to particles (Li, 2016; Opyrchał et al., 2022). PSD can be represented graphically or in a tabular form. The histogram is a descriptive way of depicting a PSD, where the width of a bar corresponds to the size class's lower or higher limit and the height of the bar refers to the quantity in that size class (Singh & Singh, 2022; Terzi, 2018).

1.2. Significance of the Research

Extensive research on the inclusion of various particle size studies has been undertaken during the previous decade. Furthermore, previous papers underlined the importance of sieving in several industries but failed to give precise information. This paper investigates all prior and present studies on the effects of various physicochemical variables on wet sieving. According to the author's understanding, this serves as another most comprehensive review studies on the subject published in recent years, offering a structured analysis, discussion, and added contribution to knowledge intended to broaden this area. This review covers (i) a trend of particle size analysis in wet and dry sieving perspectives, (ii) a variety of studies on the application of screening in different industries, (iii) major factors that affect wet sieving, and (iv) some factors that can lead to erroneous PSD analysis.

2. IMPORTANCE OF PARTICLE SIZE ANALYSIS ON MECHANICAL ALLOYING

Particle size analysis is of crucial importance in the context of mechanical alloying, which is a solid-state powder processing technique used to produce advanced materials and alloys. Mechanical alloying is a contemporary and widely used technique for producing nanostructured stainless-steel powder (Pradeep et al., 2022; Suryanarayana, 2004). Mechanical alloying involves the repeated welding, fracturing, and re-welding of powder particles in a high-energy ball mill. During this process, the size and distribution of particles play a significant role, affecting the outcome and properties of the resulting material (Shashanka & Chaira, 2015a). The main benefit of utilizing mechanical alloying for the synthesis of nanocrystalline materials is the ability to manufacture large amounts of material in solid form using basic equipment and at room temperature. Furthermore, because of the reduced processing time, mechanically alloyed powder lowers the oxidation of the constituent powders (Nayak et al., 2016; Shashanka, 2017).

Several researchers have conducted studies on the importance of particle size analysis in mechanical alloying. Suryanarayana, (2004) wrote a book on "Mechanical Alloying and Applications". This book covers the fundamentals of mechanical alloying, including particle size analysis techniques and their applications in alloy synthesis. Other books include; (Shashanka & Uzun, 2021) titled "Powder Technology" and "Properties and Technological Features of Materials" edited by (Adimule & Shashanka, 2023). Research papers and publications in materials science and metallurgy journals often include particle size analysis as a critical aspect of mechanical alloying studies. Some notable journals to explore include Materials Science and Engineering A, Journal of Alloys and Compounds, Powder Technology, and Journal of Materials Science.

Several investigations have been carried out since the mid-1980s on how particle size analysis is an important factor in mechanical alloying. Basavarajappa et al., (2023) and Pradeep et al., (2022) investigated the structural and mechanical properties of nanostructured alloy. Shashanka et al., (2023)used mechanical alloying to synthesize novel nanomaterials. This chapter mostly focuses on some insights into why particle size analysis is essential in mechanical alloying. The first and most common is to control over alloy composition, the microstructure and the milling properties (Shashanka & Chaira, 2017). Mechanical alloying involves blending different elemental powders to create an alloy (Shashanka & Chaira, 2015b). The final properties of the alloy depend on its chemical composition. Particle size analysis helps in controlling the exact proportion of each element, ensuring a uniform distribution, and preventing undesirable phase segregation (Nayak et al., 2016).

Achieving a homogeneous mixture of powders is crucial in mechanical alloying. Particle size analysis helps to ensure that the starting powders are of similar sizes or have a well-defined distribution. Homogeneity aids in the uniform distribution of elements during milling, resulting in a more consistent alloy (Shashanka, Adimule, et al., 2023). The influence on milling kinetics is one of the factors that is affected by the PSD in mechanical alloying. The PSD impacts the milling kinetics during mechanical alloying. Smaller particles have a higher surface area, which enhances their reactivity (Shashanka et al., 2022). Understanding the initial PSD can help optimize milling parameters such as milling time, rotation speed, and ball-to-powder ratio for the desired alloy properties (Shashanka et al., 2015; Shashanka & Chaira, 2015a).

The final microstructure of the alloy is closely related to the PSD. Smaller particles tend to undergo more severe plastic deformation during milling, leading to refined microstructures (Fang et al., 2018; Shashanka et al., 2018). This refinement can improve mechanical properties such as hardness and strength as well as reduction of contamination. Particle size analysis can identify any contamination in the starting powders. Contaminants can negatively affect the alloy's properties or lead to unwanted phases (Shrinivas Mahale et al., 2022). By knowing the particle size and composition of the starting powders, appropriate steps can be taken to minimize or eliminate contamination (Kale et al., 2022).

Particle size in mechanical alloying is important for the optimization of the milling process. Particle size analysis data can be used to model and optimize the milling process. Through simulation and experimentation, the ideal PSD can be determined, leading to better alloying outcomes and reduced production costs (Oğuzyer & Shashanka, 2023). Predicting and controlling the final properties of the alloy, such as grain size, crystal structure, and mechanical properties, are influenced by the starting PSD (Msomi et al., 2022). Understanding this relationship allows researchers and engineers to tailor the alloying process to achieve specific material properties for targeted applications (Shashanka & Chaira, 2014). In summary, particle size analysis is a critical aspect of mechanical alloying as it affects the composition, microstructure, and properties of the final alloy. By controlling and understanding the PSD, researchers can optimize the mechanical alloying process and produce high-quality materials with tailored properties for various industrial applications.

3. TRENDS FOR THE PAST QUARTER CENTURY (25 YEARS)

3.1. Analysis of the Journal Articles

The search was conducted, and the selection of the theme was guided by the study done by Sanchez-Suarez et al., (2022) using the selected keywords; particle size analysis; sieve analysis; wet sieving; dry sieving; particle size distribution and screening. The screening methodology was chosen based on the method stated in Sanchez-Suarez et al., (2022), which is a method widely utilized throughout history for product classification and is still highly beneficial in the industry today. Following that, a discussion of an investigative equation to get articles to extend the topic, employing the previously indicated keywords, yields the following equation: "TITLE-ABS-KEY (sieve OR screen) and (Particle AND size AND analysis) and (wet AND sieving) and (dry AND sieving)". The period was openly left to broaden the search. The given search formula produced results for publications and patents. A search on ScienceDirect was conducted using the previously mentioned keywords.

The collected list of journal articles was filtered in reading and selecting mode to narrow the ones that focused on the issue a little more. After reviewing the most relevant articles, non-related articles were removed, since they virtually had little to do with the topic. Excel was used to construct a format that includes the following elements: year and keywords. For the last 25 years, the assessment of patents associated with the searching model has been refined like journal articles. In essence, this entails locating patents that are specifically relevant to the topic of screening and identifying the patented technology or the procedure.

The authors, year, title, journal, keywords, and given abstract were collected in an Excel table to examine the fundamental data discovered in the publications. The software was responsible for identifying the numerous correlations and trends among the data from each paper that was reviewed. Each graph shows the research trends in each nation, along with the journals and publications that are read. Figure 2 and 3 shows the particle size analysis trends for the past 25 years, published articles, and patents, respectively. In Figure 2, it is apparent that similarly to dry sieving, wet sieving also receives a lot of attention, especially in mineral processing. This information can be used as input in the decision-making process of the stakeholders of this industrial chain related to mineral processing such as mining.

Figure 2. Particle size analysis trend, from 1998-2023

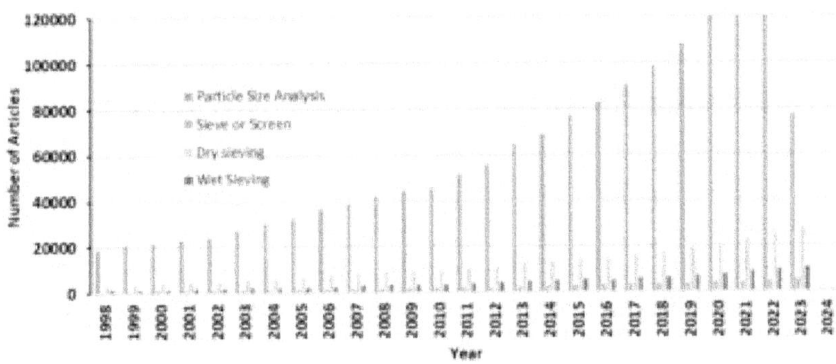

Figure 3. Patents trends, 1998-2023

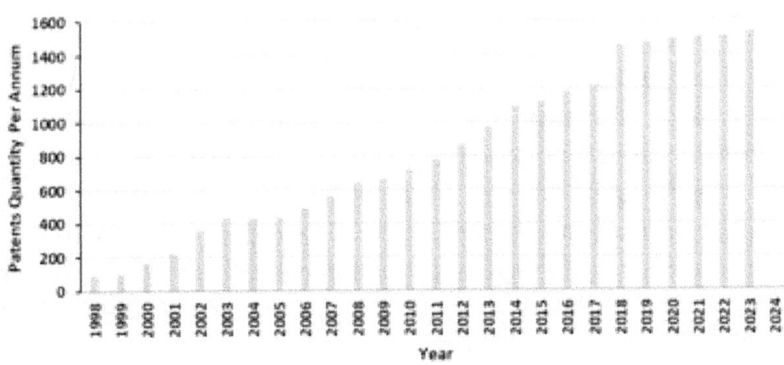

3.2. Analysis of Patents

To determine the practicality and application of sieving systems in mining and mineral processing, the prior years' applications and granted patents were examined. For this instance, the search equation: "TITLE-ABS-KEY (particle AND size AND analysis AND by AND wet AND sieving) AND PUBYEAR > 1998 AND PUBYEAR < 2024" was run again in SCOPUS, yielding a total number of patent results for all years. To narrow and update the search, we chose to focus on the last 25 years, during which a total of 21 351 patents were discovered, only 14 475 of which were directly related to some sort of sieving in mining and mineral processing, and most of which were unrelated to techniques or sieving systems. Following an examination of the seven patents directly relevant to the topic. See Figure 3 for the results presented as a graph.

4. PARTICLE SIZE ANALYSIS END-USER INDUSTRIES

Particle size analysis provides information on PSD. This can be used to calculate various particle qualities and how they would behave under different scenarios. This knowledge is crucial in industries to attain various objectives. Material characterization is critical for industrial research and development because it provides a better understanding of the relationships between material processing, structure, characteristics and performance (O'Conell, 2005). Many businesses can utilize particle size analysis to learn about the qualities of the materials they work with. This enables them to change the materials, manufacturing conditions, or products to create a better, safer, or more cost-effective product (O'Conell, 2005). Some of these industries include the pharmaceutical sector, the petroleum industry, the production of food powders and chemicals, the decorating and colour pigments industry, the construction as well as water & wastewater treatment. However, this chapter focuses on particle size analysis in mining and mineral processing.

Sieving is used in industrial processes for production safety, quality control and improved recovery in all major manufacturing sectors. Sieving is necessary to ensure that unwanted particles do not enter the production cycle with the primary product (Bertula, 2022; Sanchez-Suarez et al., 2022). This application complies with legal standards pertaining to safe food production in particular businesses, such as the food industry. The application of this technique is also important in the additive manufacturing industry to ensure the final component is as expected, i.e., 3D printing (Williams, 2015). The concept of the circular economy has prompted industries to reconsider their procedures for manufacturing and begin to invest in research for strategies that will render them sustainable in the future. In this context, sieving can be an effective technique to separate valuable products from waste and thus possibly to repurpose them into new items that can be reused or sold (Virto, 2022).

Sieving can be an interesting option for identifying and creating many useful products from a bulk component. Particle dimensions can be important in numerous industries, including food, metal powder, construction, and many others, where each product recovered from bulk can correlate to a particular market opportunity (Virto, 2022). Sieving can also be used for categorizing a powder or granular product into groups based on particle diameters, allowing different products to be identified and created. At the same time, the process will remove unwanted particles with larger dimensions, such as lumps, insects, or metal components, as well as those with smaller dimensions than the main product resulting in fine powder. Wet sieving dewaters a liquid product by separating the solid component from the water (Blaud et al., 2017; Gueli et al., 2017).

4.1. Mining and Mineral Operations

In the mineral processing plant, particle size is an important design variable. Finer grinding has been implemented to facilitate increased mineral liberation due to reduced concentrate (increased gangue) in mineral ore deposits (Donoso et al., 2013). In comminution circuits, the particle size of the feed mineral is frequently decreased to allow valuable minerals to be recovered in the subsequent downstream processing processes. Mineral liberation will be compromised, and recovery of the valuable component would be inadequate if the feed to the downstream plant is too coarse. The particle size of the resulting feed has a significant impact on several components of the process and process efficiency, hence monitoring the particle size is very important (Fang et al., 2018).

Furthermore, if the particle size is significantly reduced, this leads to increased milling energy input. The energy use in the comminution circuit will be too high, and in addition, efficiency losses will result from excessive reagent use, while filtration and thickening capacity will be reduced (Fang et al., 2018). Fang (2018) studied PSD and energy consumption during the impact crushing of single granite particles. It was evident that the comminution energy requirement is increased from 0.8 to ~1.6 units to mill or grind 18 mm coarse particle sizes and vice versa, as shown in Figure 4.

Water recovery in downstream dewatering systems is also affected by milling product particle size. Mwale et al. (2005), in addition to Fang (2018), investigated the effect of particle size on energy consumption and water recovery in comminution and dewatering systems. It has been indicated that as materials undergo grinding to finer particles, more energy is consumed, in addition, the produced material size influences water recovery (Mwale et al., 2005). The development of this energy-

Figure 4. Energy consumption per unit of the finished product versus particle size
Source: Fang et al. (2018)

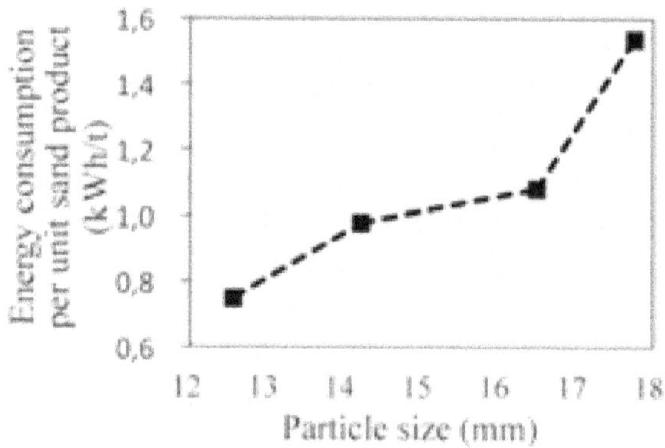

water recovery about particle size is a major variable that will aid stakeholders in integrating this during the initial planning stage. This includes mineral processing technologies on the management of downstream processes and the environment, especially salinity.

Tailings disposal is also influenced by PSD and variations in particle size cause fluctuations in the plant's downstream processing section (Lazorenko et al., 2021). Furthermore, the parameters for multiple downstream unit operations will have to be transformed according to changes in particle size. An online analysis might provide useful automated control possibilities (Pankewitz & Zeng, 2012). The demand for precise, fast, consistent, frequently measured PSD from the process stream is apparent and there is a necessity for robust online particle size analyzers.

5. VARIABLES AND PRINCIPLES GOVERNING FACTORS THAT AFFECT WET SIEVING ANALYSIS

5.1. Physicochemical Properties of the Material

5.1.1. Moisture Content

According to Allen (2003), the moisture content of a particulate material can have an impact on the material's adhesion and stickiness, which in turn affect the freedom of particle movement during the sieving process. There will always be a moisture level that allows for the most flexibility of particle movement under each screening condition and a specific particulate material. The nature of the material will, however, determine the moisture content. For instance, it may differ based on the mineralogy of the clay, the soil texture, and the air temperature during drying (Blair, 2010). Due to water-facilitated particle rearrangement, samples with higher moisture contents change in bulk density more quickly when compression is applied. However, as moisture content rises, elastic recovery declines. (Saha et al., 2022). For instance, a study by Liu, K. (2009) using a sample of soft white wheat seeds revealed that increasing the moisture level from roughly 7% to 11% seemed to improve particle mobility and increase the number of particles that passed through the sieve (Liu, 2009).

5.1.2. Surface Area

The effect of the surface area on wet sieving analysis depends on the physical properties such as particle shape or surface roughness, particle size and chemical composition of the materials. For instance, Atkinson et al. (1993), found that size class and mean specific surface area were often positively correlated, but sometimes showed a varied, nonlinear relationship (Atkinson et al., 1993). On the other hand, Atkinson et al. (1993) revealed that the materials containing organic particles had a greater mean specific surface area than those containing organic particles (Atkinson et al., 1993). In addition, the particle shape or surface roughness of the material also significantly influences the change in the surface area of a material. For instance, in the study conducted by Atkinson et al. (1992), particles with greater irregularities were found to have a greater surface area (Atkinson et al., 1992). The previous findings strongly suggest that one should not assume that just because particles are smaller, there would be more surface area accessible. For this reason, the measurement of surface area becomes much more important for small particles during sieving analysis.

5.1.3. Shape and Size

Regarding the effect of shape and size, the assumption in sieve analysis is that all particles will be square or nearly square, of similar size, and will pass through the apertures. However, particle size and shape relative to the sieve aperture are among the most significant direct elements influencing the sieving process is sieve blinding, which is influenced by several different parameters. According to a prior study by Roberts and Beddow, the mesh aperture has a significant impact on the degree of sieve blindness when the mesh aperture drops below around 100 μm, blinding increases significantly (Roberts & Beddow, 1968). Initially, it was identified that particles that were near the size of sieving apertures caused the blocking of the apertures (Liu, 2009). Apling, (1984) showed that particles with potentially blinding effects may be as insignificant as three-quarters of the size of the screen opening under specific circumstances.

Depending on the situation, an undersized particle may have a relatively significant possibility of blinding the aperture due to both its irregular shape and that of the aperture. When two or more fine particles try to pass through an aperture at once, they may also become trapped. In this regard, although near-mesh particles clog screen apertures easily, sub-mesh particles blind apertures when present singly or in conjunction with other particles (Roberts & Beddow, 1968). In addition, a sieve analysis will not produce meaningful results for elongated and flat particles since the particle size reported will presume similar sizes in three dimensions. On the other hand, researchers demonstrated that oversized particles enhanced the sieving rates by boosting the water flow of particles while simultaneously opening the apertures allowing additional particles to pass through. Consequently, sieve blinding from near-mesh and sub-mesh particles is decreased (Standish, 1985).

5.2. Operational Conditions

5.2.1. Sieving Method

Since several decades ago, sieve analysis has been used to evaluate the quality of materials based on their particle size. When the material being tested is unaffected by water, wet sieve analysis can be used. Fine material is significantly more effectively pushed through the sieve by flowing water than by shaking dry material. Regarding the effect of sieving methods, both stacked and reverse sieving procedures had significant effects on PSD. In terms of improving sieving performance and efficiency and reducing sieving loss, the reverse sieve method consistently outperformed the stacked sieve method. In addition, the experiment conducted by Liu, K. (2009), shows that using the reverse sieve method greatly increases both the chemical compositional differences among sieved fractions as well as the mass frequency of finer particle classes, when other conditions such as the milling method, flour type, tapping or non-tapping, and sieving duration are kept constant (Liu, 2009). Moreover, the improved sieve result is related to large particles' favourable influence on minimizing sieve blindness by near- or sub-sieve-sized particles. Nevertheless, reverse sieving increases operational work and necessitates greater care when performing each particle size test (Biazzi et al., 2022).

5.2.2. Tapping

In wet sieving procedures, tapping had significant effects on PSDs. To ensure that the sides of the particles connect with the sieve mesh, tapping that occurs simultaneously with shaking helps shift the orientation of the particles (Liu, 2009). Depending on the method used, the sample is moved either horizontally or vertically during sifting. Due to this, the particles and sieve move relative to one another. Depending on their size, the individual particles either pass through the sieve mesh or are trapped on the sieve surface. The downside of the horizontal movement is that some particles, especially those that are close to mesh size, tend to block some of the sieve apertures, causing sieve blinding. Tapping motion appears to have enhanced vertical movement while also assisting in dislodging particles that obstructed apertures, reducing the sieve blinding effect (Liu, 2009).

5.2.3. Sieving Duration

The duration of sieving is a major factor in obtaining an accurate measurement of particle size during the sieving process (Diaz-Zorita et al., 2007). However, the duration of sieving during a sieving operation as well as the overall energy inputs vary for each sieving system. In investigations involving soil fragments, a variety of sieving times (between 5s and 100min) have reportedly been employed (Diaz-Zorita et al., 2007). According to Liu (2009), the sieving time improves sieving efficiency significantly; this is due to the improvement in the proper orientation of the particles (Liu, 2009). Additionally, longer sieve times give particles more chances to be accurately sorted (Han et al., 2008). To pass the entire mixture through the smaller aperture sieves, the feed material and the particle fractions must be sieved for a longer period of time. The particle's mass-to-area ratio decreases as it becomes finer. In relation to the sieving time, the sieve loading rate is probably directly correlated with the sieving time (Diaz-Zorita et al., 2007).

6. CHALLENGES THAT CAN GIVE MISLEADING AND ERRONEOUS PSD RESULTS: SIEVING TECHNIQUE

In particle size analysis, inaccuracies exist in all measurements. These errors can be attributed to size and quantity. Size is generally ascribed by theory or calibration, rather than true measurement, except for imaging techniques. As a result, most of the inaccuracy often occurs in quantity measurement. Typically, quantity is measured by simple counting, resulting in a direct measurement of the number size distribution (Liu, 2009; Barry A. Wills & Finch, 2016). Direct weighing of the categorized material is frequently used to estimate the relative quantity. Considering that most accessible procedures have an ideal range of applications, size constraint is a present challenge (Hogg, 2008).

6.1. Physicochemical and Operational Associated Errors

As previously stated, Particles are categorized by sieve analysis according to their equivalent sphere diameter and sieve aperture dimension, which typically corresponds to the particles' secondary rather than their primary size (Williams, 2015). The sieving technique assumes that particles are spherical and have similar shapes. However, particles with irregular shapes, such as fibres or flakes, can become trapped in the sieve openings, leading to an inaccurate measurement of the PSD. This suggests that shape can have a significant impact on size results and potentially a source of misrepresentation. Sieve analysis, for example, may classify elongated particles as having a finer PSD than other approaches. Angular-shaped particles, on the other hand, can be characterized by the coarser PSD (Rodriguez, 2013). As demonstrated in Figure 5, the sub-angular form particle frequently becomes plugged into the screen aperture, resulting in misleading and erroneous PSD findings (Rodriguez, 2013).

Figure 5. Sub-angular shaped particles get clogged in the screen aperture resulting in PSD errors
Source: Adapted from Allen (1997) and Barbosa-Cánovas et al. (2012).

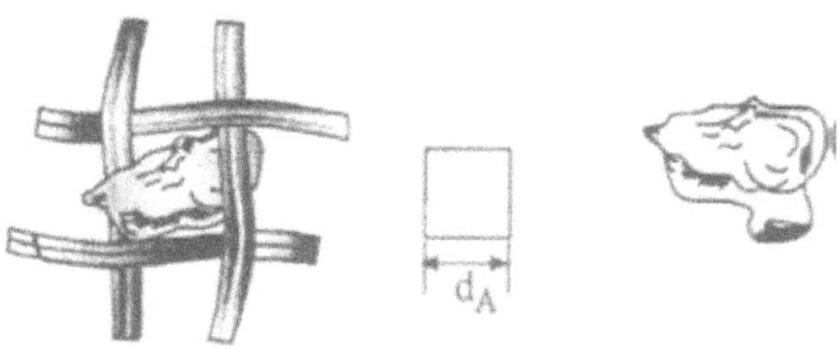

Too often, the particle density distribution is erroneously represented. Many users prefer the representation as distribution density, which is frequently and wrongly referred to as a "Gaussian curve" in shorthand (Microtrac MRB, 2023). The cumulative curve's first derivative is the distribution density, and this density distribution has a maximum when the cumulative curve increases rapidly and a minimum when the cumulative curve is flat. It is critical to note that the correct density distribution reflects the cumulative curve's slope. As a result, the measurement class quantity must be divided by the class width, see Figure 6. The number of measurement classes increases the density distribution's accuracy. The approach of connecting the histogram bars with a "balancing curve" does not necessarily produce a density distribution. With respect to the density distribution's low information content and

Figure 6. The distributions are represented as histograms for the probability density function

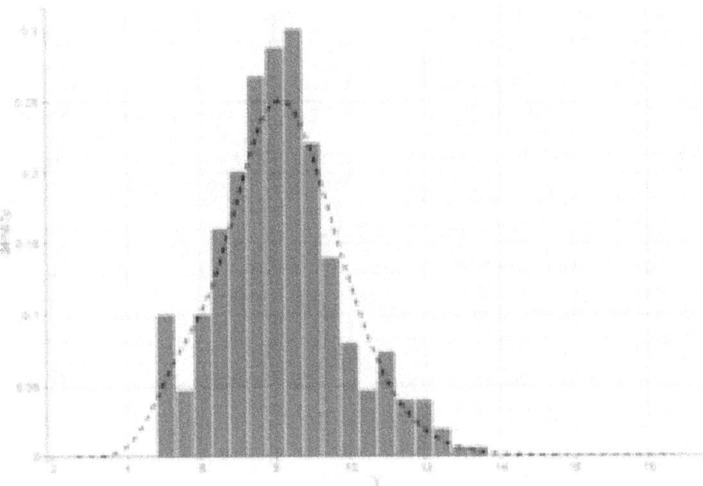

error-proneness, it should be disregarded for the benefit of a cumulative distribution (Bhrany & Brown, 1962; Microtrac MRB, 2023).

Inefficient sieving process, inadequate agitation or insufficient sample mass can lead to inefficient sieving, causing particles to remain trapped in the sieve openings. This can result in inaccurate PSD data. The orientation of particles during sieving can influence the results. If particles align in a way that allows them to pass through the sieve openings more easily, the measured PSD may not be representative of the actual distribution (Avramović et al., 2018).

6.2. Sampling Method and Feeding Technique

It is consequently critical that size analysis techniques quality control is not comprised (Wills & Finch, 2016b). This implies that it must be precise, accurate, reproducible and consistent, as substantial adjustments in plant operation may be made based on the results of these analyses. Since most particle-size instruments often require a small amount of the material, it must be ensured that the sample is a true representative of the bulk material, and thus precaution should be taken during sampling. A true representation is a subset of a larger group that accurately depicts the characteristics of the entire population (Wills & Finch, 2016b). Quality control is fundamental in every sector, and it is dependent on the quality control of particle analysis testing.

Many people are not mindful that particle testing preparation is just as significant as the actual test itself. Obtaining an accurate sample is one of the first and most critical tasks in the particle analysis process, regardless of whether the technique of analysis is a vibrating sieve or an automated particle analyzer. Any test conducted on the sample, no matter how precise or refined, is not beneficial if the sample fails to accurately represent the entire batch of material (Rajendran & Murty, 2023).

As per sampling guidelines, the standards set by the industry regulate sample size, mass, and amount required for analysis (Kongas & Saloheimo, 2014). The initial step for any particle measurement technique is to ascertain the quantity of material you need to sample based on the guidelines. Gathering a representative sample is also extremely important. It is often noticed that, when a large quantity of the test material is randomly poured out, the coarse particles are distributed on the outer edges and the fine will land in the middle. The sample taken from this pile will not be a precise representation of the bulk sample. Therefore, this is certainly not an acceptable sampling strategy for representing an entire load of material (Smith et al., 2014). When working with a substantial amount of material, the most efficient way to start accurately is to subdivide it into smaller samples using the coning and quartering method, splitter, or rotary sampler, as shown in Figure 7. Amongst them, the well-known machinery is the rotating material splitter, in Figure 7c. The rotary sampler separates the particles into several identical-size samples.

Figure 7. Different types of sampling techniques
Source: a) Coning & quartering technique (Alakangas & Impola, 2015), b) typical riffle splitter, c) Geneq Rotary splitter (Geneq, 2023)

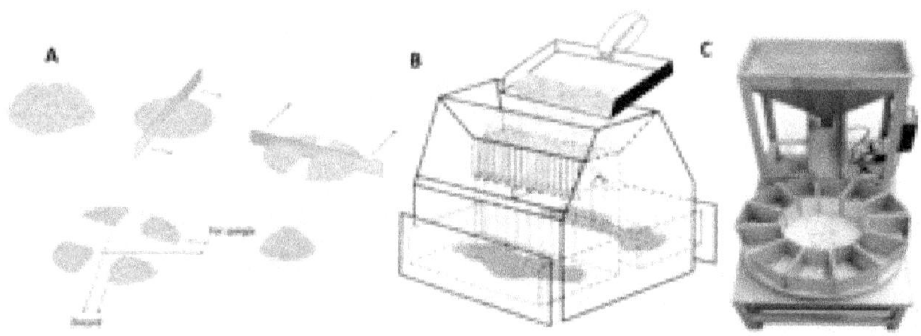

6.3. Screen Plugging and Blinding

Particle agglomeration or clumping of particles can occur during the sieving process, leading to the formation of larger particles. These agglomerates may not pass through the sieve openings, resulting in an overestimate of the PSD. Plugging occurs when near-size irregular particles become stuck in the screen apertures, as depicted in Figure 8a. This reduces the amount of open space for the material to fall through, resulting in an ineffective screening process and, in many cases, product contamination. Blinding of the screen medium occurs when the material contains high humidity or moisture and clay, causing small particles to attach and obstruct the screen surface (see Figure 13b). This can prevent fine material from going through, and the screen section's screening efficacy will progressively deteriorate as more particles plug into the screen apertures (Barlow, 2023).

According to the literature, there are three common causes or occurrences of screen plugging and blinding following analysis (Barlow, 2023). The first occurrence occurs when the crushed stone particle size is similar to the size of the screen aperture. Such particles become caught in the aperture during the screening process and cannot move through the screen smoothly, resulting in screen plugging. This type of plugging is often referred to as near-sized plugging (Barlow, 2023). The second is when flat particles absorb a substantial proportion of materials as a result of crushing or the material itself. These flat particles cannot pass smoothly through the screen. Furthermore, they spread out and congregate, preventing other materials from screening and blocking the aperture (Gerasimov et al., 2021). The final one is when the material contains a lot of moisture or needs to be flushed. One aspect is when the coarse material contains a high proportion of mud or when the material needs to be processed, the fine particles become bonded together by water. The mud blinds the screen, making it impossible for particles to pass through the apertures causing screen blinding.

Figure 8. Effect of the screen plugging and blinding: a) plugging, b) blinding
Source: Barlow (2023), Russell Finex (2023)

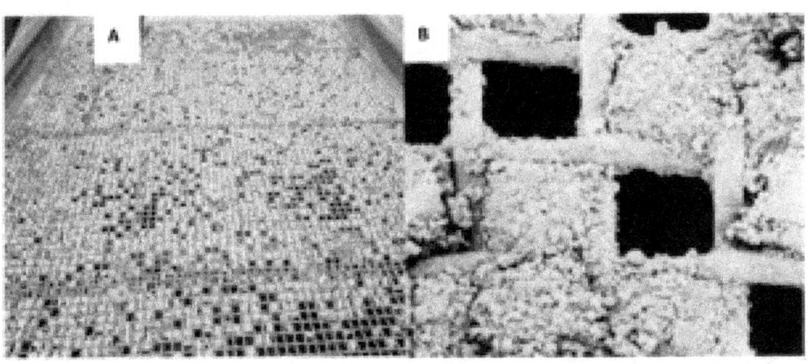

Screen blinding and plugging can be avoided by investigating the use of various screen materials, sieving vibration patterns, and shaking speeds (Liu, 2009). However, some implementations could considerably increase the cost of sieving without much improvement. The screen blinding advancement is intended to significantly enhance sieving efficacy and profitability. As a result, it is critical to pick high-efficiency screens that are designed to essentially minimize blinding and clogging. In response, screen media manufacturers have created a variety of self-cleaning screens that are intended to increase production volume and dependability with no human oversight or control (Russell Finex, 2023).

6.4. Human Error

Mistakes in the handling and operation of sieving equipment, such as incorrect sieve stacking, inconsistent shaking intensity, or improper cleaning, can introduce errors and inaccuracies in the results. To mitigate these challenges and improve the accuracy of PSD measurements using sieving techniques, it is important to carefully consider sample preparation methods, optimize the sieving process parameters, ensure the use of appropriate sieves, and employ complementary techniques such as microscopy or laser diffraction for more comprehensive analysis. Regular calibration and maintenance of sieving equipment are also essential to minimize errors.

7. MARKET ANALYSIS FOR PARTICLE SIZE ANALYSIS AND DEMAND IN 2023 AND THE FUTURE

7.1. Different Particle Size Analysers

Particle-size technologies have the potential to be tremendously successful on a global scale, but widespread endorsement and practical applications are still not as good as they ought to be (Business Wire & Research and Markets, 2022; Facts & Factors, 2023). There are various particle size measurement techniques available on the market today, with laser diffraction, dynamic light scattering (DLS), dynamic image analysis (DIA), electrozone sensing, sedimentation and sieve analysis being the most common. Different results can be obtained because distinct methods measure various dimensions of the particle using different mechanisms. There are benefits and drawbacks to each of the various measurement techniques. Particle size is estimated indirectly in laser diffraction by sensing intensity distributions of laser light scattered by particles at different angles (Malvern Instruments, 2015). A popular strategy is regarded as one of the most effective, accurate and reliable. It has several significant advantages, including the ability to measure various forms of particles, including powders, sprays, suspensions and emulsions (Mordor Intelligence, 2023). The technique has a very wide and dynamic range, allowing it to measure a whole sample; it is extremely repeatable and very fast. It provides an absolute method of particle analysis based on scientific principles, allowing measurements to be made without the requirement to calibrate any device against a reference or blank (Karlsson & Schade, 2013).

DIA involves moving a huge number of particles past a camera system and analyzing them in a completely random direction. Based on the particle images, a variety of sizes and form parameters such as breadth, length and diameter of equivalent circle are calculated (Terzi, 2018). Since DIA measures the particles in random orientation, there are systematic discrepancies in the acquired results when measuring irregularly shaped particles. DLS is a well-established and precise measuring technique for assessing particle sizes in suspensions and emulsions. It is based on particle Brownian motion in suspensions, which states that smaller particles move more quickly, while larger particles move more slowly (Mordor Intelligence, 2023). Sieving is a traditional approach, and the main problem of this particle size analysis technique is that it is impossible to measure sprays or emulsions, Also measuring dry powders becomes very challenging when measuring very small particles (powder). However, it has the advantage of being inexpensive and good for measuring big particles, notably in industries such as mining. Wet sieving, on the other hand, can quantify small particles although reproducibility and repeatability can be difficult. Materials that are cohesive and agglomerated, such as clay, are similarly difficult to measure since they cause blinding (Blaud et al., 2017).

7.2. What The Future Will Demand for Particle Size Analysis

In recent years, particle size analysis has advanced drastically beyond milling, sieving, and other material processing operations to meet industry demands. This has accelerated the particle characterization procedure while maintaining 100% accuracy every time. According to the report's analysis, the worldwide particle size analyzer market is anticipated to increase from USD 395.37 Million in 2022 to USD 695.45 Million by 2032 (Mordor Intelligence, 2023). The South Africa Particle Size Analysis (PSA) Market is estimated to expand from USD 1 995,00 to USD 3 435,00 between 2022 and 2028 (6WResearch, 2022). During the forecast period, it is expected to develop rapidly at a compound annual growth rate (CAGR) of 5.5% (Transparent Market Research, 2022). The market is expected to expand significantly because of several driving forces. The increasing investments in R&D efforts, as well as the continuous advancements in nanotechnology, are important driving forces. Furthermore, the implementation of obligatory government requirements for product quality and the growing need for improving industrial efficiency are projected to fuel market demand in the near future. Figure 9 shows that the Asia-Pacific region emerged as the largest in the global particle size analyzers market, with a 34.85% share in 2022 (Mordor Intelligence, 2023).

Figure 9. Particle size analysis, regional outlook (revenue - USD million, 2022 - 2027)
Source: Mordor Intelligence (2021)

According to a recent Mordor Intelligence research, the market penetration of particle size analyzers is very compelling and new advancements and patents are filed every year. Market incumbents with significant R&D capability or startups with top product differentiation often thrive in the market. Malvern Panalytical Ltd, Microtrac Retsch GmbH, Horiba Ltd, Agilent Technologies Inc., Beckman Coulter Inc., and Anton Paar GmbH are the leading manufacturers or developers (Business Wire & Research and Markets, 2022). Based on technology, the laser diffraction market dominates the global market with its highest revenue shares. The drawback and challenge with such devices is the high price of purchasing and maintaining the instrument (Wedd et al., 2019). This suggests that there is a need to establish a more affordable technique, which has led to why sieving has attracted a lot of attention recently.

The primary advantage of online analytic systems is that they do not, by definition, require sampling by operations individuals, hence eliminating the human source of errors. In the case of some systems' continuous measurement, they also deliver significantly more rapid results, which minimizes production delays and optimizes efficiency. Online measurement becomes essential for process control (Wills & Finch, 2016a). At the current stage, online analytical methods are scarce, and there is an urgent demand for equipment that can be employed for the automatic control of costly operations such as grinding or precipitation. Although some attempts have been made, such as the diffraction device for use in South African gold mines, more work and effort are required in this field. In the future, particle size analyzers will be expected to report on a wider range including more on the smaller particles (Wills

& Finch, 2016a). Most recent technological advancements in the field of PSA have focused on sub-sieve or the under 38 µm measurements, which are challenging to measure using existing test sieve methods (Liu, 2009).

In the global market of PSA, analysis has been divided into three types of particle dispersion: dry, moist, and spray particles. In 2021, the wet particle segment held a significant share of the global market (Business Wire & Research and Markets, 2022). The problem with wet sieving is cleaning the screens after each test. However, some progress has been achieved in sieve cleaning utilizing ultrasonic processes (Russell Finex, 2023). This procedure is far superior to the time-consuming brush and poke-cleaning technique that is commonly utilized. Since the increased emphasis on quality control programs such as ISO 9000, certification of test sieve performance has become a major demand (American Standards & Test Methods, 2017). Recent, efficient, and affordable calibration systems for use in-plant have been developed, serving as a trustworthy benchmark for the introduction of stringent quality control procedures. Following standard sieving test procedures such as the American Society for Testing and Materials (ASTM) D6913/D6913M-17 can be time-consuming, hence high-end automation equipment is required. The approach must not only minimize operator error to near zero, but it must also reduce the time required to execute a sieve. The procedure must not only reduce operator error to nearly zero but also cut the time for conducting a sieve analysis test.

7.3. Comparison of Different Particle Size Measurement Instruments on the Market

Table 1 summarises a few examples of particle size measurement techniques available on the market. The choice of method depends on factors such as the desired size range, sample characteristics, accuracy requirements, cost, and throughput. It is recommended to consider the specific needs of your application and consult with instrument manufacturers or experts to select the most suitable technique.

8. CONCLUSIONS AND RECOMMENDATIONS

- The first objective of this review article was to report the trend in the wet-sieving analysis as a particle size analysis tool for the past 25 years. The research analysis of particle size analyzers has expanded significantly. New publications are produced frequently, and patents are filed every year. This information can be used as input in the decision-making process of the stakeholders of the industrial chain related to mineral processing such as mining.

Table 1. Particle size measurement techniques available on the market in 2023

Method of Analysis	Measurement and Detection Limits	Resolution	Reproducibility	Particle Dispersion	Factors that Lead to Errors	Response Time	Ease of Use	Advantages	Disadvantages	Key Manufacturer / Developer	Reference
Laser diffraction	10 nm - 4000 µm	Medium-Low	High	Wet/Dry	Orientation Effect	Rapid	Medium operator skills	Can measure a variety of samples; liquid, dry, or aerosol Quality control: high degree of accuracy and repeatability Easy operation, rapid test cycle and wide and dynamic range	Sampling: Correct representative sample of the bulk material High equipment cost Low resolution	Malvern Mastersizer Microtrac SYNC Horiba Partica	(Malvern Instruments, 2015) (Karlsson & Schade, 2013)
Dynamic Light Scattering (DLS)	0.8 - 6500 nm	Low	High	Wet	Inevitable width-error	Medium-Rapid	Medium operator skills	Can measure zeta potential Measure in a wide range: dry, suspensions and emulsions Possibility of analyzing nm to the µm particles	Sensitive to changes in temperature, thus viscosity Closely related particles, i.e., similar sizes are not resolved presence of aggregates may affect measurements	Microtrac Nanotrac Malvern Zetasizer Horiba NanoPartica	(Karlsson & Schade, 2013)
Image Analysis (Optical & Electron Microscopy)	0.8 µm - 135 mm	High	High	Dry	Particles Matching	Medium-Long	Needs high operator skills	Can measure shape, density, and surface area in 5 minutes Simple, intuitive, high accuracy, and quality control Measure particles at a higher speed, typically about 10,000 particles/minute or greater	Sampling errors main source of unreliable measurement results Not recommended for particles with wide-range PSD Cannot analyze a sample size that is less than 1 µm or >1000 µm	Microtrac Camsizer Horiba ImageSizer Malvern Morphologi	(Karlsson & Schade, 2013)
Electrozone Sensing	0.4 - 1400 µm	High	High	Wet	Calibration	Rapid	Medium operator skills	Suitable for irregular particles Not sensitive to refractive index, conductivity, or particle opacity Can do morphology and structural analysis	Low sample representative The instrument is very expensive Easily affected by human errors	Beckman Coulter Multisizer 4e Coulter Counter Micromeritics Elzone II 5390 Sysmex XN-1000 Coulter Counter	(Karlsson & Schade, 2013)
Sedimentation	1.0 - 140 µm (gravity) 0.05 - 5 µm (centrifuge)	Medium	Medium	Wet	Dispersion Stability	Very Long	Needs medium-high skilled operator	Can operate continuously Low investment cost, high accuracy, and reproducibility Wide test size range	Long test times can take up to a few days. Complicated operations that need highly skilled operators Cannot be used for particles of size larger than 0.2 mm	CPS Disc Centrifuge Micromeritics SediGraph III Centrifugal sedimentation PSA BT-1500	(Schakowski and Fester 1988)
Sieve Analysis	20 µm- 125 mm	High	Medium	Wet/Dry	Screen blinding and plugging	Long	Low-medium operator skills	Easy handling and low instrument cost Highly robust Provide a more complete picture of the PSD	Results are greatly affected by human factors and mesh deformation Cannot accurately analyze particle size below 40 µm It is time-consuming to assemble and disassemble the sieves	W.S. Tyler RO-Tap Labotec Analysette 3 Pro Sieve Shaker Elcan Hi-Sifter	(Karlsson & Schade, 2013) https://www.telsonic.com/en/sieving/ https://www.russellfinex.com/

- Since material characterization is critical for industrial research and development, many industries such as the pharmaceutical sector, the petroleum industry, the construction and decorating industry as well as the water treatment industry. The medical and pharmaceutical industry has increased usage due to stringent quality control that has been placed by the government. This is followed by the production of additive material because of the technology advancements and then food and chemical production.
- Physicochemical properties such as moisture content shape, surface area and operational conditions such as the sieving technique, shaking or tapping as well and the sieving duration are the major governing factors that affect sieving tests, especially the wet sieving analysis.
- In particle size analysis, inaccuracies and errors can be attributed to the size and potential quantity. This is due to physicochemical properties, operation conditions and sampling technique that often leads to screen blinding and plugging. In plugging, particles become clogged in the screen aperture and small particles bond and obstruct the screen surface in blinding. All these have the potential to result in misrepresentation and erroneous PSD results.
- Market incumbents with significant R&D capability or start-ups with top product differentiation often thrive in the market. The comparison table showed some of the drawbacks and challenges with most of the particle size measurement devices. The major disadvantage of these instruments is the high price of purchasing and maintaining the instrument. As a result, a more economical technique is required, which is why sieving has recently received a lot of attention.

ACKNOWLEDGMENT

The author would like to acknowledge the authors' work team at Mintek for their dedication that facilitated the development of this review article.

REFERENCES

Adimule, V., & Shashanka, R. (2023). *Properties and Technological Features of Materials*. Academic Press.

AlakangasE.ImpolaR. (2015). Quality guidelines for wood fuels in Finland. *October*, 39. doi:10.13140/RG.2.1.3290.3127

Allen, T. (2003). *Powder sampling and particle size determination*. Elsevier.

American Standards & Test Methods. (2017). Standard Test Methods for Particle-Size Distribution (Gradation) of Soils Using Sieve Analysis. In ASTM International (Vol. 04, Issue Reapproved 2009, pp. 1–35). https://doi.org/ doi:10.1520/D6913-17.1.6

Apling, A. C. (1984). Blinding of screens by sub-sieve sized particles. *Transactions of the Institution of Mining and Metallurgy Section B. Applied Earth Science*, *93*, C92–C94.

Atkinson, C. F., Aumen, N. G., Miller, G. L., & Ward, G. M. (1992). Influence of particle shapes and size distributions on fine particulate organic matter surface area in streams. *Journal of the North American Benthological Society*, *11*(3), 261–268. doi:10.2307/1467646

Atkinson, C. F., Miller, G. L., & Aumen, N. G. (1993). Influence of Season, Size, and Collection Site on Surface Area of Fine Particular Matter in Streams. *Journal of the North American Benthological Society*, *12*(4), 410–417. doi:10.2307/1467622

Avramović, L., Bugarin, M., Milanović, D., Conić, V., Pavlović, M. M., Vuković, M., & Nikolić, N. D. (2018). The particle size distribution (PSD) as criteria for comparison of silver powders obtained by different methods of synthesis and by conditions of electrolysis. *Journal of Mining and Metallurgy, Section B. Metallurgy*, *54*(3), 291–300. doi:10.2298/JMMB171002020A

Barbosa-Cánovas, G. V., Harte, F., & Yan, H. H. (2012). Particle size distribution in food powders. *Food Engineering*, *1*, 303–328. http://www.eolss.net/sample-chapters/c10/E5-10-01-06.pdf

Barlow, B. (2023). *Screen Technology Helps Solve Blinding, Plugging Problems.* https://www.constructionequipment.com/home/article/10716519/screen-technology-helps-solve-blinding-plugging-problems

Basavarajappa, P. N., Shashanka, R., Surendranathan, A. O., & Hegde, M. M. R. (2023). Investigation of Structural and Mechanical Properties of Nanostructured TiMgSr Alloy for Biomedical applications. *Biointerface Research in Applied Chemistry, 13*(2), 1–12. doi:10.33263/BRIAC132.118

Bertula, K. (2022). *Particle size distribution analysis: Guide to method selection.* Academic Press.

Bhrany, U. N., & Brown, J. H. (1962). Minerals Beneficiation - Particle-Size. *Measurement and Control, 226.*

Biazzi, H. M., Tubin, J. S. B., Conte, R. A., Robazza, W. da S., & Paiano, D. (2022). Different sieving methods for determining the physical characteristics in ground corn. *Acta Scientiarum. Animal Sciences, 44*, e53382. Advance online publication. doi:10.4025/actascianimsci.v44i1.53382

Blair, N. (2010). The impact of soil water content and water temperature on wet aggregate stability. What answer do you want? *Proceedings of the 19th World Congress of Soil Science: Soil Solutions for a Changing World,* 1–6.

Blaud, A., Menon, M., van der Zaan, B., Lair, G. J., & Banwart, S. A. (2017). Effects of Dry and Wet Sieving of Soil on Identification and Interpretation of Microbial Community Composition. In *Advances in Agronomy* (Vol. 142, pp. 119–142). Academic Press Inc. doi:10.1016/bs.agron.2016.10.006

Business Wire & Research and Markets. (2022). *The Worldwide Particle Size Analysis Industry is Expected to Reach $609 Million by 2030.* Author.

Diaz-Zorita, M., Grove, J. H., & Perfect, E. (2007). Sieving duration and sieve loading impacts on dry soil fragment size distributions. *Soil & Tillage Research, 94*(1), 15–20. doi:10.1016/j.still.2006.06.006

Donoso, M., Robles, P. A., Gálvez, E. D., & Cisternas, L. A. (2013). Particle size effect on the efficient use of water and energy in mineral concentration processes. *Industrial & Engineering Chemistry Research, 52*(49), 17686–17690. doi:10.1021/ie402099n

Facts & Factors. (2023). *Particle Size Analysis Market Size is Projected to Grow.* https://www.fnfresearch.com/sample/particle-size-analysis-market

Fang, H. Y., Yang, J. H., & Chen, Q. (2018). Particle size distribution and energy consumption during impact crushing of single granite particles. *Journal of the Southern African Institute of Mining and Metallurgy, 118*(5), 555–561. doi:10.17159/2411-9717/2018/v118n5a12

Geneq. (2023). *Geneq Rotary Sampler Splitter*. doi:10.1108/eb052349

Gerasimov, A. M., Eremina, O. V., Cherkasova, M. V., & Dmitriev, S. V. (2021). Application of particle-size analysis in various industries. *Journal of Physics: Conference Series, 1728*(1), 012003. Advance online publication. doi:10.1088/1742-6596/1728/1/012003

Gueli, A. M., Bonfiglio, G., Pasquale, S., & Troja, S. O. (2017). Effect of particle size on pigments colour. *Color Research and Application, 42*(2), 236–243. doi:10.1002/col.22062

Han, J., Fu, J., & Schoch, R. B. (2008). Molecular sieving using nanofilters: Past, present and future. *Lab on a Chip, 8*(1), 23–33. doi:10.1039/B714128A PMID:18094759

Hogg, R. (2008). Issues_in_Particle_Size_Analysis. *Kona Powder and Particle Journal, 26*(2), 81–93. doi:10.14356/kona.2008009

Kale, S., Adimule, V., Shashanka, R., Tanaka, M., & Wei, P. (2022). *Synthesis and Modifications of Materials and its Properties*. Trans Tech Publications Ltd.

Karlsson, O., & Schade, B. E. H. (2013). Particle Analysis. In Chemistry and Technology of Emulsion Polymerisation: Second Edition (pp. 213–251). doi:10.1002/9781118638521.ch9

Kongas, M., & Saloheimo, K. (2014). *Next Generation Slurry Particle Size Analyzers*. Academic Press.

Lazorenko, G., Kasprzhitskii, A., Shaikh, F., Krishna, R. S., & Mishra, J. (2021). Utilization potential of mine tailings in geopolymers: Physicochemical and environmental aspects. *Process Safety and Environmental Protection, 147*, 559–577. doi:10.1016/j.psep.2020.12.028

Liu, K. S. (2009). Some factors affecting sieving performance and efficiency. *Powder Technology, 193*(2), 208–213. doi:10.1016/j.powtec.2009.03.027

Malvern Instruments. (2015). *A basic guide to particle characterization*. Author.

Microtrac, M. R. B. (2023). *Errors in Particle Size Analysis and How to Avoid Them*. www.microtrac.com

Mordor Intelligence. (2023). *Global Particle Size Analysis Market: Growth, Trends, Covid-19 Impact and Forecasts (2023-2028)*. https://www.mordorintelligence.com/industry-reports/particle-size-analysis-market

Msomi, V., Adimule, V., Shashanka, R., Kaloop, M., & Koutsonas, S. (2022). Materials and Production Technologies in Machinery. Trans Tech Publications Ltd.

Mwale, A. H., Musonge, P., & Fraser, D. M. (2005). The influence of particle size on energy consumption and water recovery in comminution and dewatering systems. *Minerals Engineering*, *18*(9), 915–926. doi:10.1016/j.mineng.2005.02.014

Nayak, A. K., Shashanka, R., & Chaira, D. (2016). Effect of Nanosize Yittria and Tungsten Addition to Duplex Stainless Steel during High Energy Planetary Milling. *IOP Conference Series. Materials Science and Engineering*, *115*(1), 012008. Advance online publication. doi:10.1088/1757-899X/115/1/012008

O'Conell, R. (2005). Sieve Use in the Pharmaceutical Industry. *Pharmaceutical Technology Europe*, *17*(5), 1. https://www.pharmtech.com/view/sieve-use-pharmaceutical-industry

Oğuzyer, A., & Shashanka, R. (2023). An Overview of Nanocomposites with Recent Advancements. *Journal of Metastable and Nanocrystalline Materials*, *37*, 45–56. doi:10.4028/p-VW2fCP

Pankewitz, A., & Zeng, S. (2012). Implementation of Online Particle Size and Concentration Analysis for Grinding Optimization. *XXVI International Mineral Processing Congress (IMPC)*, 4127–4136.

Pradeep, N. B., Hegde, M. M. R., Shashanka, R., & Surendranathan, A. O. (2022). Investigation of microstructure and mechanical properties of microwave consolidated TiMgSr alloy prepared by high energy ball milling. *Powder Technology*, *408*(February), 117715. doi:10.1016/j.powtec.2022.117715

Rajendran, S., & Murty, C. V. G. K. (2023). *Mineral Processing: Beneficiation Operations and Process Optimization through Modeling*. Academic Press.

Roberts, T. A., & Beddow, J. K. (1968). Some effects of particle shape and size upon blinding during sieving. *Powder Technology*, *2*(2), 121–124. doi:10.1016/0032-5910(68)80046-4

Rodriguez, J. M. (2013). Importance of the Particle Shape on Mechanical Properties of Soil Materials. Luleå University of Technology.

Russell Finex. (2023). *Ultrasonic Sieves With Deblinding Screens.* https://www.russellfinex.com/en/separation-equipment/ultrasonic-sieves/ultrasonic-mesh-screens/

Saha, N., Goates, C., Hernandez, S., Jin, W., Westover, T., & Klinger, J. (2022). Characterization of particle size and moisture content effects on mechanical and feeding behavior of milled corn (Zea mays L.) stover. *Powder Technology, 405*, 117535. doi:10.1016/j.powtec.2022.117535

Sanchez-Suarez, N., Orozco-Mendoza, G. L., Zartha-Sossa, J. W., Gafaro-Garcés, D. C., Melchor-Cahuana, L. G., & Gonzalez-Tovar, C. (2022). Trends in Sieving and Its Applications in Cereals. A Literature Review. In *Frontiers in Sustainable Food Systems* (Vol. 6). Frontiers Media S.A. doi:10.3389/fsufs.2022.902147

Shashanka, R. (2017). Synthesis of nano-structured stainless steel powder by mechanical alloying and characterization. *International Journal of Scientific and Engineering Research, 8*(June), 588–594.

Shashanka, R., Adimule, V. M., Jayaprakash, G. K., & Pandith, A. (2023). Electrochemical oxidation of methylene blue dye in wastewater using mechanically alloyed high entropy alloy modified carbon paste electrode using cyclic voltammetry. *Materials Research Express, 10*(5), 054003. Advance online publication. doi:10.1088/2053-1591/acd3d7

Shashanka, R., Basavegowda, N., Adimule, V. M., Avar, B., Somu, P., Saravana, S. K., & Baek, K. H. (2022). Assessing the Food Quality Using Carbon Nanomaterial Based Electrodes by Voltammetric Techniques. *Biosensors (Basel), 12*(12), 1–19. doi:10.3390/bios12121173 PMID:36551140

Shashanka, R., Basavegowda, N., Vinaykumar, R., Narsimhachary, D., Somu, P., & Lee, M.-J. (2023). Electrocatalytic determination of methyl orange dye using mechanically alloyed novel metallic glass modified carbon paste electrode by cyclic voltammetry. *Inorganic Chemistry Communications, 155*(June), 111010. doi:10.1016/j.inoche.2023.111010

Shashanka, R., & Chaira, D. (2014). Phase transformation and microstructure study of nano-structured austenitic and ferritic stainless steel powders prepared by planetary milling. *Powder Technology, 259*, 125–136. doi:10.1016/j.powtec.2014.03.061

Shashanka, R., & Chaira, D. (2015a). Development of nano-structured duplex and ferritic stainless steels by pulverisette planetary milling followed by pressureless sintering. *Materials Characterization, 99*, 220–229. doi:10.1016/j.matchar.2014.11.030

Shashanka, R., & Chaira, D. (2015b). Optimization of milling parameters for the synthesis of nano-structured duplex and ferritic stainless steel powders by high energy planetary milling. *Powder Technology*, *278*, 35–45. doi:10.1016/j.powtec.2015.03.007

Shashanka, R., & Chaira, D. (2017). *Ball Milled Nano-Structured Stainless Steel Powders: Fabrication and Characterization*. Educreation Publishing. https://books.google.co.za/books?id=ECvvzgEACAAJ

Shashanka, R., Chaira, D., & Kumara Swamy, B. E. (2015). Electrocatalytic response of duplex and yittria dispersed duplex stainless steel modified carbon paste electrode in detecting folic acid using cyclic voltammetry. *International Journal of Electrochemical Science*, *10*(7), 5586–5598. doi:10.1016/S1452-3981(23)17279-8

Shashanka, R., Chaira, D., & Kumara Swamy, B. E. (2018). Effect of Y2O3 nanoparticles on corrosion study of spark plasma sintered duplex and ferritic stainless steel samples by linear sweep voltammetric method. *Archives of Metallurgy and Materials*, *63*(2), 749–763. doi:10.24425/122401

Shashanka, R., & Uzun, O. (2021). Powder Metallurgy. In Fundamentals of Materials Engineering (pp. 113–140). doi:10.2174/9789811489228121010008

Shrinivas Mahale, R., Shamanth, V., Hemanth, K., Nithin, S. K., Sharath, P. C., Shashanka, R., Patil, A., & Shetty, D. (2022). Processes and applications of metal additive manufacturing. *Materials Today: Proceedings*, *54*, 228–233. doi:10.1016/j.matpr.2021.08.298

Singh, M. K., & Singh, A. (2022). Particle size analysis. *Characterization of Polymers and Fibres*, 341–358. doi:10.1016/B978-0-12-823986-5.00009-9

Smith, A., Stender, M., Pankewitz, A., & Behrens, C. (2014). The quality of ultrasonic extinction based on-line particle size distribution analysers for mineral slurries. *XXVII International Mineral Processing Congress*, 1–22.

Standish, N. (1985). The kinetics of batch sieving. *Powder Technology*, *41*(1), 57–67. doi:10.1016/0032-5910(85)85075-0

Suryanarayana, C. (2004). Mechanical Alloying and Milling (Issue 1). Marcel Dekker.

Terzi, M. (2018). Particle Size Distribution Analysis in Aggregate Processing Plants Using Digital Image Processing Methods. *Revista Română de Materiale / Romanian. Journal of Materials*, *48*(4), 514–521.

Transparent Market Research. (2022). Particle Size Analyzer Market. *Outlook*, 2022–2031. https://transparencymarketresearch.com/particle-size-analyzer-market.html

Virto. (2022). *Why Sieving?* https://www.virtogroup.com/why-sieving/

Wedd, M., Ward-Smith, S., & Rawle, A. (2019). Particle size analysis. Encyclopedia of Analytical Science, 144–157. doi:10.1016/B978-0-12-409547-2.14522-6

Williams, N. (2015). The Magazine for the Metal Additive Manufacturing Industry. *Additive Manufacturing*, *1*(3), 84.

Wills, B. A., & Finch, J. A. (2016a). Particle Size Analysis. In *Wills' Mineral Processing Technology* (pp. 91–107). Elsevier., doi:10.1016/B978-0-08-097053-0.00004-2

Wills, B. A., & Finch, J. A. (2016b). *Wills' mineral processing technology : an introduction to the practical aspects of ore treatment and mineral recovery*. Academic Press.

WResearch. (2022). *South Africa Particle Size Analysis Market Outlook.* https://www.6wresearch.com/industry-report/nigeria-denature-alcohol-market-outlook

Compilation of References

(2021). Mechanical and Tribological Characteristics of Aluminium 2618 Matrix Composite Reinforced with Boron Carbide. *Biointerface Research in Applied Chemistry*, *12*(4), 4544–4556. doi:10.33263/BRIAC124.45444556

Abderrazak, H., Schoenstein, F., Abdellaoui, M., & Jouini, N. (2011, March 1). Spark plasma sintering consolidation of nanostructured TiC prepared by mechanical alloying. *International Journal of Refractory & Hard Metals*, *29*(2), 170–176. doi:10.1016/j.ijrmhm.2010.10.003

AboulFotouh, K., Zhang, Y., Maniruzzaman, M., Williams, R. O. III, & Cui, Z. (2020, September 25). Amorphous solid dispersion dry powder for pulmonary drug delivery: Advantages and challenges. *International Journal of Pharmaceutics*, *587*, 119711. doi:10.1016/j.ijpharm.2020.119711 PMID:32739389

Adamczyk-Cieślak, B., Koralnik, M., Kuziak, R., Majchrowicz, K., & Mizera, J. (2021). Studies of bainitic steel for rail applications based on carbide-free, low-alloy steel. *Metallurgical and Materials Transactions. A, Physical Metallurgy and Materials Science*, *52*(12), 5429–5442. doi:10.1007/s11661-021-06480-6

Adeyemi, I. A. (1983, July 1). Dry-milling of sorghum for ogi manufacture. *Journal of Cereal Science*, *1*(3), 221–227. doi:10.1016/S0733-5210(83)80024-1

Adimule, V., & Shashanka, R. (2023). Dielectric, Supercapacitive, Photoluminescent Properties of Hybrid Metal Organic Frameworks. Academic Press.

Adimule, V., & Shashanka, R. (2023). *Properties and Technological Features of Materials*. Academic Press.

Adimule, V., Keri, R., Rajendrachari, S., Kendrekar, P., & Kulkarni, C. V. (2023, February). Highly photoluminescence and wide band gap insulating metal hybrid nanoparticles array of samarium-doped SrO: CoO: synthesis, characterizations and sensor characteristics. *Journal of Materials Science Materials in Electronics*, *34*(5), 442. doi:10.1007/s10854-023-09899-2

Adimule, V., Koutsonas, S. S., Kaloop, M., & Shashanka, R. (2022). *Materials and Production Technologies in Machinery*. Trans Tech Publications. doi:10.4028/b-sG2xM9

Affolter, C., Thorwarth, G., Arabi-Hashemi, A., Müller, U., & Weisse, B. (2019). Ductile Compressive Behavior of Biomedical Alloys. *Metals*, *10*(1), 60. doi:10.3390/met10010060

Agimelen, O. S., Svoboda, V., Ahmed, B., Cardona, J., Dziewierz, J., Brown, C. J., McGlone, T., Cleary, A., Tachtatzis, C., Michie, C., Florence, A. J., Andonovic, I., Mulholland, A. J., & Sefcik, J. (2018, December 1). Multi-sensor inline measurements of crystal size and shape distributions during high shear wet milling of crystal slurries. *Advanced Powder Technology*, *29*(12), 2987–2995. doi:10.1016/j.apt.2018.09.003

Aglan, H. A., & Fateh, M. (2006). Fatigue damage tolerance of bainitic and pearlitic rail steels. *International Journal of Damage Mechanics*, *15*(4), 393–410. doi:10.1177/1056789506060775

Aglan, H. A., Liu, Z. Y., Hassan, M. F., & Fateh, M. (2004). Mechanical and fracture behavior of bainitic rail steel. *Journal of Materials Processing Technology*, *151*(1-3), 268–274. doi:10.1016/j.jmatprotec.2004.04.073

Aglan, H., & Fateh, M. (2007). Fracture and fatigue crack growth analysis of rail steels. *Journal of Mechanics of Materials and Structures*, *2*(2), 335–346. doi:10.2140/jomms.2007.2.335

Ahmed, M. M. Z., Ataya, S., El-Sayed Seleman, M. M., Mahdy, A. M. A., Alsaleh, N. A., & Ahmed, E. (2020). Heat Input and Mechanical Properties Investigation of Friction Stir Welded AA5083/AA5754 and AA5083/AA7020. *Metals*, *11*(1), 68. doi:10.3390/met11010068

Ahuir-Torres, J. I., Kotadia, H. R., Öpoz, T. T., & Sharp, M. C. (2023, February 28). A Study on the Corrosion Behaviour of Laser Textured Pure Aluminium in Saltwater. *Processes (Basel, Switzerland)*, *11*(3), 721. doi:10.3390/pr11030721

Aizawa, T., Kurihara, T., & Sakayori, H. (2022). Manufacturing of Aluminum Alloy Parts from Recycled Feedstock by PIG Die-Casting and Hot Stamping. *Lubricants (Basel, Switzerland)*, *11*(1), 13. doi:10.3390/lubricants11010013

Akrami, S., Edalati, P., Fuji, M., & Edalati, K. (2021). High-entropy ceramics: Review of principles, production and applications. *Materials Science and Engineering R Reports*, *146*, 100644. doi:10.1016/j.mser.2021.100644

AlakangasE.ImpolaR. (2015). Quality guidelines for wood fuels in Finland. *October*, 39. doi:10.13140/RG.2.1.3290.3127

Alexandra, B., Hidalgo, A., & Frykholm, R. (2017). *Powder Metallurgy Strategies to Improve Properties and Processing of Titanium Alloys: A Review*. doi:10.1002/adem.201600743

Al-Hajji, L. (2019, May). A comparative study on the zinc metatitanate microstructure by ball milling and solvothermal approaches. *Journal of Structural Chemistry*, *60*(5), 830–837. doi:10.1134/S0022476619050172

Ali, H. S., York, P., Ali, A. M., & Blagden, N. (2011, January 20). Hydrocortisone nanosuspensions for ophthalmic delivery: A comparative study between microfluidic nanoprecipitation and wet milling. *Journal of Controlled Release*, *149*(2), 175–181. doi:10.1016/j.jconrel.2010.10.007 PMID:20946923

Alipal, J., Lee, T. C., Koshy, P., Abdullah, H. Z., & Idris, M. I. (2021). Evolution of anodised titanium for implant applications. *Heliyon, 7*(7), e07408. doi:10.1016/j.heliyon.2021.e07408 PMID:34296002

Alizadeh, M., Garkani Nejad, F., Dourandish, Z., Tajik, S., Karimi, F., Mohammadzadeh Jahani, P., Aghaei Afshar, A., Zaimbashi, R., Sheikhshoaie, I., & Beitollahi, H. (2022). A brief review on the recent achievements in electrochemical detection of folic acid. In Journal of Food Measurement and Characterization (Vol. 16, Issue 5). doi:10.1007/s11694-022-01421-2

Allen, T. (2003). *Powder sampling and particle size determination*. Elsevier.

American Standards & Test Methods. (2017). Standard Test Methods for Particle-Size Distribution (Gradation) of Soils Using Sieve Analysis. In ASTM International (Vol. 04, Issue Reapproved 2009, pp. 1–35). https://doi.org/ doi:10.1520/D6913-17.1.6

Anis, M. A., Srivastava, J. P., Duhan, N. R., & Sarkar, P. K. (2018, June). Rolling contact fatigue and wear in rail steels: An overview. *IOP Conference Series. Materials Science and Engineering, 377*(1), 012098. doi:10.1088/1757-899X/377/1/012098

Antherjanam, S., & Saraswathyamma, B. (2021, September). Simultaneous electrochemical determination of hydrazine and hydroxylamine on a thiadiazole derivative modified pencil graphite electrode. *Materials Chemistry and Physics, 275*, 125223. doi:10.1016/j.matchemphys.2021.125223

Apling, A. C. (1984). Blinding of screens by sub-sieve sized particles. *Transactions of the Institution of Mining and Metallurgy Section B. Applied Earth Science, 93*, C92–C94.

Aravindan, N., & M.V., S. (2017). Differential pulse voltammetry as an alternate technique for over oxidation of polymers: Application of electrochemically synthesized over oxidized poly (Alizarin Red S) modified disposable pencil graphite electrodes for simultaneous detection of hydroquinone and catechol. *Journal of Electroanalytical Chemistry (Lausanne, Switzerland), 789*. Advance online publication. doi:10.1016/j.jelechem.2017.02.037

Arendarchuck, B. E., Mayer, A. R., de Oliveira, W. R., Pukasiewicz, A. G. M., Lourençato, L. A., Fals, H. D. C., & Martínez-Cámara, E. (2022). Effect of Ti-B Grain Refiners on Wear and Corrosion of the A332 Alloy with Sr Modification. *Applied Sciences (Basel, Switzerland), 13*(1), 430. doi:10.3390/app13010430

Arndt, O. R., Baggio, R., Adam, A. K., Harting, J., Franceschinis, E., & Kleinebudde, P. (2018, December 1). Impact of different dry and wet granulation techniques on granule and tablet properties: A comparative study. *Journal of Pharmaceutical Sciences, 107*(12), 3143–3152. doi:10.1016/j.xphs.2018.09.006 PMID:30244008

Ashwini, R., Kumar, M. K. P., Rekha, M. Y., Santosh, M. S., & Srivastava, C. (2022). Optimization of NiFeCrCoCu high entropy alloy nanoparticle – graphene (HEA-G) composite for the enhanced electrochemical sensitivity towards urea oxidation. *Journal of Alloys and Compounds, 903*, 163846. Advance online publication. doi:10.1016/j.jallcom.2022.163846

Atazadeh, N., Heydari, M. S., Baharvandi, H. R., & Ehsani, N. (2016, December 1). Reviewing the effects of different additives on the synthesis of the Ti3SiC2 MAX phase by mechanical alloying technique. *International Journal of Refractory & Hard Metals*, *61*, 67–78. doi:10.1016/j.ijrmhm.2016.08.003

Atkinson, C. F., Aumen, N. G., Miller, G. L., & Ward, G. M. (1992). Influence of particle shapes and size distributions on fine particulate organic matter surface area in streams. *Journal of the North American Benthological Society*, *11*(3), 261–268. doi:10.2307/1467646

Atkinson, C. F., Miller, G. L., & Aumen, N. G. (1993). Influence of Season, Size, and Collection Site on Surface Area of Fine Particular Matter in Streams. *Journal of the North American Benthological Society*, *12*(4), 410–417. doi:10.2307/1467622

Avramović, L., Bugarin, M., Milanović, D., Conić, V., Pavlović, M. M., Vuković, M., & Nikolić, N. D. (2018). The particle size distribution (PSD) as criteria for comparison of silver powders obtained by different methods of synthesis and by conditions of electrolysis. *Journal of Mining and Metallurgy, Section B. Metallurgy*, *54*(3), 291–300. doi:10.2298/JMMB171002020A

Awotunde, M. A., Adegbenjo, A. O., Obadele, B. A., Okoro, M., Shongwe, B. M., & Olubambi, P. A. (2019, April 1). Influence of sintering methods on the mechanical properties of aluminium nanocomposites reinforced with carbonaceous compounds: A review. *Journal of Materials Research and Technology*, *8*(2), 2432–2449. doi:10.1016/j.jmrt.2019.01.026

Baghayeri, M., Maleki, B., & Zarghani, R. (2014). Voltammetric behavior of tiopronin on carbon paste electrode modified with nanocrystalline Fe50Ni50 alloys. *Materials Science and Engineering C*, *44*, 175–182. Advance online publication. doi:10.1016/j.msec.2014.08.023 PMID:25280694

Balazic, M., Kopac, J., Jackson, M. J., & Ahmed, W. (2007). Review: Titanium and titanium alloy applications in medicine. *Int J Nano Biomater*, *1*(1), 3–34. doi:10.1504/IJNBM.2007.016517

Ballupete Nagaraju, S., Kodigarahalli Somashekara, M., Puttegowda, M., Manjulaiah, H., Kini, C. R., & Channarayapattana Venkataramaiah, V. (2022, December 15). Effect of B4C/Gr on Hardness and Wear Behavior of Al2618 Based Hybrid Composites through Taguchi and Artificial Neural Network Analysis. *Catalysts*, *12*(12), 1654. doi:10.3390/catal12121654

Balordi, M., Santucci de Magistris, G., Casali, A., Pini, F., Cammi, A., Lacavalla, M., & Rotella, V. (2023, March 14). The Anti-Snow Behaviour of Icephobic Coatings: Laboratory and In-Field Testing. *Coatings*, *13*(3), 616. doi:10.3390/coatings13030616

Barbacki, A. (1995). The role of bainite in shaping mechanical properties of steels. *Journal of Materials Processing Technology*, *53*(1-2), 57–63. doi:10.1016/0924-0136(95)01961-D

Barbarino, S., Flores, E. S., Ajaj, R. M., Dayyani, I., & Friswell, M. I. (2014). A review on shape memory alloys with applications to morphing aircraft. *Smart Materials and Structures*, *23*(6), 063001. doi:10.1088/0964-1726/23/6/063001

Barbosa-Cánovas, G. V., Harte, F., & Yan, H. H. (2012). Particle size distribution in food powders. *Food Engineering*, *1*, 303–328. http://www.eolss.net/sample-chapters/c10/E5-10-01-06.pdf

Barlow, B. (2023). *Screen Technology Helps Solve Blinding, Plugging Problems*. https://www.constructionequipment.com/home/article/10716519/screen-technology-helps-solve-blinding-plugging-problems

Barro, Ó., Arias-González, F., Lusquiños, F., Comesaña, R., del Val, J., Riveiro, A., Badaoui, A., Gómez-Baño, F., & Pou, J. (2020). Improved Commercially Pure Titanium Obtained by Laser Directed Energy Deposition for Dental Prosthetic Applications. *Metals*, *11*(1), 70. doi:10.3390/met11010070

Bartos, C., Szabó-Révész, P., Bartos, C., Katona, G., Jójárt-Laczkovich, O., & Ambrus, R. (2016, April 21). The effect of an optimized wet milling technology on the crystallinity, morphology and dissolution properties of micro-and nanonized meloxicam. *Molecules (Basel, Switzerland)*, *21*(4), 507. doi:10.3390/molecules21040507 PMID:27110752

Basavarajappa, P. N., Hegde, M. M. R., Rajendrachari, S., & Surendranathan, A. O. (2023). Investigation of structural and Mechanical properties of Nanostructured TiMgSr Alloy for Biomedical applications, *Biointerface Research in Applied Chemistry*, *13*(2), 1-12.

Basavarajappa, P. N., Hegde, M. M. R., Rajendrachari, S., & Surendranathan, A. O. (2023). Investigation of Structural and Mechanical Properties of Nanostructured TiMgSr Alloy for Biomedical applications. *Biointerface Research in Applied Chemistry*, *13*, 1–12.

Bas, J. A., Calero, J. A., & Dougan, M. J. (2003, January 1). Sintered soft magnetic materials. Properties and applications. *Journal of Magnetism and Magnetic Materials*, *254*, 391–398. doi:10.1016/S0304-8853(02)00934-4

Bedolla-Becerril, E., Garcia-Guerra, J., Lopez-Morelos, V. H., Garcia-Renteria, M. A., Falcon-Franco, L. A., Martinez-Landeros, V. H., García-Villarreal, S., & Flores-Villaseñor, S. E. (2022). Tribological Behaviour of Al-2024/TiC Metal Matrix Composites. *Coatings*, *13*(1), 77. doi:10.3390/coatings13010077

Been, J., Faller, K., & Eli, T. (1999). Using Ti-5111 for Marine Fastener Applications. *Journal of the Minerals Metals & Materials Society*, *51*(6), 21–24. doi:10.1007/s11837-999-0088-5

Ben Said, L., Chabchoub, A. K., & Wali, M. (2023, April 22). Mathematical Model Describing the Hardening and Failure Behaviour of Aluminium Alloys: Application in Metal Shear Cutting Process. *Mathematics*, *11*(9), 1980. doi:10.3390/math11091980

Benjamin, J. S. (1970). Dispersion Strengthened Superalloys by Mechanical Alloying. *Metallurgical Transactions*, *1*(10), 2943–2951. doi:10.1007/BF03037835

Benjamin, J. S., & Arzt, E. (1989). *New materials by mechanical alloying techniques*. DGM Informationgesellschaft.

Benjamin, J. S., & Co, A. (1990). Mechanical Alloying - A Perspective. *Metal Powder Report*, *45*(2), 122–127. doi:10.1016/S0026-0657(10)80124-9

Bertula, K. (2022). *Particle size distribution analysis: Guide to method selection*. Academic Press.

Beynon, J. H., & Perez-Unzueta, A. J. (1993). Microstructure and wear resistance of pearlitic rail steels. *Wear, 162/164*, 173–182. doi:10.1016/0043-1648(93)90498-B

Bhadeshia, H. K. D. H. (1997). Martensite and bainite in steels: transformation mechanism & mechanical properties. *J. Phys. IV France, 7*, C5-367-C5-376.

Bhadeshia, H. K. D. H. (2000). Mechanically Alloyed Metals. *Materials Science and Technology, 16*(11-12), 1404–1411. doi:10.1179/026708300101507361

Bhadeshia, H. K. D. H. (2001). *Bainite in steels* (2nd ed.). University Press.

Bhatt, B., Murthy, T. S. R. C., Limaye, P. K., Nagaraj, A., Singh, K., Sonber, J. K., Sairam, K., Sashanka, A., Rao, G. V. S. N., & Rao, T. S. (2016). Tribological studies of monolithic chromium diboride against cemented tungsten carbide (WC–Co) under dry condition. *Ceramics International, 42*(14), 15536–15546. doi:10.1016/j.ceramint.2016.06.208

Bhirud, N. L., & Gawande, R. R. (2017, December). Measurement and prediction of cutting temperatures during dry milling: Review and discussions. *Journal of the Brazilian Society of Mechanical Sciences and Engineering, 39*(12), 5135–5158. doi:10.1007/s40430-017-0869-7

Bhrany, U. N., & Brown, J. H. (1962). Minerals Beneficiation - Particle-Size. *Measurement and Control, 226.*

Biazzi, H. M., Tubin, J. S. B., Conte, R. A., Robazza, W. da S., & Paiano, D. (2022). Different sieving methods for determining the physical characteristics in ground corn. *Acta Scientiarum. Animal Sciences, 44*, e53382. Advance online publication. doi:10.4025/actascianimsci.v44i1.53382

Blair, N. (2010). The impact of soil water content and water temperature on wet aggregate stability. What answer do you want? *Proceedings of the 19th World Congress of Soil Science: Soil Solutions for a Changing World,* 1–6.

Blaud, A., Menon, M., van der Zaan, B., Lair, G. J., & Banwart, S. A. (2017). Effects of Dry and Wet Sieving of Soil on Identification and Interpretation of Microbial Community Composition. In *Advances in Agronomy* (Vol. 142, pp. 119–142). Academic Press Inc. doi:10.1016/bs.agron.2016.10.006

Bogdanoff, T., Tiryakioğlu, M., Liljenfors, T., Jarfors, A. E. W., Seifeddine, S., & Ghassemali, E. (2023, March 15). On the Effectiveness of Rotary Degassing of Recycled Al-Si Alloy Melts: The Effect on Melt Quality and Energy Consumption for Melt Preparation. *Sustainability (Basel), 15*(6), 5189. doi:10.3390/su15065189

Bordia, R. K., Kang, S. J., & Olevsky, E. A. (2017, June). Current understanding and future research directions at the onset of the next century of sintering science and technology. *Journal of the American Ceramic Society, 100*(6), 2314–2352. doi:10.1111/jace.14919

Braginsky, M., Tikare, V., & Olevsky, E. (2005, January 1). Numerical simulation of solid state sintering. *International Journal of Solids and Structures, 42*(2), 621–636. doi:10.1016/j.ijsolstr.2004.06.022

Braun, J. L., Rost, C. M., Lim, M., Giri, A., Olson, D. H., Kotsonis, G. N., Stan, G., Brenner, D. W., Maria, J.-P., & Hopkins, P. E. (2018). Charge-induced disorder controls the thermal conductivity of entropy-stabilized oxides. *Advanced Materials*, *30*(51), 1805004. doi:10.1002/adma.201805004 PMID:30368943

Buchczik, D., Wegehaupt, J., & Krauze, O. (2017). Indirect measurements of milling product quality in the classification system of electromagnetic mill. In *2017 22nd International Conference on Methods and Models in Automation and Robotics (MMAR)* (pp. 1039-1044). IEEE. 10.1109/MMAR.2017.8046973

Business Wire & Research and Markets. (2022). *The Worldwide Particle Size Analysis Industry is Expected to Reach $609 Million by 2030*. Author.

Caballero, F. G., Santofimia, M. J., García-Mateo, C., Chao, J., & De Andres, C. G. (2009). Theoretical design and advanced microstructure in super high strength steels. *Materials & Design*, *30*(6), 2077–2083. doi:10.1016/j.matdes.2008.08.042

Camarano, A., Giuranno, D., & Narciso, J. (2019). SiC-IrSi3 for High Oxidation Resistance. *Materials (Basel)*, *13*(1), 98. doi:10.3390/ma13010098 PMID:31878174

Cañellas, F. M., Al-Rifai, N., Padrela, L., Tajber, L., Khamiakova, T., Otava, M., & Geertman, R. (2023, July 21). A comparative study of dry and wet top-down milling approaches for the preparation of microparticle suspensions. *Powder Technology*, *428*, 118829. doi:10.1016/j.powtec.2023.118829

Cantor, B. (2014). Multicomponent and high entropy alloys. *Entropy (Basel, Switzerland)*, *16*(9), 4749–4768. doi:10.3390/e16094749

Cantor, B., Chang, I. T. H., Knight, P., & Vincent, A. J. B. (2004). Microstructural development in equiatomic multicomponent alloys. *Materials Science and Engineering A*, *375–377*, 213–218. doi:10.1016/j.msea.2003.10.257

Cao, W., Chen, P., Lin, R., Li, F., Ge, B., Song, D., Cheng, Z., & Wang, C. (2023). Boosting energy-storage performance in lead-free ceramics via polyphase engineering in the superparaelectric state. *Composites. Part B, Engineering*, *255*, 110630. doi:10.1016/j.compositesb.2023.110630

Cao, X., Zhu, J., Gao, F., & Gao, Z. (2020). Influence of Fine Grains on the Bending Fatigue Behavior of Two Implant Titanium Alloys. *Materials (Basel)*, *14*(1), 171. doi:10.3390/ma14010171 PMID:33396522

Cardarelli, F. (2008). *Less Common Nonferrous Metals*. Springer.

Castle, E., Csanádi, T., Grasso, S., Dusza, J., & Reece, M. (2018). Processing and Properties of High-Entropy Ultra-High Temperature Carbides. *Scientific Reports*, *8*(1), 8609. doi:10.1038/s41598-018-26827-1 PMID:29872126

Castro, R., & Van Benthem, K. (Eds.). (2012 Sep 13). *Sintering: mechanisms of convention nanodensification and field assisted processes*. Springer Science & Business Media.

Chaira, D. (2014, June 1). Phase transformation and microstructure study of nano-structured austenitic and ferritic stainless steel powders prepared by planetary milling. *Powder Technology*, *259*, 125–136. doi:10.1016/j.powtec.2014.03.061

Chaira, D. (2015, January 1). Development of nano-structured duplex and ferritic stainless steels by pulverisette planetary milling followed by pressure less sintering. *Materials Characterization*, *99*, 220–229. doi:10.1016/j.matchar.2014.11.030

Chang, L. C. (2005). The rolling/sliding wear performance of high silicon carbide-free bainitic steels. *Wear*, *258*(5-6), 730–743. doi:10.1016/j.wear.2004.09.064

Chaoyang, C., Guangjie, G., Lingfei, L., Fei, F., & Lin, L. (2020). Effect of Two-Stage Homogenization Heat Treatment on Microstructure and Mechanical Properties of AA2060 Alloy. *Crystals*, *11*(1), 40. doi:10.3390/cryst11010040

Chatterjee, A., Sen, S., Paul, S., Roy, P., Seikh, A. H., Alnaser, I. A., Das, K., Sutradhar, G., & Ghosh, M. (2023, March 13). Fabrication and Characterization of SiC-reinforced Aluminium Matrix Composite for Brake Pad Applications. *Metals*, *13*(3), 584. doi:10.3390/met13030584

Chausov, M., Pylypenko, A., Maruschak, P., Zasimchuk, V., Brezinová, J., Brezina, J., & Viňáš, J. (2022). Impact of the Initial Phase Composition of Alloys on the Effects Manifested by Yield Sites That Occur on Sheet Aluminum Alloys Subjected to Impact-Oscillatory Loading. *Materials (Basel)*, *16*(1), 249. doi:10.3390/ma16010249 PMID:36614587

Chen, H., Guo, H., & Shang, C. J. (2014). Study on Wear Behaviours of 400 and 450 Grade Wear Resistant Steels. *Materials Research Innovations*, *18*(sup4), S4-270. . doi:10.1179/1432891714Z.000000000691

Cheng, Y.-L., Lee, C.-Y., & Huang, Y.-L. (2016). Alloying Elements of Magnesium Alloys: A Literature Review. Intech.

Chen, G., Xu, G., Zurob, H. S., Hu, H., & Wan, X. (2019). Effect of strain rate on the bainitic transformation in Fe-C-Mn-Si medium-carbon bainitic steels. *Metallurgical and Materials Transactions. A, Physical Metallurgy and Materials Science*, *50*(2), 573–580. doi:10.1007/s11661-018-5051-z

Cheng, B., Lou, H., Sarkar, A., Zeng, Z., Zhang, F., Chen, X., Tan, L., Glazyrin, K., liermann, H.-P., Yan, J., Wang, L., Djenadic, R., Hahn, H., & Zeng, Q. (2020). Lattice distortion and stability of (Co0.2Cu0.2Mg0.2Ni0.2Zn0.2)O high-entropy oxide under high pressure. *Materials Today. Advances*, *8*, 100102. doi:10.1016/j.mtadv.2020.100102

Cheng, C., Zhang, F., Cheng, F., Hou, J., Ren, B., Miao, Y., Xin, L., & Wang, X. (2023). The effect of plasma-assisted ball milling on preparation and sintering behavior of (Zr0.1429Hf0.1429Ce0.1429Y0.2857La0.2857)O2-δ high entropy fluorite oxide. *Ceramics International*, *49*(8), 13092–13101. doi:10.1016/j.ceramint.2022.12.185

Cheng, J., Liu, P., Peng, T., Liu, Q., Chen, W., Liu, B., Yuan, Y., Zhang, W., Song, F., Gu, J., & Zhang, D. (2020). Mechanically alloyed NiTiO3/transition metal heterostructures: Introducing oxygen vacancies for exceptionally enhanced hydrogen evolution reaction activity. *Journal of Materials Chemistry. A, Materials for Energy and Sustainability*, *8*(30), 14908–14914. Advance online publication. doi:10.1039/D0TA05030J

Chen, H., Fu, J., Zhang, P., Peng, H., Abney, C. W., Jie, K., Liu, X., Chi, M., & Dai, S. (2018). Entropy-stabilized metal oxide solid solutions as CO oxidation catalysts with high-temperature stability. *Journal of Materials Chemistry. A, Materials for Energy and Sustainability*, *6*(24), 11129–11133. doi:10.1039/C8TA01772G

Chen, H., Xiang, H., Dai, F. Z., Liu, J., & Zhou, Y. (2019). Porous high entropy (Zr0.2Hf0.2Ti0.2Nb0.2Ta0.2)B2: A novel strategy towards making ultrahigh temperature ceramics thermal insulating. *Journal of Materials Science and Technology*, *35*(10), 2404–2408. doi:10.1016/j.jmst.2019.05.059

Chen, H., Xiang, H., Dai, F.-Z., Liu, J., Lei, Y., Zhang, J., & Zhou, Y. (2019). High porosity and low thermal conductivity high entropy (Zr0.2Hf0.2Ti0.2Nb0.2Ta0.2)C. *Journal of Materials Science and Technology*, *35*(8), 1700–1705. doi:10.1016/j.jmst.2019.04.006

Chen, H., Zhao, D., Wang, Q., Qiang, Y., & Qi, J. (2017). Effects of Impact Energy on the Wear Resistance and Work Hardening Mechanism of Medium Manganese Austenitic Steel. *Friction*, *5*(4), 447–454. doi:10.1007/s40544-017-0158-6

Chen, M., Jiang, L., Guo, G., & An, Q. (2011, August). Experimental and FEM study of coated and uncoated tools used for dry milling of compacted graphite cast iron. *Transactions of Tianjin University.*, *17*(4), 235–241. doi:10.1007/s12209-011-1609-1

Chen, S., Ye, H., & Lin, X. (2017). Effect of Rare Earth and Alloying Elements on the Thermal Conductivity of Austenitic Medium Manganese Steel. *International Journal of Minerals Metallurgy and Materials*, *24*(6), 670–674. doi:10.1007/s12613-017-1449-7

Chicardi, E., García-Garrido, C., & Gotor, F. J. (2019). Low temperature synthesis of an equiatomic (TiZrHfVNb)C5 high entropy carbide by a mechanically-induced carbon diffusion route. *Ceramics International*, *45*(17), 21858–21863. doi:10.1016/j.ceramint.2019.07.195

Chiu, Y.-C., Pan, T.-A., Chen, G.-M., Jiang, X.-C., Bor, H.-Y., Tzeng, Y.-C., & Lee, S.-L. (2020). The Effects of a Trace Amount of Manganese and the Homogenization on the Recrystallization of Al–7Mg–0.15Ti Alloys. *Molecules (Basel, Switzerland)*, *26*(1), 168. doi:10.3390/molecules26010168 PMID:33396531

Chlanda, A., Witkowska, J., Morgiel, J., Nowińska, K., Choińska, E., Swieszkowski, W., & Wierzchoń, T. (2018). Multi-scale characterization and biological evaluation of composite surface layers produced under glow discharge conditions on NiTi shape memory alloy for potential cardiological application. *Micron*, *114*, 14–22. doi:10.1016/j.micron.2018.07.009

Chopade, J., Barjibhe, R., & Kumar, B. (2013). Critical review on shape memory alloy. *International Journal of Engineering*, *2*(2), 1–6.

Clermont, B., & De Haas, B. (2010, March). Optimization of mill performance by using online ball and pulp measurements. *Journal of the Southern African Institute of Mining and Metallurgy*, *110*(3), 133–140.

Cologna, M., Rashkova, B., & Raj, R. (2010). Flash sintering of nanograin zirconia in <5 s at 850°C. *Journal of the American Ceramic Society*, *93*(11), 3556–3559. doi:10.1111/j.1551-2916.2010.04089.x

Contuzzi, N., Casalino, G., Boccaccio, A., Ballini, A., Charitos, I. A., Bottalico, L., & Santacroce, L. (2022). Metals Biotribology and Oral Microbiota Biocorrosion Mechanisms. *Journal of Functional Biomaterials*, *14*(1), 14. doi:10.3390/jfb14010014 PMID:36662061

Corker, D. L., Whatmore, R. W., Ringgaard, E., & Wolny, W. W. (2000, November 1). Liquid-phase sintering of PZT ceramics. *Journal of the European Ceramic Society*, *20*(12), 2039–2045. doi:10.1016/S0955-2219(00)00089-3

Cross, J. O., Opila, R. L., Boyd, I. W., & Kaufmann, E. N. (2015, December). Materials characterization and the evolution of materials. *MRS Bulletin*, *40*(12), 1019–1034. doi:10.1557/mrs.2015.271

Cui, C., Uhlenwinkel, V., Schulz, A., & Zoch, H.-W. (2019). Austenitic Stainless Steel Powders with Increased Nitrogen Content for Laser Additive Manufacturing. *Metals*, *10*(1), 61. doi:10.3390/met10010061

Daskalakis, E., Scott, A., & Jha, A. (2023). Enhancing the ignitability of the Al–TiO2–B2O3 powder mixture through intensive vibratory ball milling (C:P12). *Ceramics International*, *49*(14), 24142–24155. Advance online publication. doi:10.1016/j.ceramint.2022.12.242

Dastanpoor, E., & Enayati, M. H. (2017). Role of impurities in mechanical alloying amorphization of Cu-Zr and Cu-Zr-Al systems. *Indian Journal of Engineering and Materials Sciences*, *24*(1).

de Fatima Ulbrich, K., Winiarski, J. P., Jost, C. L., & de Campos, C. E. M. (2020). Green and facile solvent-free synthesis of NiTe2 nanocrystalline material applied to voltammetric determination of antioxidant morin. *Materials Today. Communications*, *25*, 101251. Advance online publication. doi:10.1016/j.mtcomm.2020.101251

de Fatima Ulbrich, K., Winiarski, J. P., Jost, C. L., & Maduro de Campos, C. E. (2020). Mechanochemical synthesis of a Ni3-xTe2 nanocrystalline composite and its application for simultaneous electrochemical detection of dopamine and adrenaline. *Composites. Part B, Engineering*, *183*, 107649. Advance online publication. doi:10.1016/j.compositesb.2019.107649

de Oro Calderon, R., Gierl-Mayer, C., & Danninger, H. (2022). Fundamentals of sintering: liquid phase sintering. *Encyclopedia of Materials: Metals and Alloys, 3*, 481-92.

De Oro Calderón, R., Bernardo, E., Campos, M., Gierl-Mayer, C., Danninger, H., & Torralba, J. M. (2016, January 1). Tailoring master alloys for liquid phase sintering: Effect of introducing oxidation-sensitive elements. *Powder Metallurgy*, *59*(1), 31–40. doi:10.1080/00325899.2016.1148897

Demirskyi. (2020). Synthesis of medium-entropy (Zr1/3Hf1/3Ta1/3)B2 using the spark plasma consolidation of diboride powders. *Journal of the Ceramic Society of Japan*, *128*, 977–980. doi:10.2109/jcersj2.20151

Di Egidio, G., Tonelli, L., Morri, A., Boromei, I., Shashkov, P., & Martini, C. (2022). Influence of Anodizing by Electro-Chemical Oxidation on Fatigue and Wear Resistance of the EV31A-T6 Cast Magnesium Alloy. *Coatings*, *13*(1), 62. doi:10.3390/coatings13010062

Diantonio, C.B., & Ewsuk, K.G. (2010). Master sintering curve and its application in sintering of electronic ceramics. In *Sintering of Advanced Materials* (pp. 130-161). Woodhead Publishing.

Diaz-Zorita, M., Grove, J. H., & Perfect, E. (2007). Sieving duration and sieve loading impacts on dry soil fragment size distributions. *Soil & Tillage Research*, *94*(1), 15–20. doi:10.1016/j.still.2006.06.006

Ding, Z., Long, W., Jiu, Y., Zhong, S., Jiang, D., Liu, C., Yang, J., & Qiao, J. (2022). Formation and Morphological Transition of Diversified Eutectic Structures in Ag-Based Brazing Filler Metals with Sn Addition. *Crystals*, *13*(1), 68. doi:10.3390/cryst13010068

Dippo, O. F., Mesgarzadeh, N., Harrington, T. J., Schrader, G. D., & Vecchio, K. S. (2020). Bulk high-entropy nitrides and carbonitrides. *Scientific Reports*, *10*(1), 1–11. doi:10.1038/s41598-020-78175-8 PMID:33277546

Dobrzański, L.A., Dobrzańska-Danikiewicz, A.D., Achtelik-Franczak, A., Dobrzański, L.B., Hajduczek, E., & Matula, G. (2017). Fabrication technologies of the sintered materials including materials for medical and dental application. In *Powder Metallurgy–Fundamentals and Case Studies*. InTech.

Donoso, M., Robles, P. A., Gálvez, E. D., & Cisternas, L. A. (2013). Particle size effect on the efficient use of water and energy in mineral concentration processes. *Industrial & Engineering Chemistry Research*, *52*(49), 17686–17690. doi:10.1021/ie402099n

Dubrovinskaia, N., Dubrovinsky, L. & Solozhenko, V. L. (2007). Comment on 'synthesis of ultra-incompressible superhard rhenium diboride at ambient pressure'. *Science, 318*.

Du, M., Liu, B., Liu, Y., & Yang, Y. (2022). Dynamic Behavior of Additively Manufactured FeCoCrNi High Entropy Alloy. *Metals*, *13*(1), 75. doi:10.3390/met13010075

Dusza, J., Švec, P., Girman, V., Sedlák, R., Castle, E. G., Csanádi, T., Kovalčíková, A., & Reece, M. J. (2018). Microstructure of (Hf-Ta-Zr-Nb)C high-entropy carbide at micro and nano/atomic level. *Journal of the European Ceramic Society*, *38*(12), 4303–4307. doi:10.1016/j.jeurceramsoc.2018.05.006

Dwivedi, A., Khan, M. M., & Pali, H. S. (2023). A comprehensive review of thermal enhancement techniques in microchannel heat exchangers and heat sinks. *Journal of Thermal Analysis and Calorimetry*, *148*(23), 1–43. doi:10.1007/s10973-023-12451-3

Ekberg, A. (1996, July). Rolling contact fatigue of railway wheels–computer modeling and in-field data. In *Proceedings of 2nd mini conf. Contact mechanics and wear of railway systems* (pp. 154-163). Academic Press.

Elahinia, M. H., Hashemi, M., Tabesh, M., & Bhaduri, S. B. (2012). Progress in Materials Science Manufacturing and processing of NiTi implants : A review. *Progress in Materials Science*, *57*(5), 911–946. doi:10.1016/j.pmatsci.2011.11.001

Ertug, B. (Ed.). (2013 Feb 6). *Sintering applications*. BoD–Books on Demand. doi:10.5772/56064

Esamael, S.K., & Fatalla, A.A. (n.d.). *Analysis of Effect of Alloying Magnesium by Powder Metallurgy on Mechanical Properties: Review*. Academic Press.

Euchner, H., & Mayrhofer, P. H. (2015). Designing thin film materials-Ternary borides from first principles. *Thin Solid Films*, *583*, 46–49. doi:10.1016/j.tsf.2015.03.035 PMID:26082562

Exarchos, D. A., Dalla, P. T., Tragazikis, I. K., Dassios, K. G., Zafeiropoulos, N. E., Karabela, M. M., De Crescenzo, C., Karatza, D., Musmarra, D., Chianese, S., & Matikas, T. E. (2018). Development and Characterization of High Performance Shape Memory Alloy Coatings for Structural Aerospace Applications. *Materials (Basel)*, *11*(5), 832. doi:10.3390/ma11050832 PMID:29783626

Exner, H. E. & Arzt, E. (1996). Sintering Processes. *Phys. Metall.*, 2627–2662. doi:10.1016/B978-044489875-3/50036-3

Eylon, D., Fujishiro, S., Postans, P. J., & Froes, F. H. (1984). High-Temperature Titanium Alloys—A Review. *JOM*, *36*(11), 55–62. doi:10.1007/BF03338617

Eymann, K., Riedl, T., Bram, A., Ruhnow, M., Boucher, R., Kirchner, A., & Kieback, B. (2012, September 15). Consolidation of mechanically alloyed nanocrystalline Cu–Nb–ZrO2 powder by spark plasma sintering. *Journal of Alloys and Compounds*, *535*, 62–69. doi:10.1016/j.jallcom.2012.04.102

Facts & Factors. (2023). *Particle Size Analysis Market Size is Projected to Grow*. https://www.fnfresearch.com/sample/particle-size-analysis-market

Failla, S., Galizia, P., Fu, S., Grasso, S., & Sciti, D. (2020). Formation of high entropy metal diborides using arc-melting and combinatorial approach to study quinary and quaternary solid solutions. *Journal of the European Ceramic Society*, *40*(3), 588–593. doi:10.1016/j.jeurceramsoc.2019.10.051

Fang, Z.Z., & Wang, H. (2013). Sintering of ultrafine and nanosized ceramic and metallic particles. In *Ceramic Nanocomposites* (pp. 431-473). Woodhead Publishing.

Fang, H. Y., Yang, J. H., & Chen, Q. (2018). Particle size distribution and energy consumption during impact crushing of single granite particles. *Journal of the Southern African Institute of Mining and Metallurgy*, *118*(5), 555–561. doi:10.17159/2411-9717/2018/v118n5a12

Fang, Z. Z. (Ed.). (2010 Sep 27). *Sintering of advanced materials*. Elsevier. doi:10.1533/9781845699949

Fan, X., Mi, Z., Yang, L., & Su, H. (2022). Application of DFT Simulation to the Investigation of Hydrogen Embrittlement Mechanism and Design of High Strength Low Alloy Steel. *Materials (Basel), 16*(1), 152. doi:10.3390/ma16010152 PMID:36614491

Farahani, E., Liberati, A. C., Moreau, C., Dolatabadi, A., & Stoyanov, P. (2023, February 27). Comparative Evaluation of the Shear Adhesion Strength of Ice on PTFE Solid Lubricant. *Lubricants (Basel, Switzerland), 11*(3), 105. doi:10.3390/lubricants11030105

Feng, L., Fahrenholtz, W. G., & Hilmas, G. E. (2019). Low-temperature sintering of single-phase, high-entropy carbide ceramics. *Journal of the American Ceramic Society, 102*(12), 7217–7224. doi:10.1111/jace.16672

Feng, L., Fahrenholtz, W. G., Hilmas, G. E., & Curtarolo, S. (2023). Boro/carbothermal reduction co-synthesis of dual-phase high-entropy boride-carbide ceramics. *Journal of the European Ceramic Society, 43*(6), 2708–2712. doi:10.1016/j.jeurceramsoc.2022.12.056

Feng, L., Lee, S. H., & Kim, H. N. (2017). Effects of high-energy ball milling and reactive spark plasma sintering on the densification of HfC-SiC composites. *Journal of the European Ceramic Society, 37*(5), 1891–1898. doi:10.1016/j.jeurceramsoc.2016.12.036

Fernandes, F. B., Mahesh, K. K., & dos Santos Paula, A. (2013). Thermomechanical treatments for Ni-Ti alloys. *Shape Memory Alloys-Processing, Characterization and Applications*, 3-26.

Ferreira, P. P., Carvalho, F. M., Ariza-Echeverri, E. A., Delfino, P. M., Bauri, L. F., Ferreira, A. M., Braga, A. P., Eleno, L. T. F., Goldenstein, H., & Tschiptschin, A. P. (2022). Synergism between B and Nb Improves Fire Resistance in Microalloyed Steels. *Metals, 13*(1), 84. doi:10.3390/met13010084

Fourmont, A., Le Gallet, S., Politano, O., Desgranges, C., & Baras, F. (2020). Effects of planetary ball milling on AlCoCrFeNi high entropy alloys prepared by Spark Plasma Sintering: Experiments and molecular dynamics study. *Journal of Alloys and Compounds, 820*, 153448. doi:10.1016/j.jallcom.2019.153448

Frensemeier, M., Kaiser, J. S., Frick, C. P., Schneider, A. S., Arzt, E., Fertig, R. S. III, & Kroner, E. (2015). Temperature-Induced Switchable Adhesion using Nickel-Titanium-Polydimethylsiloxane Hybrid Surfaces. *Advanced Functional Materials, 25*(20), 3013–3021. doi:10.1002/adfm.201500437 PMID:26120295

Friák, M., Golian, M., Holec, D., Koutná, N., & Šob, M. (2019). An Ab Initio Study of Magnetism in Disordered Fe-Al Alloys with Thermal Antiphase Boundaries. *Nanomaterials (Basel, Switzerland), 10*(1), 44. doi:10.3390/nano10010044 PMID:31878105

Froes, F. H., & Eylon, D. (1990). Powder -metallurgy of titanium alloys. *International Materials Reviews, 35*(1), 162–184. doi:10.1179/095066090790323984

Fuerstenau, D. W., & Abouzeid, A. Z. (2002, November 1). The energy efficiency of ball milling in comminution. *International Journal of Mineral Processing, 67*(1-4), 161–185. doi:10.1016/S0301-7516(02)00039-X

Fujiki, A. (2001, January 15). Present state and future prospects of powder metallurgy parts for automotive applications. *Materials Chemistry and Physics*, *67*(1-3), 298–306. doi:10.1016/S0254-0584(00)00455-7

Gao, B., Tan, Z., Liu, Z., Gao, G., Zhang, M., Zhang, G., & Bai, B. (2019). Influence of non-uniform microstructure on rolling contact fatigue behavior of high-speed wheel steels. *Engineering Failure Analysis*, *100*, 485–491. doi:10.1016/j.engfailanal.2019.03.002

Gao, G., Liu, R., Fan, Y., Qian, G., Gui, X., Misra, R. D. K., & Bai, B. (2022). Mechanism of subsurface microstructural fatigue crack initiation during high and very-high cycle fatigue of advanced bainitic steels. *Journal of Materials Science and Technology*, *108*, 142–157. doi:10.1016/j.jmst.2021.08.060

Gao, X., Geng, K., Sun, C., Zhang, S., Zhou, J., Wu, J., Zhang, X., & Wang, X. (2022). Effects of Graphite Particle Content and Holding Time on the Microstructure and Mechanical Properties of the Graphite/AZ91D Composite. *Metals*, *13*(1), 57. doi:10.3390/met13010057

Gao, Y., Yang, Z., Xiao, H., Lei, Q., Liu, B., & Liu, Y. (2022). Effect of Welding Current on Wear Behavior of PTA-Welded Cu35Ni25Co25Cr15 HEA Coating. *Coatings*, *13*(1), 52. doi:10.3390/coatings13010052

Garcia-Mateo, C., & Caballero, F. G. (2007). Design of carbide-free low-temperature ultra-high strength bainitic steels. *International Journal of Materials Research*, *98*(2), 137–143. doi:10.3139/146.101440

Garnham, J. E., & Davis, C. L. (2008). The role of deformed rail microstructure on rolling contact fatigue initiation. *Wear*, *265*(9-10), 1363–1372. doi:10.1016/j.wear.2008.02.042

Gaur, A., Chauhan, V. S., & Vaish, R. (2023). Planetary ball milling induced piezocatalysis for dye degradation using BaTiO3 ceramics. *Environmental Science. Advances*, *2*(3), 462–472. doi:10.1039/D2VA00210H

Geambazu, L. E., Cotruţ, C. M., Miculescu, F., & Csaki, I. (2021). Mechanically alloyed CoCrFeNiMo0.85 high-entropy alloy for corrosion resistance coatings. *Materials (Basel)*, *14*(14), 3802. Advance online publication. doi:10.3390/ma14143802 PMID:34300721

Geetha, M., Singh, A. K., Asokamani, R., & Gogia, A. K. (2009). Ti based biomaterials, the ultimate choice for orthopaedic implants – A review. *Progress in Materials Science*, *54*(3), 397–425. doi:10.1016/j.pmatsci.2008.06.004

Geneq. (2023). *Geneq Rotary Sampler Splitter*. doi:10.1108/eb052349

Gerasimov, A. M., Eremina, O. V., Cherkasova, M. V., & Dmitriev, S. V. (2021). Application of particle-size analysis in various industries. *Journal of Physics: Conference Series*, *1728*(1), 012003. Advance online publication. doi:10.1088/1742-6596/1728/1/012003

German, R. (2014). *Sintering: From empirical observations to scientific principles*. Butterworth-Heinemann.

German, R.M. (1996). *Sintering theory and practice*. Academic Press.

German, R. M., Suri, P., & Park, S. J. (2009, January). Liquid phase sintering. *Journal of Materials Science, 44*(1), 1–39. doi:10.1007/s10853-008-3008-0

Ge, S., Wang, Q., & Wang, J. (2017). The Impact Wear-Resistance Enhancement Mechanism of Medium Manganese Steel and Its Applications in Mining Machines. *Wear, 376–377*, 1097–1104. doi:10.1016/j.wear.2017.01.015

Ghodrati, M., Ahmadian, M., & Mirzaeifar, R. (2018). Modeling of rolling contact fatigue in rails at the microstructural level. *Wear, 406*, 205–217. doi:10.1016/j.wear.2018.04.016

Ghosh, M., Hendy, M., Raush, J., & Momeni, K. (2022). A Phase-Field Model for In-Space Manufacturing of Binary Alloys. *Materials (Basel), 16*(1), 383. doi:10.3390/ma16010383 PMID:36614722

Gibbs, P. J., De Cooman, B. C., Brown, D. W., Clausen, B., Schroth, J. G., Merwin, M. J., & Matlock, D. K. (2014). Strain Partitioning in Ultra-Fine Grained Medium-Manganese Transformation Induced Plasticity Steel. *Materials Science and Engineering A, 609*, 323–333. doi:10.1016/j.msea.2014.03.120

Gild, J. (2019). A high-entropy silicide: (Mo0.2Nb0.2Ta0.2Ti0.2W0.2)Si2. *Journal of Materials, 5*, 337–343.

Gild, J., Kaufmann, K., Vecchio, K., & Luo, J. (2019). Reactive flash spark plasma sintering of high-entropy ultrahigh temperature ceramics. *Scripta Materialia, 170*, 106–110. doi:10.1016/j.scriptamat.2019.05.039

Gild, J., Wright, A., Quiambao-Tomko, K., Qin, M., Tomko, J. A., Shafkat bin Hoque, M., Braun, J. L., Bloomfield, B., Martinez, D., Harrington, T., Vecchio, K., Hopkins, P. E., & Luo, J. (2020). Thermal conductivity and hardness of three single-phase high-entropy metal diborides fabricated by borocarbothermal reduction and spark plasma sintering. *Ceramics International, 46*(5), 6906–6913. doi:10.1016/j.ceramint.2019.11.186

Gild, J., Zhang, Y., Harrington, T., Jiang, S., Hu, T., Quinn, M. C., Mellor, W. M., Zhou, N., Vecchio, K., & Luo, J. (2016). High-Entropy Metal Diborides: A New Class of High-Entropy Materials and a New Type of Ultrahigh Temperature Ceramics. *Scientific Reports, 6*(1), 37946. doi:10.1038/srep37946 PMID:27897255

Gilman, P. S., & Benjamin, J. S. (1983). Mechanical Alloying. *Annual Review of Materials Science, 13*(1), 279–300. doi:10.1146/annurev.ms.13.080183.001431

Giridasappa, A., Kiran, K. K., & Rajendrachari, S. (2023). Nanotechnology and their Evaluation of Bi-Functional Applications. *J Metastable Nanocrystalline Mater, 37*, 33–43. doi:10.4028/p-dKORv2

Girsch, G., & Heyder, R. (2006, June). Advanced pearlitic and bainitic high-strength rails promise to improve rolling contact fatigue resistance. In *7th World Congress on Railway Research (WCRR2006)*, Montreal, Canada.

Go, S. H., Kim, K. S., Kim, J. S. & Cheon, C. (2023). Effects of attrition milling on the microstructure and piezoelectric properties of BiFeO3–BaTiO3 ceramics. *J. Korean Ceram. Soc.* doi:10.1007/s43207-023-00291-8

Gökçe, H., & Öveçoğlu, M.L. (2023). Microstructure-property evolution of mechanically alloyed Al-20 wt% Si matrix powders and sintered composites reinforced withTiB2 particulates. *Engineering Science and Technology, an International Journal, 39.*

Golla, B. R., Bhandari, T., Mukhopadhyay, A., & Basu, B. (2014). Titanium Diboride. *Ultra-High Temp. Ceram. Mater. Extrem. Environ. Appl., 9781118700*, 316–360.

Gong, M., & Bassi, A. (2016, December 1). Carotenoids from microalgae: A review of recent developments. *BiotechnologyAdvances, 34*(8), 1396–1412. doi:10.1016/j.biotechadv.2016.10.005 PMID:27816618

Gong, W., Li, J., Chu, Z., Yang, D., Subhan, S., Li, J., Huang, M., Zhang, H., & Zhao, Z. (2022). A low-cost high-entropy porous CrO/CrN/C biosensor for highly sensitive simultaneous detection of dopamine and uric acid. *Microchemical Journal, 175*, 107188. Advance online publication. doi:10.1016/j.microc.2022.107188

Gontijo, M., Hoflehner, C., Ilie, S., Six, J., & Sommitsch, C. (2020). Holding Time Influence on the Hot Ductility Behavior of a Continuously Cast Low Alloy Steel. *Metals, 11*(1), 64. doi:10.3390/met11010064

Gonzaga, S., Molina, A., Guardián, R., Martínez, H., Vélez, E. V., & Tapia, J. S.-O. (2023). Synthesis of Magnesium-Based Alloys by Mechanical Alloying for Implant Applications. *Coatings, 13*(2), 1–15. doi:10.3390/coatings13020260

Gorsse, S., & Senkov, O. N. (2018). About the reliability of CALPHAD predictions in multicomponent systems. *Entropy (Basel, Switzerland), 20*(12), 899. doi:10.3390/e20120899 PMID:33266623

Goryczka, T. (2013). Ni25Ti50Cu25 shape memory alloy produced by nonconventional techniques. *Shape Memory Alloys: Processing, Characterization and Applications, Dr. Francisco Manuel Braz Fernandes*, 53-76.

Green, M. R., Rainforth, W. M., Frolish, M. F., & Beynon, J. H. (2007). The effect of microstructure and composition on the rolling contact fatigue behaviour of cast bainitic steels. *Wear, 263*(1-6), 756–765. doi:10.1016/j.wear.2007.01.070

Guan, D. (2015). *Processing, microstructure and mechanical behavior of bulk nanostructured Mg alloy.* Academic Press.

Guan, C., Yu, T., Zhao, Y., Chen, L., & Chen, Y. (2022). Repair of Gear by Laser Cladding Ni60 Alloy Powder: Process, Microstructure and Mechanical Performance. *Applied Sciences (Basel, Switzerland), 13*(1), 319. doi:10.3390/app13010319

Gueli, A. M., Bonfiglio, G., Pasquale, S., & Troja, S. O. (2017). Effect of particle size on pigments colour. *Color Research and Application, 42*(2), 236–243. doi:10.1002/col.22062

Gui, X., Wang, K., Gao, G., Misra, R. D. K., Tan, Z., & Bai, B. (2016). Rolling contact fatigue of bainitic rail steels: The significance of microstructure. *Materials Science and Engineering A*, *657*, 82–85. doi:10.1016/j.msea.2016.01.052

Guo, C., Chen, W., & Deng, T. (2023). Fundamental study on the preparation of insulating ceramics via the phase reconstruction of phosphate tailings. *Ceramics International*, *49*(6), 9419–9431. doi:10.1016/j.ceramint.2022.11.107

Gupta, S., Shashanka, R., & Chaira, D. (2015). Synthesis of nano-structured duplex and ferritic stainless steel powders by planetary milling: An experimental and simulation study. In *IOP conference series: materials science and engineering* (Vol. 75, No. 1, p. 012033). IOP Publishing.

Gupta, S., Shashanka, R., & Chaira, D. (2015). Synthesis of nano-structured duplex and ferritic stainless steel powders by planetary milling: An experimental and simulation study. In *IOP conference series: Materials science and engineering* (Vol. 75, No. 1, p. 012033). IOP Publishing.

Gupta, S., Shashanka, R., & Chaira, D. (2015). Synthesis of nanostructured duplex and ferritic stainless steel powders by planetary milling: An experimental and simulation study. *IOP Conference Series. Materials Science and Engineering*, *75*(1), 012033. Advance online publication. doi:10.1088/1757-899X/75/1/012033

Gupta, V. K., Khosravi, S., Karimi-Maleh, H., Alizadeh, M., & Sharafi, S. (2015). A voltammetric sensor for determination of methyldopa in the presence of hydrochlorothiazide using fe: Co Nanoalloy modified carbon paste electrode. *International Journal of Electrochemical Science*, *10*(4).

Gurudas, M., Kumar, T.N., & Ghosh. (2017). Enhancement of Mechanical Properties in Bainitic Steel Processed from Different Austenitization Temperatures. *Steel Research International, 87*, 1700259 (1-9). doi:10.1002/srin.201700259.

Hafnium Boride Powder (HfB2 Powder), CAS 12007-23-7. (n.d.). Heeger Materials. https://heegermaterials.com/boride-powders/1244-hafnium-boride-hfb2-powder-cas-12007-23-7.html

Hajizad, O., Kumar, A., Li, Z., Petrov, R. H., Sietsma, J., & Dollevoet, R. (2019). Influence of microstructure on mechanical properties of bainitic steels in railway applications. *Metals, 9*(7), 1-19.

Halakoei, H., Ghalkhani, M., Sobhani-Nasab, A., & Rahimi-Nasrabadi, M. (2021). An efficient electrochemical sensor based on CeVO4-CuWO4nanocomposite for methyldopa. *Materials Research Express*, *8*(8), 085001. Advance online publication. doi:10.1088/2053-1591/ac16f1

Hamweendo, A., & Botef, I. (2019). Modern Production Methods for Titanium Alloys: A review. Titanium Alloys-Novel Aspects of Their Manufacturing and Processing, 1–17.

Hamweendo, A., Malama, T., & Botef, I. (2016). Titanium-Nickel Alloys for Bone Tissue Engineering Application via Cold Titanium-Nickel Alloys for Bone Tissue Engineering Application. *International Conference on Competitive Manufacturing COMA16*, 273–279.

Han, J. (2022). Microstructural evolution of chemically vapor-deposited tantalum carbide at elevated temperature. *Journal of Ceramic Processing Research, 23*, 751–757.

Han, J., Fu, J., & Schoch, R. B. (2008). Molecular sieving using nanofilters: Past, present and future. *Lab on a Chip*, *8*(1), 23–33. doi:10.1039/B714128A PMID:18094759

Han, Z. D., Luan, H. W., Liu, X., Chen, N., Li, X. Y., Shao, Y., & Yao, K. F. (2018). Microstructures and mechanical properties of TixNbMoTaW refractory high-entropy alloys. *Materials Science and Engineering A*, *712*, 380–385. Advance online publication. doi:10.1016/j.msea.2017.12.004

Harrington, T. J., Gild, J., Sarker, P., Toher, C., Rost, C. M., Dippo, O. F., McElfresh, C., Kaufmann, K., Marin, E., Borowski, L., Hopkins, P. E., Luo, J., Curtarolo, S., Brenner, D. W., & Vecchio, K. S. (2019). Phase stability and mechanical properties of novel high entropy transition metal carbides. *Acta Materialia*, *166*, 271–280. doi:10.1016/j.actamat.2018.12.054

Hasan, I. M. A., Abd-Elsabur, K. M., Assaf, F. H., & Abd-Elsabour, M. (2022). Folic Acid Determination in Food Samples Using Green Synthesized Copper Oxide Nanoparticles and Electro-Poly (Methyl Orange) Sensor. *Electrocatalysis (New York)*, *13*(6), 759–772. Advance online publication. doi:10.1007/s12678-022-00756-0

Hasan, S. M., Chakrabarti, D., & Singh, S. B. (2018). Dry rolling/sliding wear behaviour of pearlitic rail and newly developed carbide-free bainitic rail steels. *Wear*, *408*, 151–159. doi:10.1016/j.wear.2018.05.006

Hasan, S. M., Ghosh, M., Chakrabarti, D., & Singh, S. B. (2020). Development of continuously cooled low-carbon, low-alloy, high strength carbide-free bainitic rail steels. *Materials Science and Engineering A*, *771*, 138590. doi:10.1016/j.msea.2019.138590

Hasan, S., Chakrabarti, D., & Singh, S. B. (2021). Thermomechanical Treatment of Carbide-Free Bainitic Steel. *Metal Science and Heat Treatment*, *63*(7), 354–362. doi:10.1007/s11041-021-00695-5

Hassan, S. S., Nafady, A., Sirajuddin, Solangi, A. R., Kalhoro, M. S., Abro, M. I., & Sherazi, S. T. H. (2015). Ultra-trace level electrochemical sensor for methylene blue dye based on nafion stabilized ibuprofen derived gold nanoparticles. *Sensors and Actuators. B, Chemical*, *208*, 320–326. Advance online publication. doi:10.1016/j.snb.2014.11.021

He, Z.-M., Jiang, Q.-C., Fu, S.-B., & Xie, J.-P. (1987). Improved Work-Hardening Ability and Wear Resistance of Austenitic Manganese Steel under Non-Severe Impact-Loading Conditions. *Wear*, *120*(3), 305–319. doi:10.1016/0043-1648(87)90024-X

Hînganu, D., Scutariu, M. M., & Hînganu, M. V. (2018). The existence of labial SMAS - Anatomical, imaging and histological study. *Annals of Anatomy = Anatomischer Anzeiger: Official Organ of the Anatomische Gesellschaft*, *218*, 271–275. doi:10.1016/j.aanat.2018.04.009

Hlavatý, I., Sigmund, M., Krejčí, L., & Mohyla, P. (2009). The Bainitic Steels for Rails Applications. *Materials Engineering*, *16*(4), 44–50.

Hogg, R. (2008). Issues_in_Particle_Size_Analysis. *Kona Powder and Particle Journal*, *26*(2), 81–93. doi:10.14356/kona.2008009

Hong, W., Chen, F., Shen, Q., Han, Y.-H., Fahrenholtz, W. G., & Zhang, L. (2019). Microstructural evolution and mechanical properties of (Mg,Co,Ni,Cu,Zn)O high-entropy ceramics. *Journal of the American Ceramic Society*, *102*(4), 2228–2237. doi:10.1111/jace.16075

Hrapkowicz, B., Lesz, S., Karolus, M., Garbiec, D., Wiśniewski, J., Rubach, R., Gołombek, K., Kremzer, M., & Popis, J. (2022, February 22). Microstructure and Mechanical Properties of Spark Plasma Sintered Mg-Zn-Ca-Pr Alloy. *Metals*, *12*(3), 375. doi:10.3390/met12030375

Hrapkowiczk, B., Lesz, S., & Kremzer, M. (2021). Mechanical alloying of Mg-Zn-Ca-Er alloy. *Bulletin of the Polish Academy of Sciences. Technical Sciences*, *69*, 1–8. doi:10.24425/bpasts.2021.137587

Huang, Y. B., Shi, L. B., Zhao, X. J., Cai, Z. B., Liu, Q. Y., & Wang, W. J. (2018). On the formation and damage mechanism of rolling contact fatigue surface cracks of wheel/rail under the dry condition. *Wear*, *400*, 62–73. doi:10.1016/j.wear.2017.12.020

Hua, P., Xia, M., Onuki, Y., & Sun, Q. (2021). Nanocomposite NiTi shape memory alloy with high strength and fatigue resistance. *Nature Nanotechnology*, *16*(4), 409–413. doi:10.1038/s41565-020-00837-5 PMID:33479541

Hulanicki, A., Glab, S., & Ingman, F. (1991). Chemical sensors definitions and classification. *Pure and Applied Chemistry*, *63*(9), 1247–1250. Advance online publication. doi:10.1351/pac199163091247

Huo, S., Chen, L., Liu, X., Kong, Q., Wang, Y., Gu, H., & Zhou, Y. (2022). Reactive sintering of dual-phase high-entropy ceramics with superior mechanical properties. *Journal of Materials Science and Technology*, *129*, 223–227. doi:10.1016/j.jmst.2022.03.035

Hussain, I., Lamiel, C., Ahmad, M., Chen, Y., Shuang, S., Javed, M. S., Yang, Y., & Zhang, K. (2021). High entropy alloys as electrode material for supercapacitors: A review. In Journal of Energy Storage (Vol. 44). doi:10.1016/j.est.2021.103405

Ibrahim, H., Aziz, A., & Rahmat, A. (2014, March 1). Enhanced liquid-phase sintering of W–Cu composites by liquid infiltration. *International Journal of Refractory & Hard Metals*, *43*, 222–226. doi:10.1016/j.ijrmhm.2013.12.004

Imshinetskiy, I., Kashepa, V., Nadaraia, K., Mashtalyar, D., Suchkov, S., Zadorozhny, P., Ustinov, A., Sinebryukhov, S., & Gnedenkov, S. (2022). PEO Coatings Modified with Halloysite Nanotubes: Composition, Properties, and Release Performance. *International Journal of Molecular Sciences*, *24*(1), 305. doi:10.3390/ijms24010305 PMID:36613748

Ishida, M. (2013). Rolling contact fatigue (RCF) defects of rails in Japanese railways and its mitigation strategies. *Electronic Journal of Structural Engineering*, *13*(1), 67–74. doi:10.56748/ejse.131621

Ito, A., & Matsumoto, S. (2023). Chemically vapor deposited oxide-based thick film scintillators. *Japanese Journal of Applied Physics*, *62*(1), 010612. doi:10.35848/1347-4065/aca249

Ivasishin, O. M., Anokhin, V. M., Demidik, A. N., & Savvakin, D. G. (2000). *Cost-effective blended elemental powder metallurgy of titanium alloys for transportation application. In Key Engineering Materials* (Vol. 188). Trans Tech Publications Ltd.

Ivošević, Š., & Rudolf, R. (2019). Materials with Shape Memory Effect for Applications in Maritime. *Maritime Technical Journal, 218*(3), 25–41. doi:10.2478/sjpna-2019-0016

Iwan, S., Lin, C.-M., Perreault, C., Chakrabarty, K., Chen, C.-C., Vohra, Y., Hrubiak, R., Shen, G., & Velisavljevic, N. (2022). High-Entropy Borides under Extreme Environment of Pressures and Temperatures. *Materials (Basel), 15*(9), 3239. doi:10.3390/ma15093239 PMID:35591574

Jani, J. M., Leary, M., & Subic, A. (2014). Shape memory alloys in automotive applications. *Applied Mechanics and Materials, 663*, 248–253. doi:10.4028/www.scientific.net/AMM.663.248

Jiao, J., Xu, J., Jing, C., Cheng, X., Wu, D., Ru, H., Zeng, K., & Sheng, L. (2022). Effect of Laser Micro-Texturing on Laser Joining of Carbon Fiber Reinforced Thermosetting Composites to TC4 Alloy. *Materials (Basel), 16*(1), 270. doi:10.3390/ma16010270 PMID:36614608

Johnson, D. R., & Becker, W. T. (1993). Toughness of tempered upper and lower bainitic microstructures in a 4150 steel. *Journal of Materials Engineering and Performance, 2*(2), 255–263. doi:10.1007/BF02660294

Jost, N., & Schmidt, I. (1986). Friction-Induced Martensitic Transformation in Austenitic Manganese Steels. *Wear, 111*(4), 377–389. doi:10.1016/0043-1648(86)90134-1

Kale, S., Adimule, V., Shashanka, R., Tanaka, M., & Wei, P. (2022). Synthesis and Modifications of Materials and its Properties. *Trans Tech Publ. Ltd. Switz., 134*, 1–134.

Kale, S., Adimule, V., Shashanka, R., Tanaka, M., & Wei, P. (2022). *Synthesis and Modifications of Materials and its Properties*. Trans Tech Publications Ltd.

Kanaginahal, G. M., & Murthy, M. (2017). Microstructural Study and Evaluation of Few Mechanical Properties of Hybrid Composites. *Adv Mater Manuf Charact, 7*, 7–14. https://doi.org/http://dx.doi.org/10.11127/ijammc2017.04.02C

Kanaginahal, G. M., Mahale, R. S., & Vasanth, S. (2023) Studies on Dielectric, Super-Capacitive and Photoluminescence Properties of Metal Organic Frameworks Subjected to Varying Temperature and Frequencies. *Nano Hybrids Compos, 39*. https://doi.org/https://doi.org/10.4028/p-si6964

Kanaginahal, G. M., Kumar Muniraju, A., & Madhav Murthy, M. (2018). Coatings for Enhancement of Properties of Polymer Matrix Composites: A review. *Materials Today: Proceedings, 5*(1), 2462–2465. doi:10.1016/j.matpr.2017.11.026

Kanaginahal, G. M., Mahale, R. S., Vasanth, S., Sharath, P. C., & Shashanka, R. (2023, June 30). Studies on Dielectric, Super-Capacitive and Photoluminescence Properties of Metal Organic Frameworks Subjected to Varying Temperature and Frequencies. *Nano Hybrids and Composites., 39*, 17–25. doi:10.4028/p-si6964

Kanaginahal, G. M., Suresh Hebbar, H., & Shahapurkar, K. (2023). Leverage of Weave Pattern and Composite Thickness on Dynamic Mechanical Analysis, Water Absorption and Flammability Response of Bamboo Fabric/Epoxy Composites. *Heliyon*, *9*(1), 1–13. doi:10.1016/j.heliyon.2023.e12950 PMID:36820186

Kane, S. N., Mishra, A., & Dutta, A. K. (2016). Microstructure and mechanical properties of the as-cast Mg-Zn-Mn-Ca alloys. *Journal of Physics: Conference Series*, *755*. Advance online publication. doi:10.1088/1742-6596/755/1/011001

Kang, S.J. (2004). *Sintering: Densification, grain growth and microstructure*. Elsevier.

Kang, S.L. (2010). Liquid phase sintering. In *Sintering of advanced materials* (pp. 110-129). Woodhead Publishing.

Kang, J., Zhang, F. C., Yang, X. W., Lv, B., & Wu, K. M. (2017). Effect of tempering on the microstructure and mechanical properties of a medium carbon bainitic steel. *Materials Science and Engineering A*, *686*, 150–159. doi:10.1016/j.msea.2017.01.044

Kapito, A., Stumpf, W., & Papo, M. J. (2012). On the development of bainitic alloys for railway wheel applications. *Journal of the Southern African Institute of Mining and Metallurgy*, *7A*, 539–544.

Karlsson, O., & Schade, B. E. H. (2013). Particle Analysis. In Chemistry and Technology of Emulsion Polymerisation: Second Edition (pp. 213–251). doi:10.1002/9781118638521.ch9

Kauffman, G. B., & Mayo, I. (1993). The metal with a memory. *American Heritage of Invention & Technology*, *9*(2), 18–23.

Kavak, S., Bayrak, K. G., Bellek, M., Mertdinç, S., Muhaffel, F., Gökçe, H., Ayas, E., Derin, B., Öveçoğlu, M. L., & Ağaoğulları, D. (2022). Synthesis and characterization of (HfMoTiWZr)C high entropy carbide ceramics. *Ceramics International*, *48*(6), 7695–7705. doi:10.1016/j.ceramint.2021.11.317

Ke, B., Ji, W., Zou, J., Wang, W., & Fu, Z. (2023). Densification mechanism, microstructure and mechanical properties of ZrC ceramics prepared by high-pressure spark plasma sintering. *Journal of the European Ceramic Society*, *43*(8), 3053–3061. doi:10.1016/j.jeurceramsoc.2023.02.038

Keerthivasan, N. & Anandakrishnan, V. (2022). Effect of recycled TiC addition in the compaction and sintering behavior of Al-Mg-Cu-Zn + r-TiC powder metallurgy composites. *Part. Sci. Technol.* . doi:10.1080/02726351.2022.2104187

Kejić, D., Žuvelek, V., & Valčić, A. (2023, March 9). Parametric Numerical Study of Welded Aluminium Beam-to-Column Joints. *Buildings*, *13*(3), 718. doi:10.3390/buildings13030718

Kelsey, R. G., & Shafizadeh, F. (1980, May). Enhancement of cellulose accessibility and enzymatic hydrolysis by simultaneous wet milling. *Biotechnology and Bioengineering*, *22*(5), 1025–1036. doi:10.1002/bit.260220511

Kenfaui, D., Guillemet-Fritsch, S., Valdez-Nava, Z., Laudebat, L., Tenailleau, C., Dufour, P., & Locatelli, M.-L. (2023). Tuning electrical conductivity in AlN-based ceramics by incorporating graphene. *Journal of the European Ceramic Society*, *43*(5), 1887–1896. doi:10.1016/j.jeurceramsoc.2022.12.039

Khot, M., Raut, G., Ghosh, D., Alarcón-Vivero, M., Contreras, D., & Ravikumar, A. (2020, January 1). Lipid recovery from oleaginous yeasts: Perspectives and challenges for industrial applications. *Fuel*, *259*, 116292. doi:10.1016/j.fuel.2019.116292

Kim, Y.G., & Merrick, H.F. (1979). *Characterization of an oxide dispersion strengthened superalloy, MA-6000E, for turbine blade applications*. Academic Press.

Kim, D., Hong, K., Sim, J., Lee, J., & Lee, W. (2021). Shape Memory and Mechanical Properties of Cold Rolled and Annealed Fe-17Mn-5Si-5Cr-4Ni-1Ti-0.3C Alloy. *Materials (Basel)*, *14*(2), 255. doi:10.3390/ma14020255 PMID:33419156

Klaasen, H., Kollo, L., & Kübarsepp, J. (2007). Mechanical properties and wear performance of compression sintered TiC based cermets. *Powder Metallurgy*, *50*(2), 132–136. doi:10.1179/174329007X153305

Kocich, R., Szurman, I., & Kursa, M. (2013). The methods of preparation of Ti–Ni–X alloys and their forming. *Shape Memory Alloys-Processing, Characterization and Applications*, 28-35.

Kongas, M., & Saloheimo, K. (2014). *Next Generation Slurry Particle Size Analyzers*. Academic Press.

Kong, X., Dang, Z., Liu, X., Wang, M., & Hou, N. (2022). Simulation and Experimental Analysis of Tool Wear and Surface Roughness in Laser Assisted Machining of Titanium Alloy. *Crystals*, *13*(1), 40. doi:10.3390/cryst13010040

Kostryzhev, A., Singh, N., Chen, L., Killmore, C., & Pereloma, E. (2018). Comparative effect of Mo and Cr on microstructure and mechanical properties in NbV-microalloyed bainitic steels. *Metals*, *8*(2), 1-19.

Kovalev, D. Y., Kochetov, N. A., & Chuev, I. I. (2021). Fabrication of high-entropy carbide (TiZrHfTaNb)C by high-energy ball milling. *Ceramics International*, *47*(23), 32626–32633. doi:10.1016/j.ceramint.2021.08.158

Krishnan, R. G., Rejithamol, R., & Saraswathyamma, B. (2020). Non-enzymatic electrochemical sensor for the simultaneous determination of adenosine, adenine and uric acid in whole blood and urine. *Microchemical Journal*, *155*, 104745. Advance online publication. doi:10.1016/j.microc.2020.104745

Krishnan, R. G., & Saraswathyamma, B. (2021). Disposable electrochemical sensor for coumarin induced milk toxicity in raw milk samples. *Measurement*, *170*, 108709. Advance online publication. doi:10.1016/j.measurement.2020.108709

Kshama, S. S., Swamynathan, K., & Shashanka, R. (2023). Electrochemical Devices for Soil Analysis. In *ACS Symposium Series* (Vol. 1437). 10.1021/bk-2023-1437.ch006

Kumar, A., Kim, J.-W., Yadav, N., Vij, A., Kumar, S., Sharma, M. K., Kumari, K., Huh, S.-H., & Koo, B. H. (2023). Structural insights into the rare-earth modified La1.4-xSmxSr1.6Mn2O7 layered perovskites and probing their complex magnetic and magnetocaloric properties. *Materials Chemistry and Physics*, *302*, 127695. doi:10.1016/j.matchemphys.2023.127695

Kumar, A., & Singh, A. (2021). Mechanical properties of nanostructured bainitic steels. *Materialia*, *15*, 101034. doi:10.1016/j.mtla.2021.101034

Kumar, A., Singh, A., & Suhane, A. (2022). Mechanically alloyed high entropy alloys: Existing challenges and opportunities. *Journal of Materials Research and Technology*, *17*, 2431–2456. Advance online publication. doi:10.1016/j.jmrt.2022.01.141

Kumar, P., Skotnicova, K., Mallick, A., Gupta, M., Cegan, T., & Jurica, J. (2020). Mechanical Characterization of Graphene Nanoplatelets-Reinforced Mg-3Sn Alloy Synthesized by Powder Metallurgy. *Metals*, *11*(1), 62. doi:10.3390/met11010062

K, V., Gorle, R., S, A., & Bakshi, S. R. (2021). Novel single phase (Ti0.2W0.2Ta0.2Mo0.2V0.2) C0.8 high entropy carbide using ball milling followed by reactive spark plasma sintering. *Journal of the European Ceramic Society*, *41*(13), 6756–6762. doi:10.1016/j.jeurceramsoc.2021.06.009

Lai, G. Y., Wood, W. E., Clark, R. A., Zackay, V. F., & Parker, E. R. (1974). The effect of austenitizing temperature on the microstructure and mechanical properties of as-quenched 4340 steel. *Metallurgical and Materials Transactions. B, Process Metallurgy and Materials Processing Science*, *5*, 1663–1670.

Lakshmi Priya, M., Rana, D. & Ramalingam, M. (2017). Ceramic nanofiber composites. *Nanofiber Compos. Biomed. Appl.* 33–54. doi:10.1016/B978-0-08-100173-8.00030-2

Laszkiewicz-Łukasik, J., Jaworska, L., Putyra, P., Klimczyk, P., & Garzeł, G. (2016). The influence of SPS heating rates on the synthesis reaction of tantalum diboride. *Boletín de la Sociedad Española de Cerámica y Vidrio*, *55*(4), 159–168. doi:10.1016/j.bsecv.2016.03.001

Latypov, M. I., Shin, S., De Cooman, B. C., & Kim, H. S. (2016). Micromechanical Finite Element Analysis of Strain Partitioning in Multiphase Medium Manganese TWIP+TRIP Steel. *Acta Materialia*, *108*, 219–228. doi:10.1016/j.actamat.2016.02.001

Laydi, M. R., & Lexcellent, C. (2015). Determination of phase transformation surfaces around crack tip in shape memory alloys. In *MATEC Web of Conferences* (Vol. 33, p. 02010). EDP Sciences. 10.1051/matecconf/20153302010

Lazorenko, G., Kasprzhitskii, A., Shaikh, F., Krishna, R. S., & Mishra, J. (2021). Utilization potential of mine tailings in geopolymers: Physicochemical and environmental aspects. *Process Safety and Environmental Protection*, *147*, 559–577. doi:10.1016/j.psep.2020.12.028

Lee, K. M., & Polycarpou, A. A. (2005). Wear of conventional pearlitic and improved bainitic rail steels. *Wear*, *259*(1-6), 391–399. doi:10.1016/j.wear.2005.02.058

Lee, W. E., Gilbert, M., Murphy, S. T., & Grimes, R. W. (2013). Opportunities for advanced ceramics and composites in the nuclear sector. *Journal of the American Ceramic Society, 96*(7), 2005–2030. doi:10.1111/jace.12406

Lexcellent, C. (2013). *Shape-memory alloys handbook*. John Wiley & Sons. doi:10.1002/9781118577776

Li, K., & Chen, W. (2021). Recent progress in high-entropy alloys for catalysts: synthesis, applications, and prospects. In Materials Today Energy (Vol. 20). doi:10.1016/j.mtener.2021.100638

Li, Y., & Milbourn, D. (2011). Vanadium in bainitic steels: A review of recent developments. *Advanced Steels: The Recent Scenario in Steel Science and Technology*, 303-308.

Li, Y., Yuan, Y., & Wang, J. (2023). Controllable degradation behavior of Mg-Sr-Y alloys for the bio-applications. *NPJ Mater Degrad, 7*. doi:10.1038/s41529-023-00366-1

Li, B., Hu, Y., Li, J., Liu, M., Kong, L., Hu, Y., & Kang, L. (2016). Mechanical alloying synthesis of Co9S8 particles as materials for supercapacitors. *Metals, 6*(6), 142. Advance online publication. doi:10.3390/met6060142

Li, F., Zhou, L., Liu, J. X., Liang, Y., & Zhang, G. J. (2019). High-entropy pyrochlores with low thermal conductivity for thermal barrier coating materials. *J. Adv. Ceram., 8*(4), 576–582. doi:10.1007/s40145-019-0342-4

Li, H. F., Qiu, K. J., Zhou, F. Y., Li, L., & Zheng, Y. F. (2016). Design and development of novel antibacterial Ti-Ni-Cu shape memory alloys for biomedical application. *Scientific Reports, 6*(1), 37475. doi:10.1038/srep37475 PMID:27897182

Li, H., Jiang, C., Yu, Z., Wan, Y., Ma, Y., & Yang, Z. (2022). Numerical Study on the Progressive Damage Behavior of the Interfacial Debonding between Shape Memory Alloy and Polymer Matrix. *Materials (Basel), 16*(1), 168. doi:10.3390/ma16010168 PMID:36614508

Li, J. C. M., Feng, C. R., & Rath, B. B. (2020). Emission of Dislocations from Grain Boundaries and Its Role in Nanomaterials. *Crystals, 11*(1), 41. doi:10.3390/cryst11010041

Li, M., Wang, X., Cheng, Y., Su, L., Xu, P., & Li, X. (2021). High-Entropy Alloy Nanoparticles As Catalyst for Nanomolar-Level Detection of Neurotransmitter Serotonin in Serum. *Proceedings of the IEEE International Conference on Micro Electro Mechanical Systems (MEMS), 2021-January*. 10.1109/MEMS51782.2021.9375269

Lin, Y., & Zhou, Q. (2018, February). The effect of V Ni microalloying on microstructure and properties of bainite steel. In *2017 3rd International Forum on Energy, Environment Science and Materials (IFEESM 2017)* (pp. 1656-1659). Atlantis Press. 10.2991/ifeesm-17.2018.302

Lin, F. L., Liu, B., Hu, C. C., & Song, K. X. (2023). Novel high-entropy microwave dielectric ceramics Sr(La0.2Nd0.2Sm0.2Eu0.2Gd0.2)AlO4 with excellent temperature stability and mechanical properties. *Journal of the European Ceramic Society, 43*(6), 2506–2512. doi:10.1016/j.jeurceramsoc.2023.01.028

Li, Q., Huang, X., & Huang, W. (2019). Fatigue property and microstructure deformation behavior of multiphase microstructure in a medium-carbon bainite steel under rolling contact condition. *International Journal of Fatigue*, *125*, 381–393. doi:10.1016/j.ijfatigue.2019.04.019

Liu, M., Fan, Y., Gui, X., Hu, J., Wang, X., & Gao, G. (2022). Relationship between Microstructure and Properties of 1380 MPa Grade Bainitic Rail Steel Treated by Online Bainite-Based Quenching and Partitioning Concept. *Metals, 12*(2), 1-15.

Liu, B., Zhao, J., Liu, Y., Xi, J., Li, Q., Xiang, H., & Zhou, Y. (2021). Application of high-throughput first-principles calculations in ceramic innovation. *Journal of Materials Science and Technology*, *88*, 143–157. doi:10.1016/j.jmst.2021.01.071

Liu, C. P., Pan, J. Z., Liu, P. T., Ren, R. M., & Zhao, X. J. (2020). Influence of original microstructure on rolling contact fatigue property of D2 wheel steel. *Wear*, *456*, 203380. doi:10.1016/j.wear.2020.203380

Liu, C., Ren, Z., Xu, Y., Pang, S., Zhao, X., & Zhao, Y. (2018). Biodegradable Magnesium Alloys Developed as Bone Repair Materials: A Review. *Scanning*, *2018*, 1–15. Advance online publication. doi:10.1155/2018/9216314 PMID:29725492

Liu, J., Li, Y., Zhang, Y., Hu, Y., Shi, L., Ding, H., Wang, W., Liu, F., Zhou, S., & Shi, T. (2020). Dry rolling/sliding wear of bainitic rail steels under different contact stresses and slip ratios. *Materials (Basel)*, *13*(4678), 1–15. doi:10.3390/ma13204678 PMID:33092302

Liu, J., Shao, G., Liu, D., Chen, K., Wang, K., Ma, B., Ren, K., & Wang, Y. (2020). Design and synthesis of chemically complex ceramics from the perspective of entropy. *Materials Today. Advances*, *8*, 100114. doi:10.1016/j.mtadv.2020.100114

Liu, K. S. (2009). Some factors affecting sieving performance and efficiency. *Powder Technology*, *193*(2), 208–213. doi:10.1016/j.powtec.2009.03.027

Liu, L., Gou, Y., Gao, X., Zhang, P., Chen, W., Feng, S., Hu, F., & Li, Y. (2014). Electrochemically reduced graphene oxide-based electrochemical sensor for the sensitive determination of ferulic acid in A. sinensis and biological samples. *Materials Science and Engineering C*, *42*, 227–233. Advance online publication. doi:10.1016/j.msec.2014.05.045 PMID:25063114

Liu, Q., Ding, D., & Ning, C. (2014). Anodic Fabrication of Ti-Ni-O Nanotube Arrays on Shape Memory Alloy. *Materials (Basel)*, *7*(4), 3262–3273. doi:10.3390/ma7043262 PMID:28788616

Liu, X., Su, Y., Zhang, G., Wang, R., & Cai, X. (2022). Effect of Longitudinal Magnetic Field on the Microstructure and Properties of A-TIG Welding with Different TiO2 Coating Amounts. *Crystals*, *13*(1), 66. doi:10.3390/cryst13010066

Li, Y., Kang, G., Wang, C., Dou, P., & Wang, J. (2006). Vertical short-crack behavior and its application in rolling contact fatigue. *International Journal of Fatigue*, *28*(7), 804–811. doi:10.1016/j.ijfatigue.2005.06.047

Li, Z., Chen, L., Chang, F., Hong, C., Zhao, X., Fang, Y., & Dai, P. (2022). Synthesis, microstructure and properties of Ti(C,N)-(HfZrTaNbTi)C5-HEA high-entropy cermets by high-energy ball milling and spark plasma sintering. *Ceramics International*, *48*(20), 30826–30837. doi:10.1016/j.ceramint.2022.07.036

Loh, C. S., Yokoi, H., & Arai, T. (2006, January). New shape memory alloy actuator: design and application in the prosthetic hand. In *2005 IEEE Engineering in Medicine and Biology 27th Annual Conference* (pp. 6900-6903). IEEE.

Loh, Z. H., Samanta, A. K., & Heng, P. W. (2015, July 1). Overview of milling techniques for improving the solubility of poorly water-soluble drugs. *Asian Journal of Pharmaceutical Sciences*, *10*(4), 255–274. doi:10.1016/j.ajps.2014.12.006

Lonardelli, I., Bortolotti, M., van Beek, W., Girardini, L., Zadra, M., & Bhadeshia, H. K. D. H. (2012). Powder metallurgical nanostructured medium carbon bainitic steel: Kinetics, structure, and in situ thermal stability studies. *Materials Science and Engineering A*, *555*, 139–147. doi:10.1016/j.msea.2012.06.043

Long, X. Y., Branco, R., Zhang, F. C., Berto, F., & Martins, R. F. (2020). Influence of Mn addition on cyclic deformation behaviour of bainitic rail steels. *International Journal of Fatigue*, *132*, 105362. doi:10.1016/j.ijfatigue.2019.105362

Long, X. Y., Kang, J., Lv, B., & Zhang, F. C. (2014). Carbide-free bainite in medium carbon steel. *Materials & Design*, *64*, 237–245. doi:10.1016/j.matdes.2014.07.055

Long, X. Y., Zhang, F. C., Kang, J., Lv, B., & Shi, X. B. (2014). Low-temperature bainite in low-carbon steel. *Materials Science and Engineering A*, *594*, 344–351. doi:10.1016/j.msea.2013.11.089

Lu, L. (2019). Nanoporous noble metal-based alloys: a review on synthesis and applications to electrocatalysis and electrochemical sensing. In Microchimica Acta (Vol. 186, Issue 9). doi:10.1007/s00604-019-3772-3

Lü, L., & Lai, M. O. (1998). Introduction to Mechanical Alloying. In Mechanical Alloying. doi:10.1007/978-1-4615-5509-4_1

Luo, G., Zhang, G., Zhang, Y., Li, A., Sun, Y., Tu, R., & Shen, Q. (2023). Wide temperature range of stable dielectric properties in relaxor BaTiO3-based ceramics by co-doping synergistic engineering. *Materials Chemistry and Physics*, *127629*. Advance online publication. doi:10.1016/j.matchemphys.2023.127629

Madhu, K. S., Sharath, B. N., Venkatesh, C. V., & Pradeep, D. G. (2021). Evaluation of mechanical properties of ceramic reinforced aluminium-7029 hybrid composite. In *IOP Conference Series: Materials Science and Engineering* (Vol. 1189, No. 1, p. 012019). IOP Publishing. 10.1088/1757-899X/1189/1/012019

Madhu, K. S., Venkatesh, C. V., Sharath, B. N., & Karthik, S. (2022, November 2). Characterization and Evaluation of Mechanical Properties of Al-Zn Based Hybrid Metal Matrix Composites. *Applied Science and Engineering Progress.*, *16*(1), 5804. doi:10.14416/j.asep.2022.03.008

Maged, A., Abu El-Magd, S. A., Radwan, A. E., Kharbish, S., & Zamzam, S. (2023). Evaluation insight into Abu Zenima clay deposits as a prospective raw material source for ceramics industry: Remote Sensing and Characterization. *Scientific Reports*, *13*(1), 58. doi:10.1038/s41598-022-26484-5 PMID:36593265

Magel, E., Sroba, P., Sawley, K., & Kalousek, J. (2005). *Control of rolling contact fatigue of rails.* Center for Surface Transportation Technology, National Research Council Canada.

Mahale, R. S., & Shashanka, R. (2022). Mechanical testing of spark plasma sintered materials: A review. In *AIP Conference Proceedings* (Vol. 2469, No. 1). AIP Publishing. 10.1063/5.0080183

Mahale, R. S., Shashanka, R., Vasanth, S., & Vinaykumar, R. (2022). Voltammetric determination of various food azo dyes using different modified carbon paste electrodes. In Biointerface Research in Applied Chemistry (Vol. 12, Issue 4). doi:10.33263/BRIAC124.45574566

Mahale, R. S., Kanaginahal, G. M., & Vasanth, S. (2023). Applications of Fused Deposition Modeling in Dentistry. In R. Keshavamurthy, V. Tambrallimath, & J. P. Davim (Eds.), *Development, Properties, and Industrial Applications of 3D Printed Polymer Composites* (pp. 211–219). IGI Global. doi:10.4018/978-1-6684-6009-2.ch012

Mahale, R. S., Rajendrachari, S., Vasanth, S., Krishna, H., Kapanigowda, N. S., Chikkegowda, S. P., & Patil, A. (2022). Technology and challenges in additive manufacturing of duplex stainless steels. *Biointerface Research in Applied Chemistry*, *12*(1), 1110–1119.

Mahale, R. S., Shamanth, V., & Sharath, P. C. (2020). *Review on Processing and Characterization of Duplex Stainless Steels.* Lecture Notes in Mechanical Engineering., doi:10.1007/978-981-15-7557-0_20

Mahale, R. S., Shamanth, V., Sharath, P. C., Shashanka, R., & Hemanth, K. (2020). A review on spark plasma sintering of duplex stainless steels. *Materials Today: Proceedings*, *45*, 138–144. doi:10.1016/j.matpr.2020.10.357

Mahale, R. S., Vasanth, S., Chikkegouda, S. P., Rajendrachari, S., Narsimhachary, D., & Basavegowda, N. (2023, August 9). Electrochemical Determination of Ascorbic Acid by Mechanically Alloyed Super Duplex Stainless Steel Powders. *Metals*, *13*(8), 1430. doi:10.3390/met13081430

Mahale, R. S., Vasanth, S., & Krishna, H. (2022). Electrochemical sensor applications of nanoparticle modified carbon paste electrodes to detect various neurotransmitters: A review. *Applied Mechanics and Materials*, *908*, 908. doi:10.4028/p-mizm85

Malagutti, M. A., de Fátima Ulbrich, K., Winiarski, J. P., Paes, V. Z. C., Geshev, J., Jost, C. L., & Maduro de Campos, C. E. (2022). Mechanochemical synthesis of γ-CoTe2 nanocrystals and their application for determination of ferulic acid. *Materials Today. Communications*, *31*, 103481. Advance online publication. doi:10.1016/j.mtcomm.2022.103481

Malamatari, M., Taylor, K. M., Malamataris, S., Douroumis, D., & Kachrimanis, K. (2018, March 1). Pharmaceutical nanocrystals: Production by wet milling and applications. *Drug Discovery Today, 23*(3), 534–547. doi:10.1016/j.drudis.2018.01.016 PMID:29326082

Malvern Instruments. (2015). *A basic guide to particle characterization*. Author.

Manogar, B., Yang, F., & Bolzoni, L. (2022, September 15). Effect of Nb Addition on the Phase Stability, Microstructure, and Mechanical Properties of Powder Metallurgy Ti-5Fe-xNb Alloys. *Metals, 12*(9), 1528. doi:10.3390/met12091528

Mantovani, D. (2000). Shape memory alloys: Properties and biomedical applications. *Journal of the Minerals Metals & Materials Society, 52*(10), 36–44. doi:10.1007/s11837-000-0082-4

Mao, F., Wang, F., Shuang, Y., Hu, J., & Chen, J. (2019). Deformation Behavior and Experiments on a Light Alloy Seamless Tube via a Tandem Skew Rolling Process. *Metals, 10*(1), 59. doi:10.3390/met10010059

Martins, A. R., Cunha, J. F., Patrício, J., & Caravana, J. (2016). Familial superior mesenteric artery syndrome. *BMJ Case Reports, 2016*. Advance online publication. doi:10.1136/bcr-2016-214784 PMID:27126095

Marušáková, D., Corrêa, C. A., Aparicio, C., Libera, O., Berka, J., Vilémová, M., & Gávelová, P. (2022). The Influence of High-Temperature Helium and the Amount of Revert Material on the Material Properties of Inconel 738. *Coatings, 13*(1), 45. doi:10.3390/coatings13010045

Masoumi, M., Tressia, G., Centeno, D. M. A., & Goldenstein, H. (2021). Improving the mechanical properties and wear resistance of a commercial pearlitic rail steel using a two-step heat treatment. *Metallurgical and Materials Transactions. A, Physical Metallurgy and Materials Science, 52*(11), 4888–4906. doi:10.1007/s11661-021-06432-0

McCormick, P. G. (1995). Application of mechanical alloying to chemical refining (overview). *Materials Transactions, JIM, 36*(2), 161-9.

Mehta, K., & Gupta, K. (2019). *Fabrication and processing of shape memory alloys*. Springer International Publishing. doi:10.1007/978-3-319-99307-2

Meisenheimer, P. B., Kratofil, T. J., & Heron, J. T. (2017). Giant enhancement of exchange coupling in entropy-stabilized oxide heterostructures. *Scientific Reports, 7*(1), 13344. doi:10.1038/s41598-017-13810-5 PMID:29042610

Mekki, O. B., & Auricchio, F. (2011). Performance evaluation of shape-memory-alloy superelastic behavior to control a stay cable in cable-stayed bridges. *International Journal of Non-linear Mechanics, 46*(2), 470–477. doi:10.1016/j.ijnonlinmec.2010.12.002

Meng, J., Chen, X., Jiang, J., & Liu, L. (2022). Investigation on Anisotropic Mechanical Behavior of Ti-6Al-4V Alloy via Schmid Factor and Kernel Average Misorientation Distribution. *Metals, 13*(1), 89. doi:10.3390/met13010089

Meng, X., Xu, J., Zhu, J., Zhang, P., Rydosz, A., Reece, M. J., & Gao, F. (2021). Pyrochlore-fluorite dual-phase high-entropy ceramic foams with extremely low thermal conductivity from particle-stabilized suspension. *Scripta Materialia, 194*, 113714. doi:10.1016/j.scriptamat.2020.113714

Miao, Fan, Gui, Hu, Wang, & Gao. (2022). Relationship between Microstructure and Properties of 1380 MPa Grade Bainitic Rail Steel Treated by Online Bainite-Based Quenching and Partitioning Concept. *Metals, 12*(330), 1–15.

Miao, Q., Ding, W., Kuang, W., Zhou, B., Hao, T., Dai, C., & Yin, Z. (2022). Microstructure Evolution and Formation of Gradient Structures in Single Crystal Nickel-Based Superalloy by Surface Mechanical Creep-Feed Grinding Treatment. *Materials (Basel), 16*(1), 321. doi:10.3390/ma16010321 PMID:36614660

Microtrac, M. R. B. (2023). *Errors in Particle Size Analysis and How to Avoid Them.* www.microtrac.com

Miracle, D. B., Miller, J., Senkov, O., Woodward, C., Uchic, M., & Tiley, J. (2014). Exploration and development of high entropy alloys for structural applications. *Entropy (Basel, Switzerland), 16*(1), 494–525. doi:10.3390/e16010494

Mirzaeifar, R., DesRoches, R., & Yavari, A. (2011). A combined analytical, numerical, and experimental study of shape-memory-alloy helical springs. *International Journal of Solids and Structures, 48*(3-4), 611–624. doi:10.1016/j.ijsolstr.2010.10.026

Mogale, N. F., & Matizamhuka, W. R. (2020, August 11). Spark plasma sintering of titanium aluminides: A progress review on processing, structure-property relations, alloy development and challenges. *Metals, 10*(8), 1080. doi:10.3390/met10081080

Mohd Jani, J., Leary, M., Subic, A., & Gibson, M. A. (2014). A review of shape memory alloy research, applications and opportunities. *Materials & Design, 56*, 1078-1113. doi:10.1016/j.matdes.2013.11.084

Moon, A. P., Sangal, S., Layek, S., Giribaskar, S., & Mondal, K. (2015). Corrosion behavior of high-strength bainitic rail steels. *Metallurgical and Materials Transactions. A, Physical Metallurgy and Materials Science, 46*(4), 1500–1518. doi:10.1007/s11661-014-2732-0

Mordor Intelligence. (2023). *Global Particle Size Analysis Market: Growth, Trends, Covid-19 Impact and Forecasts (2023-2028).* https://www.mordorintelligence.com/industry-reports/particle-size-analysis-market

Moskovskikh, D., Vorotilo, S., Buinevich, V., Sedegov, A., Kuskov, K., Khort, A., Shuck, C., Zhukovskyi, M., & Mukasyan, A. (2020). Extremely hard and tough high entropy nitride ceramics. *Scientific Reports, 10*(1), 19874. doi:10.1038/s41598-020-76945-y PMID:33199795

Motors, G. (2013). *Chevrolet debuts lightweight "smart material" on corvette.* Corporate Newsroom. Available at: http://media. gm. com/media/us/en/gm/home. detail. html/content/Pages/news/us/en/2013/Feb/0212-corvette. html

Msomi, V., Adimule, V., Shashanka, R., Kaloop, M., & Koutsonas, S. (2022). Materials and Production Technologies in Machinery. Trans Tech Publications Ltd.

Mukesh, Y. B., Naik, P. K., & Raghavendra Rao, R. (2021). Impact of heat treatment on the mechanical performance of hot extruded Al6061-BN reinforced metal matrix composites. *Bulletin of the Polish Academy of Sciences. Technical Sciences*, *69*(3), 1–6.

Murakami, Y., & Ritchie, R.O. (2012). Effects of hydrogen on fatigue-crack propagation in steels. In *Gaseous Hydrogen Embrittlement of Materials in Energy Technologies* (pp. 379-417). Woodhead Publishing.

Mwale, A. H., Musonge, P., & Fraser, D. M. (2005). The influence of particle size on energy consumption and water recovery in comminution and dewatering systems. *Minerals Engineering*, *18*(9), 915–926. doi:10.1016/j.mineng.2005.02.014

Nacházel, J., Palán, J., Dlouhý, J., Sláma, P., & Nový, Z. (2019). Effects of Rotary Swaging Parameters and Artificial Ageing on Mechanical Properties and Microstructure of 2024 Precipitation-Hardenable Aluminium Alloy. *Materials (Basel)*, *13*(1), 143. doi:10.3390/ma13010143 PMID:31905879

Nahirny, E. P., Bergamini, M. F., & Marcolino-Junior, L. H. (2020). Improvement in the performance of an electrochemical sensor for ethanol determination by chemical treatment of graphite. *Journal of Electroanalytical Chemistry (Lausanne, Switzerland)*, *877*, 114659. Advance online publication. doi:10.1016/j.jelechem.2020.114659

Nayak, A. K., Shashanka, R., & Chaira, D. (2016). Effect of nanosize yittria and tungsten addition to duplex stainless steel during high energy planetary milling. In *IOP Conference Series: Materials Science and Engineering* (Vol. 115, No. 1, p. 012008). IOP Publishing. 10.1088/1757-899X/115/1/012008

Nayak, A. K., Shashanka, R., & Chaira, D. (2016). Effect of nanosizeyittria and tungsten addition to duplex stainless steel during high energy planetary milling. In *IOP Conference Series: Materials Science and Engineering* (Vol. 115, No. 1, p. 012008). IOP Publishing.

Nayak, A. K., Shashanka, R., & Chaira, D. (2016). Effect of NanosizeYittria and Tungsten Addition to Duplex Stainless Steel During High Energy Planetary Milling. *IOP Conf. Series. Materials Science and Engineering*, *115*, 012008.

Nečas, D., Marek, I., Pinc, J., Vojtěch, D., & Kubásek, J. (2022). Advanced Zinc–Magnesium Alloys Prepared by Mechanical Alloying and Spark Plasma Sintering. *Materials (Basel)*, *15*(15), 5272. Advance online publication. doi:10.3390/ma15155272 PMID:35955207

Nisar, A., Bajpai, S., Khan, M. M., & Balani, K. (2020). Wear damage tolerance and high temperature oxidation behavior of HfB2:ZrB2–SiC composites. *Ceramics International*, *46*(13), 21689–21698. doi:10.1016/j.ceramint.2020.05.276

Nisar, A., Khan, M. M., Bajpai, S., & Balani, K. (2019). Processing, microstructure and mechanical properties of HfB 2 -ZrB 2 -SiC composites: Effect of B 4 C and carbon nanotube reinforcements. *International Journal of Refractory & Hard Metals*, *81*, 111–118. doi:10.1016/j.ijrmhm.2019.02.014

Nisar, A., Khan, M. M., & Balani, K. (2019). Enhanced thermo-mechanical damage tolerance of functionally graded ZrB2-20SiC ceramic reinforced with carbon nanotubes. *Ceramics International*, *45*(5), 6198–6208. doi:10.1016/j.ceramint.2018.12.097

Nithin, S. K., Hemanth, K., Shamanth, V., Shrinivas Mahale, R., Sharath, P. C., & Patil, A. (2022). Importance of Condition Monitoring in Mechanical Domain. *Materials Today: Proceedings*, *54*, 234–239. doi:10.1016/j.matpr.2021.08.299

Nykamp, G., Carstensen, U., & Müller, B. W. (2002, August 21). Jet milling—A new technique for microparticle preparation. *International Journal of Pharmaceutics*, *242*(1-2), 79–86. doi:10.1016/S0378-5173(02)00150-3 PMID:12176228

O'Conell, R. (2005). Sieve Use in the Pharmaceutical Industry. *Pharmaceutical Technology Europe, 17*(5), 1. https://www.pharmtech.com/view/sieve-use-pharmaceutical-industry

Ogonowski, S., Wołosiewicz-Głąb, M., Ogonowski, Z., Foszcz, D., & Pawełczyk, M. (2018, March 29). Comparison of wet and dry grinding in electromagnetic mill. *Minerals (Basel)*, *8*(4), 138. doi:10.3390/min8040138

Oğuzyer, A., & Rajendrachari, S. (2023). An Overview of Nanocomposites with Recent Advancements. *J. Metastable Nanocrystalline Mater.*, *37*, 45–56. doi:10.4028/p-VW2fCP

Okada, S., Kudou, K., & Lundström, T. (1995). Preparations and some properties of w2b, δ-wb and wb2crystals from high-temperature metal solutions. *Japanese Journal of Applied Physics*, *34*(1R), 226–231. doi:10.1143/JJAP.34.226

Oko, D. N., Garbarino, S., Zhang, J., Xu, Z., Chaker, M., Ma, D., Guay, D., & Tavares, A. C. (2015). Dopamine and ascorbic acid electrooxidation on Au, AuPt and Pt nanoparticles prepared by pulse laser ablation in water. *Electrochimica Acta*, *159*, 174–183. Advance online publication. doi:10.1016/j.electacta.2015.01.192

Omarov, S., Nauryz, N., Talamona, D., & Perveen, A. (2022). Surface Modification Techniques for Metallic Biomedical Alloys: A Concise Review. *Metals*, *13*(1), 82. doi:10.3390/met13010082

Onal, O., Canadinc, D., Sehitoglu, H., Verzal, K., & Jiang, Y. (2012). Investigation of rolling contact crack initiation in bainitic and pearlitic rail steels. *Fatigue & Fracture of Engineering Materials & Structures*, *35*(11), 985–997. doi:10.1111/j.1460-2695.2012.01686.x

Oro, R., Campos, M., & Torralba, J. M. (2012, July 1). Study of high temperature wetting and infiltration for optimising liquid phase sintering in low alloy steels. *Powder Metallurgy*, *55*(3), 180–190. doi:10.1179/1743290111Y.0000000007

Oses, C., Toher, C., & Curtarolo, S. (2020). High-entropy ceramics. *Nature Reviews. Materials*, *5*(4), 295–309. doi:10.1038/s41578-019-0170-8

Pacyna, J. (2008). The microstructure and properties of the new bainitic rail steels. *Journal of Achievements in Materials and Manufacturing Engineering*, *28*(1), 19–22.

Panek, J., Kubisztal, J., & Bierska-Piech, B. (2014). Ni50Mo40Ti10 alloy prepared by mechanical alloying as electroactive material for hydrogen evolution reaction. *Surface and Interface Analysis*, *46*(10–11), 716–720. Advance online publication. doi:10.1002/sia.5501

Pankewitz, A., & Zeng, S. (2012). Implementation of Online Particle Size and Concentration Analysis for Grinding Optimization. *XXVI International Mineral Processing Congress (IMPC)*, 4127–4136.

Park, B. Y., Kim, Y., & Lee, K. (2020). Optimization Studies of AC4CH Material in the Cylinder Block of a Diesel Engine Application. *Processes (Basel, Switzerland)*, *9*(1), 70. doi:10.3390/pr9010070

Park, C., Hwang, T., Kim, G.-D., Nam, H., & Kang, N. (2022). Effect of the Initial Grain Size on Laser Beam Weldability for High-Entropy Alloys. *Crystals*, *13*(1), 65. doi:10.3390/cryst13010065

Pavitra, V., Praveen, B. M., Nagaraju, G., & Shashanka, R. (2022, January 8). Energy storage, Photocatalytic and Electrochemical nitrite sensing of ultrasound-assisted stable Ta2O5 nanoparticles. *Topics in Catalysis*, 1–4. doi:10.1007/s11244-021-01553-7

Pereloma, E., & Edmonds, D. V. (Eds.). (2012). *Phase transformations in steels: Diffusionless transformations, high strength steels, modelling and advanced analytical techniques*. Elsevier.

Peters, B. M., Kumpfert, J., Ward, C. H., & Leyens, C. (2003). Titanium Alloys for Aerospace Applications. *Advanced Engineering Materials*, *5*(6), 419–427. doi:10.1002/adem.200310095

Peterson, G. I., Dobrynin, A. V., & Becker, M. L. (2017). Biodegradable Shape Memory Polymers in Medicine. *Advanced healthcare materials*, *6*(21). doi:10.1002/adhm.201700694

Phutela, C., Aboulkhair, N. T., Tuck, C. J., & Ashcroft, I. (2019). The Effects of Feature Sizes in Selectively Laser Melted Ti-6Al-4V Parts on the Validity of Optimised Process Parameters. *Materials (Basel)*, *13*(1), 117. doi:10.3390/ma13010117 PMID:31887981

Pio, E., Medina, A., & Mart, C. (2021). Evolution of Face-Centered Cubic Ti Alloys Transformation by X-ray Diffraction Profile Analysis in Mechanical Alloying. *Metals*, *11*(11), 1–17. doi:10.3390/met11111841

Plata-Torres, M., Torres-Huerta, A. M., Domínguez-Crespo, M. A., Arce-Estrada, E. M., & Ramírez-Rodríguez, C. (2007). Electrochemical performance of crystalline Ni-Co-Mo-Fe electrodes obtained by mechanical alloying on the oxygen evolution reaction. *International Journal of Hydrogen Energy*, *32*(17), 4142–4152. Advance online publication. doi:10.1016/j.ijhydene.2007.05.023

Pointner, P. (2008). High strength rail steels-The importance of material properties in contact mechanics problems. *Wear*, *265*(9-10), 1373–1379. doi:10.1016/j.wear.2008.03.015

Pomeroy, M. (Ed.). (2021 May 24). *Encyclopedia of materials: technical ceramics and glasses*. Elsevier.

Post, B., Glaserf, F. W. & Moskowitz, D. (n.d.). Transition Metal Diborides. *Acta Metall.*

Prabhu, B., Suryanarayana, C., An, L., & Vaidyanathan, R. (2006). Synthesis and Characterization of High Volume Fraction Al–Al2O3 Nanocomposite Powders by High-Energy Milling. *Materials Science and Engineering A*, *425*(1–2), 192–200. doi:10.1016/j.msea.2006.03.066

Pradeep, N. B., Venkatesha, B. K., Parameshwara, S., Raghavendra Rao, R., & Raviprakash, M. (2022). Synthesis of nanostructured ternary Ti based alloy for bio-medical applications. *Materials Today: Proceedings*, *54*(2), 372-377. doi:10.1016/j.matpr.2021.09.429

Pradeep, N. B., Hegde, M. R., Rajendrachari, S., & Surendranathan, A. O. (2022, August 1). Investigation of microstructure and mechanical properties of microwave consolidated TiMgSr alloy prepared by high energy ball milling. *Powder Technology*, *408*, 117715. doi:10.1016/j.powtec.2022.117715

Protonotariou, S., Ritzoulis, C., & Mandala, I. (2021, April 1). Jet milling conditions impact on wheat flour particle size. *Journal of Food Engineering*, *294*, 110418. doi:10.1016/j.jfoodeng.2020.110418

Purushothama, H. T., & Arthoba Nayaka, Y. (2017). Electrochemical study of hydrochlorothiazide on electrochemically pre-treated pencil graphite electrode as a sensor. *Sensing and Bio-Sensing Research*, *16*, 12–18. Advance online publication. doi:10.1016/j.sbsr.2017.09.004

Qiao, L., Liu, Y., Gao, Y., Bi, J., Li, Y., Liu, C., Gao, J., Wang, W., & Qian, Z. (2022). First-principles prediction, fabrication and characterization of (Hf0.2Nb0.2Ta0.2Ti0.2Zr0.2)B2 high-entropy borides. *Ceramics International*, *48*(12), 17234–17245. doi:10.1016/j.ceramint.2022.02.281

Qin, M., Shivakumar, S., Lei, T., Gild, J., Hessong, E. C., Wang, H., Vecchio, K. S., Rupert, T. J., & Luo, J. (2022). Processing-dependent stabilization of a dissimilar rare-earth boride in high-entropy (Ti0.2Zr0.2Hf0.2Ta0.2Er0.2)B2 with enhanced hardness and grain boundary segregation. *Journal of the European Ceramic Society*, *42*(12), 5164–5171. doi:10.1016/j.jeurceramsoc.2022.05.034

Qin, M., Yan, Q., Liu, Y., Wang, H., Wang, C., Lei, T., Vecchio, K. S., Xin, H. L., Rupert, T. J., & Luo, J. (2021). Bulk high-entropy hexaborides. *Journal of the European Ceramic Society*, *41*(12), 5775–5781. doi:10.1016/j.jeurceramsoc.2021.05.027

Qi, P., Wang, J., Wang, X., Wang, X., Wang, Z., Xu, H., Di, S., Wang, Q., & Wang, X. (2018). Sensitive determination of fenitrothion in water samples based on an electrochemical sensor layered reduced graphene oxide, molybdenum sulfide (MoS2)-Au and zirconia films. *Electrochimica Acta*, *292*, 667–675. Advance online publication. doi:10.1016/j.electacta.2018.09.187

Qiu, J., Zhang, M., Tan, Z., Gao, G., & Bai, B. (2019). Research on the microstructures and mechanical properties of bainite/martensite rail treated by the controlled-cooling process. *Materials (Basel)*, *12*(3061), 1–9. doi:10.3390/ma12193061 PMID:31547092

Qiu, S., Xiang, H., Dai, F.-Z., Wang, H., Huang, M., Wan, C., Meng, Q., Li, J., Wang, X., & Zhou, Y. (2022). Medium-entropy (Me,Ti)0.1(Zr,Hf,Ce)0.9O2 (Me = Y and Ta): Promising thermal barrier materials for high-temperature thermal radiation shielding and CMAS blocking. *Journal of Materials Science and Technology*, *123*, 144–153. doi:10.1016/j.jmst.2022.01.019

Raghavan, J., Bartkiewicz, T., Boyko, S., Kupriyanov, M., Rajapakse, N., & Yu, B. (2010). Damping, tensile, and impact properties of superelastic shape memory alloy (SMA) fiber-reinforced polymer composites. *Composites. Part B, Engineering*, *41*(3), 214–222. doi:10.1016/j.compositesb.2009.10.009

Rahaman, M.N. (2017). *Ceramic processing and sintering*. CRC Press.

Rahman, M., Senthil Kumar, A., & Not Available, N. A. (2001, August). Evaluation of minimal of lubricant in end milling. *International Journal of Advanced Manufacturing Technology*, *18*(4), 235–241. doi:10.1007/s001700170063

Rajendrachari, S. (2022b). New Advances in Powder Technology. *New Adv. Powder Technol.* . doi:10.5772/intechopen.102106

Rajendrachari, S. (2023). New Advances in Powder Technology. Academic Press.

Rajendrachari, S., Basavegowda, N., Adimule, V. M., Avar, B., Somu, P., Saravana, S. K., & Baek, K. H. (2022). Assessing the Food Quality Using Carbon Nanomaterial Based Electrodes by Voltammetric Techniques. In Biosensors (Vol. 12, Issue 12). doi:10.3390/bios12121173

Rajendrachari, S., Somashekharappa, K. K., & Mahale, R. S. (2022). A Review on Cyclic Voltammetric Investigation of Toxic Heavy Metals. Frontiers in Voltammetry, 1–10.

Rajendrachari, S. (2018). Effect of sintering temperature on the pitting corrosion of ball milled duplex stainless steel by using linear sweep voltammetry. *Anal. Bioanal. Electrochem.*, *10*, 349–361.

Rajendrachari, S. (2018). Effect of sintering temperature on the pitting corrosion of ball milled duplex stainless steel by using linear sweep voltammetry. *Analytical and Bioanalytical Electrochemistry*, *10*(3), 349–361.

Rajendrachari, S. (2018, January 1). Effect of sintering temperature on the pitting corrosion of ball milled duplex stainless steel by using linear sweep voltammetry. *Analytical and Bioanalytical Electrochemistry.*, *10*(3), 349–361.

Rajendrachari, S. (2022, June 21). An overview of high-entropy alloys prepared by mechanical alloying followed by the characterization of their microstructure and various properties. *Alloys.*, *1*(2), 116–132. doi:10.3390/alloys1020008

Rajendrachari, S., Adimule, V. M., Jayaprakash, G. K., & Pandith, A. (2023, May 17). Electrochemical oxidation of methylene blue dye in wastewater using mechanically alloyed high entropy alloy modified carbon paste electrode using cyclic voltammetry. *Materials Research Express*, *10*(5), 054003. doi:10.1088/2053-1591/acd3d7

Rajendrachari, S., Adimule, V., Gulen, M., Khosravi, F., & Somashekharappa, K. K. (2022). Synthesis and Characterization of High Entropy Alloy 23Fe-21Cr-18Ni-20Ti-18Mn for Electrochemical Sensor Applications. *Materials (Basel)*, *15*(21), 7591. Advance online publication. doi:10.3390/ma15217591 PMID:36363181

Rajendrachari, S., Basavegowda, N., Vinaykumar, R., Narsimhachary, D., Somu, P., & Lee, M. J. (2023, June 30). Electrocatalytic determination of methyl orange dye using mechanically alloyed novel metallic glass modified carbon paste electrode by cyclic voltammetry. *Inorganic Chemistry Communications*, *155*, 111010. doi:10.1016/j.inoche.2023.111010

Rajendrachari, S., Kamacı, Y., Taş, R., Ceylan, Y., Bülbül, A. S., Uzun, O., & Karaoğlanlı, A. C. (2019, December 1). Antimicrobial investigation of CuO and ZnO nanoparticles prepared by a rapid combustion method. *Physical Chemistry Research.*, *7*(4), 799–812.

Rajendrachari, S., Swamy, B. K., Reddy, S., & Chaira, D. (2013). Synthesis of silver nanoparticles and their applications. *Analytical and Bioanalytical Electrochemistry*, *5*(4), 455–466.

Rajendran, S., & Murty, C. V. G. K. (2023). *Mineral Processing: Beneficiation Operations and Process Optimization through Modeling*. Academic Press.

Ramakrishna, D., Saravu, S., & Rajendrachari, S. (2022). A Palladium Complex Dispersed in Ionic Liquid as an Efficient Catalytic Combination for the Synthesis of Benzazoles. *Topics in Catalysis*, *65*, 1–12. doi:10.1007/s11244-021-01554-6

Ramakrishnan, P. (2013). Automotive applications of powder metallurgy. In *Advances in powder metallurgy* (pp. 493-519). Woodhead Publishing.

Rausch, K.D., Hummel, D., Johnson, L.A., & May, J.B. (2019). Wet milling: the basis for corn biorefineries. In *Corn* (pp. 501-535). AACC International Press.

Ravikanth Reddy, C., Satya Prasad, K., & Srinivasarao, B. (2023). Microstructure and mechanical properties of Mg–Ni–Gd alloy synthesised by powder metallurgy. *Powder Metallurgy*, *66*(3), 270–277. Advance online publication. doi:10.1080/00325899.2022.2163553

Ray, A. (2017). Niobium microalloying in rail steels. *Materials Science and Technology*, *33*(14), 1584–1600. doi:10.1080/02670836.2017.1309111

Rendón, J., & Olsson, M. (2009). Abrasive Wear Resistance of Some Commercial Abrasion Resistant Steels Evaluated by Laboratory Test Methods. *Wear*, *267*(11), 2055–2061. doi:10.1016/j.wear.2009.08.005

Ren, X., & Otsuka, K. (1997). Origin of rubber-like behaviour in metal alloys. *Nature*, *389*(6651), 579–582. doi:10.1038/39277

Rezende, A. B., Fonseca, S. T., Fernandes, F. M., Miranda, R. S., Grijalba, F. A. F., Farina, P. F. S., & Mei, P. R. (2020). Wear behavior of bainitic and pearlitic microstructures from microalloyed railway wheel steel. *Wear*, *456*, 203377. doi:10.1016/j.wear.2020.203377

Rezende, A. B., Fonseca, S. T., Miranda, R. S., Fernandes, F. M., Grijalba, F. A. F., Farina, P. F. S., & Mei, P. R. (2021). Effect of niobium and molybdenum addition on the wear resistance and the rolling contact fatigue of railway wheels. *Wear, 466*, 203571. doi:10.1016/j.wear.2020.203571

Richardson, H., Bopp, C., Ha, B., Thomas, R., & Santhanam, K. S. V. (2020). Electrodeposition from a Graphene Bath: A Sustainable Copper Composite Alloy in a Graphene Matrix. *Journal of Composites Science, 5*(1), 9. doi:10.3390/jcs5010009

Rivas-Cantu, R. C., Jones, K. D., & Mills, P. L. (2013, April). A citrus waste-based biorefinery as a source of renewable energy: Technical advances and analysis of engineering challenges. *Waste Management & Research, 31*(4), 413–420. doi:10.1177/0734242X13479432 PMID:23439875

Robert, W., & Haasen, P. (n.d.). *Physical Metallurgy* (4th ed.). Academic Press.

Roberts, T. A., & Beddow, J. K. (1968). Some effects of particle shape and size upon blinding during sieving. *Powder Technology, 2*(2), 121–124. doi:10.1016/0032-5910(68)80046-4

Rodriguez, J. M. (2013). Importance of the Particle Shape on Mechanical Properties of Soil Materials. Luleå University of Technology.

Rodriguez, L., Paris, J. Y., Denape, J., & Delbé, K. (2023, February 6). Micro-Arcs Oxidation Layer Formation on Aluminium and Coatings Tribological Properties—A Review. *Coatings, 13*(2), 373. doi:10.3390/coatings13020373

Rosentrater, K.A., & Bucklin, R. (2022). Structural, physical, and engineering properties of cereal grains and grain products. In *Storage of Cereal Grains and Their Products* (pp. 135-178). Woodhead Publishing.

Russell Finex. (2023). *Ultrasonic Sieves With Deblinding Screens*. https://www.russellfinex.com/en/separation-equipment/ultrasonic-sieves/ultrasonic-mesh-screens/

Ryu, H. J., Hong, S. H., & Baek, W. H. (2000, October 31). Microstructure and mechanical properties of mechanically alloyed and solid-state sintered tungsten heavy alloys. *Materials Science and Engineering A, 291*(1-2), 91–96. doi:10.1016/S0921-5093(00)00968-0

S, V., U, M. I., & Ghani, J. A. (2023, February 17). A Study on the Effects of Hybridized Metal Oxide and Carbonaceous Nano-Cutting Fluids in the End Milling of AA6082 Aluminum Alloy. *Lubricants, 11*(2), 87. https://doi.org/ doi:10.3390/lubricants11020087

Saengvong, P., Boonlakhorn, J., Jumpatam, J., Chanlek, N., Prasongkit, J., Putasaeng, B., Moontragoon, P., Srepusharawoot, P., & Thongbai, P. (2023). Effects of sintering condition on giant dielectric and nonlinear current-voltage properties of Na1/2Y1/2Cu3Ti3.975Ta0.025O12 ceramics. *Heliyon, 9*(1), e12946. doi:10.1016/j.heliyon.2023.e12946 PMID:36704279

Saha, S., & Islam, A. (2018). Shape memory material-concepts, recent trends and future directions. In *Student Conference on Electronics, Informatics and Vision 2018* (pp. 1-6). Academic Press.

Saha, N., Goates, C., Hernandez, S., Jin, W., Westover, T., & Klinger, J. (2022). Characterization of particle size and moisture content effects on mechanical and feeding behavior of milled corn (Zea mays L.) stover. *Powder Technology*, *405*, 117535. doi:10.1016/j.powtec.2022.117535

Saha, S., & Roy, S. (2022). Metallic Dental Implants Wear Mechanisms, Materials, and Manufacturing Processes: A Literature Review. *Materials (Basel)*, *16*(1), 161. doi:10.3390/ma16010161 PMID:36614500

Saheb, N., Khan, M. S., & Hakeem, A. S. (2016, January 1). Effect of processing on mechanically alloyed and spark plasma sintered Al-A12O3 nanocomposites. *Journal of Nanomaterials*, *16*(1), 377.

Sairam, K., Sonber, J. K., Murthy, T. S. R. C., Subramanian, C., Fotedar, R. K., & Hubli, R. C. (2014). Reaction spark plasma sintering of niobium diboride. *International Journal of Refractory & Hard Metals*, *43*, 259–262. doi:10.1016/j.ijrmhm.2013.12.011

Sakthivel, R., Kubendhiran, S., Chen, S. M., & Kumar, J. V. (2019). Rational design and facile synthesis of binary metal sulfides VS2-SnS2 hybrid with functionalized multiwalled carbon nanotube for the selective detection of neurotransmitter dopamine. *Analytica Chimica Acta*, *1071*, 98–108. Advance online publication. doi:10.1016/j.aca.2019.04.058 PMID:31128761

Salleh, E. M., Ramakrishnan, S., & Hussain, Z. (2016). Synthesis of Biodegradable Mg-Zn Alloy by Mechanical Alloying: Effect of Milling Time. *Procedia Chemistry*, *19*, 525–530. doi:10.1016/j.proche.2016.03.048

Samek, L., de Moor, E., Penning, J., & De Cooman, B. C. (2006). Influence of Alloying Elements on the Kinetics of Strain-Induced Martensitic Nucleation in Low-Alloy, Multiphase High-Strength Steels. *Metallurgical and Materials Transactions. A, Physical Metallurgy and Materials Science*, *37A*(1), 109–124. doi:10.1007/s11661-006-0157-0

Sanchez-Suarez, N., Orozco-Mendoza, G. L., Zartha-Sossa, J. W., Gafaro-Garcés, D. C., Melchor-Cahuana, L. G., & Gonzalez-Tovar, C. (2022). Trends in Sieving and Its Applications in Cereals. A Literature Review. In *Frontiers in Sustainable Food Systems* (Vol. 6). Frontiers Media S.A. doi:10.3389/fsufs.2022.902147

Santhy, A., Beena, S., Veena, & Shahina, S. (2020). Electrochemical quantification of vitamin B9 on poly tyrosine modified pencil graphite electrode. *IOP Conference Series. Materials Science and Engineering*, *872*(1), 012128. Advance online publication. doi:10.1088/1757-899X/872/1/012128

Santhy, A., Saraswathyamma, B., Parvathy Krishnan, A., & Luscious, L. (2020). Nanomaterials incorporated electrochemical sensors for the monitoring of pyridoxine: A mini review. *Materials Today: Proceedings*, *46*, 2998–3004. Advance online publication. doi:10.1016/j.matpr.2020.12.703

Santra, D.K., & Schoenlechner, R. (2017). Amaranth part 2—sustainability, processing, and applications of amaranth. In *Sustainable protein sources* (pp. 257-264). Academic Press.

Sarkar, A., Djenadic, R., Usharani, N. J., Sanghvi, K. P., Chakravadhanula, V. S. K., Gandhi, A. S., Hahn, H., & Bhattacharya, S. S. (2017). Nanocrystalline multicomponent entropy stabilised transition metal oxides. *Journal of the European Ceramic Society*, *37*(2), 747–754. doi:10.1016/j.jeurceramsoc.2016.09.018

Sarkar, N., Ghosh, S. K., Bannerjee, S., & Aikat, K. (2012, January 1). Bioethanol production from agricultural wastes: An overview. *Renewable Energy*, *37*(1), 19–27. doi:10.1016/j.renene.2011.06.045

Sarker, P., Harrington, T., Toher, C., Oses, C., Samiee, M., Maria, J.-P., Brenner, D. W., Vecchio, K. S., & Curtarolo, S. (2018). High-entropy high-hardness metal carbides discovered by entropy descriptors. *Nature Communications*, *9*(1), 4980. doi:10.1038/s41467-018-07160-7 PMID:30478375

Sarraf, M., Rezvani, E., & Saeid, G. (2022). A state - of - the - art review of the fabrication and characteristics of titanium and its alloys for biomedical applications. *Bio-Design and Manufacturing*, *5*(2), 371–395. doi:10.1007/s42242-021-00170-3 PMID:34721937

Satoh, Y., & Iwafuchi, K. (2005). Effect of Rail/Wheel Rolling Contact on Microstructure in Surface Layer of Rail. *Quarterly Report of RTRI*, *46*(3), 178–183. doi:10.2219/rtriqr.46.178

Satoh, Y., Tatsumi, M., Kasiwaya, K., Ueda, M., & Yokoyama, H. (1999). Development of anti-darkspot bainitic steel rail. *Quarterly Report of RTRI*, *40*(2), 86–91. doi:10.2219/rtriqr.40.86

Satyam, S., Goutam, M., & Totten G. (2009). Overview of Pearlitic Rail Steel: Accelerated Cooling, Quenching, Microstructure, and Mechanical Properties. *Journal of ASTM International*, *6*(7), 1–26.

Sawley, K., & Kristan, J. (2003). Development of bainitic rail steels with potential resistance to rolling contact fatigue. *Fatigue & Fracture of Engineering Materials & Structures*, *26*(10), 1019–1029. doi:10.1046/j.1460-2695.2003.00671.x

Seth, P. P., Parkash, O., & Kumar, D. (2020). Structure and mechanical behavior of in situdeveloped Mg2Si phase in magnesium and aluminum alloys - a review. *RSC Advances*, *10*(61), 37327–37345. doi:10.1039/D0RA02744H PMID:35521263

Shang, C., Axinte, E., Sun, J., Li, X., Li, P., Du, J., Qiao, P., & Wang, Y. (2017). CoCrFeNi(W1 – xMox) high-entropy alloy coatings with excellent mechanical properties and corrosion resistance prepared by mechanical alloying and hot pressing sintering. *Materials & Design*, *117*, 193–202. Advance online publication. doi:10.1016/j.matdes.2016.12.076

Shanghai Greenearth Chemicals Co. (n.d.). *Excellent properties of Vanadium diboride*. https://www.greenearthchem.com/excellent-properties-of-vanadium-diboride.html

Sharabash, A. M., & Andrawes, B. O. (2009). Application of shape memory alloy dampers in the seismic control of cable-stayed bridges. *Engineering Structures*, *31*(2), 607–616. doi:10.1016/j.engstruct.2008.11.007

Sharath, B. N., Venkatesh, C. V., Afzal, A., Aslfattahi, N., Aabid, A., Baig, M., & Saleh, B. (2021, May 28). Multi ceramic particles inclusion in the aluminium matrix and wear characterization through experimental and response surface-artificial neural networks. *Materials (Basel), 14*(11), 2895. doi:10.3390/ma14112895 PMID:34071305

Sharath, B. N., Venkatesh, C. V., Afzal, A., Baig, M. A., & Kumar, A. P. (2021, January 1). Study on effect of ceramics on dry sliding wear behaviour of Al-Cu-Mg based metal matrix composite at different temperature. *Materials Today: Proceedings, 46*, 8723–8733. doi:10.1016/j.matpr.2021.04.034

Sharma, N., Alam, S. N. & Ray, B. C. (2019). Fundamentals of Spark Plasma Sintering (SPS): An Ideal Processing Technique for Fabrication of Metal Matrix Nanocomposites. *Spark Plasma Sinter. Mater.*, 21–59. . doi:10.1007/978-3-030-05327-7_2

Sharma, A. S., Yadav, S., Biswas, K., & Basu, B. (2018). High-entropy alloys and metallic nanocomposites: Processing challenges, microstructure development and property enhancement. *Materials Science and Engineering R Reports, 131*, 1–42. doi:10.1016/j.mser.2018.04.003

Sharma, S. K., Saxena, K. K., Malik, V., Mohammed, K. A., Prakash, C., Buddhi, D., & Dixit, S. (2022). Significance of Alloying Elements on the Mechanical Characteristics of Mg-Based Materials for Biomedical Applications. *Crystals, 12*(8), 1138. Advance online publication. doi:10.3390/cryst12081138

Sharma, Y., Musico, B. L., Gao, X., Hua, C., May, A. F., Herklotz, A., Rastogi, A., Mandrus, D., Yan, J., Lee, H. N., Chisholm, M. F., Keppens, V., & Ward, T. Z. (2018). Single-crystal high entropy perovskite oxide epitaxial films. *Physical Review Materials, 2*(6), 060404. doi:10.1103/PhysRevMaterials.2.060404

Shashanka, R. (2016). *Fabrication of nano-structured duplex and ferritic stainless steel by planetary milling followed by consolidation* [Doctoral dissertation]. NIT, Rourkela, India.

Shashanka, R. (2022). Carbon composite voltammetric sensors for food quality assessment. Electrochemical Sensors Based on Carbon Composite Materials: Fabrication, properties and applications, 1–13. doi:10.1088/978-0-7503-5127-0ch8

Shashanka, R., & Chaira, D. (2017). *Ball Milled Nano-Structured Stainless Steel Powders: Fabrication and Characterization.* Educreation Publishing. https://books.google.co.za/books?id=ECvvzgEACAAJ

Shashanka, R., & Chaira, D. (n.d.-a). Optimization of Milling Parameters. *Ball Milled Nano-Structured Stainless Steel Powders: Fabrication and Characterization, 32.*

Shashanka, R., & Chaira, D. (n.d.-b). Fabrication of NANO-Y2O3 Dispersed Stainless Steel By Conventional And Spark Plasma Sintering. *Ball Milled Nano-Structured Stainless Steel Powders: Fabrication and Characterization, 54.*

Shashanka, R., & Uzun, O. (2021). Powder Metallurgy. In Fundamentals of Materials Engineering (pp. 113–140). doi:10.2174/9789811489228121010008

Shashanka, R. (2017). Synthesis of nano-structured stainless steel powder by mechanical alloying and characterization. *International Journal of Scientific and Engineering Research*, *8*(June), 588–594.

Shashanka, R. (2017). Synthesis of nanostructured stainless steel powder by mechanical alloying-an overview. *International Journal of Scientific and Engineering Research*, *8*(6).

Shashanka, R. (2017). Synthesis of nano-structured stainless steel powder by mechanical alloying-an overview. *International Journal of Scientific and Engineering Research*, *8*, 588–594.

Shashanka, R. (2019). Non-lubricated dry sliding wear behavior of spark plasma sintered nano-structured stainless steel. *Journal of Materials & Environmental Sciences*, *10*(8), 767–777.

Shashanka, R., & Chaira, D. (2015, July 1). Optimization of milling parameters for the synthesis of nano-structured duplex and ferritic stainless steel powders by high energy planetary milling. *Powder Technology*, *278*, 35–45. doi:10.1016/j.powtec.2015.03.007

Shashanka, R., & Chaira, D. (2016, January). Effects of nano-Y 2 O 3 and sintering parameters on the fabrication of PM duplex and ferritic stainless steels. *Acta Metallurgica Sinica*, *29*(1), 58–71. doi:10.1007/s40195-015-0362-1

Shashanka, R., & Chaira, D. (2017a, March 4). Effect of sintering temperature and atmosphere on nonlubricated sliding wear of nano-yttria-dispersed and yttria-free duplex and ferritic stainless steel fabricated by powder metallurgy. *Tribology Transactions*, *60*(2), 324–336. doi:10.1080/10402004.2016.1168897

Shashanka, R., & Chaira, D. (2017b). *Ball Milled Nano-Structured Stainless Steel Powders: Fabrication and Characterization*. Educreation Publishing.

Shashanka, R., Chaira, D., & Chakravarty, D. (2016). Fabrication of nano-yttria dispersed duplex and ferritic stainless steels by planetary milling followed by spark plasma sintering and non-lubricated sliding wear behaviour study. *J Mater Sci Eng B.*, *6*, 111–125.

Shashanka, R., Chaira, D., & Chakravarty, D. (2016, June 28). Fabrication of Nano-Yttria Dispersed Duplex and Ferritic Stainless Steels by Planetary Milling Followed by Spark Plasma Sintering and Non-Lubricated Sliding Wear Behaviour Study. *Journal of Materials Science and Engineering B*, *6*(3). Advance online publication. doi:10.17265/2161-6221/2016.5-6.001

Shashanka, R., Chaira, D., & Kumara Swamy, B. E. (2015a). Electrocatalytic response of duplex and yittria dispersed duplex stainless steel modified carbon paste electrode in detecting folic acid using cyclic voltammetry. *International Journal of Electrochemical Science*, *10*(7), 5586–5598. doi:10.1016/S1452-3981(23)17279-8

Shashanka, R., Chaira, D., & Kumara Swamy, B. E. (2015b). Electrochemical investigation of duplex stainless steel at carbon paste electrode and its application to the detection of dopamine, ascorbic and uric acid. *International Journal of Scientific and Engineering Research*, *6*(9).

Shashanka, R., Chaira, D., & Kumara Swamy, B. E. (2018). Effect of Y2O3 nanoparticles on corrosion study of spark plasma sintered duplex and ferritic stainless steel samples by linear sweep voltammetric method. *Archives of Metallurgy and Materials*, *63*(2). Advance online publication. doi:10.24425/122401

Shashanka, R., Chaira, D., & Kumara Swamy, B. K. (2016). Fabrication of yttria dispersed duplex stainless steel electrode to determine dopamine, ascorbic and uric acid electrochemically by using cyclic voltammetry. *International Journal of Scientific and Engineering Research*, *7*(3), 1275–1285.

Shashanka, R., Esgin, H., Yilmaz, V. M., & Caglar, Y. (2020). Fabrication and Characterization of Green Synthesized ZnO Nanoparticle Based Dye-Sensitized Solar Cells. *Journal of Science. Advanced Materials and Devices*, *5*(2), 185–191. doi:10.1016/j.jsamd.2020.04.005

Shashanka, R., & Kumara Swamy, B. E. (2020a). Biosynthesis of silver nanoparticles using leaves of Acacia melanoxylon and their application as dopamine and hydrogen peroxide sensors. *Physical Chemistry Research*, *8*(1). Advance online publication. doi:10.22036/pcr.2019.205211.1688

Shashanka, R., & Kumara Swamy, B. E. (2020b). Simultaneous electro-generation and electro-deposition of copper oxide nanoparticles on glassy carbon electrode and its sensor application. *SN Applied Sciences*, *2*(5), 956. Advance online publication. doi:10.1007/s42452-020-2785-1

Shashanka, R., Uzun, O., & Chaira, D. (2020). Synthesis of nano-structured duplex and ferritic stainless steel powders by dry milling and its comparison with wet milling. *Archives of Metallurgy and Materials*, *65*(1).

Shashanka, R., Uzun, O., & Chaira, D. (2020). Synthesis of nanostructured duplex and ferritic stainless steel powders by dry milling and its comparison with wet milling. *Archives of Metallurgy and Materials*, *65*(1). Advance online publication. doi:10.24425/amm.2019.131091

Shi, Y., Li, R., & Lei, Z. (2022). Influences of Synthetic Parameters on Morphology and Growth of High Entropy Oxide Nanotube Arrays. *Coatings*, *13*(1), 46. doi:10.3390/coatings13010046

Shrinivas Mahale, R., Shamanth, V., Hemanth, K., Nithin, S., Sharath, P., Shashanka, R., Patil, A., & Shetty, D. (2022). Processes and applications of metal additive manufacturing. *Materials Today: Proceedings*, *54*, 228–233. doi:10.1016/j.matpr.2021.08.298

Shur, E. A., Bychkova, N. Y., & Trushevsky, S. M. (2005). Physical metallurgy aspects of rolling contact fatigue of rail steels. *Wear*, *258*(7-8), 1165–1171. doi:10.1016/j.wear.2004.03.027

Sidambe, A. T. (2014). Biocompatibility of Advanced Manufactured Titanium Implants—A Review. *Materials (Basel)*, *7*(12), 8168–8188. doi:10.3390/ma7128168 PMID:28788296

Simoncini, M., Costa, A., Fichera, S., & Forcellese, A. (2020). Experimental Analysis and Optimization to Maximize Ultimate Tensile Strength and Ultimate Elongation of Friction Stir Welded AA6082 Aluminum Alloy. *Metals*, *11*(1), 69. doi:10.3390/met11010069

Singh, M. K., & Singh, A. (2022). Particle size analysis. *Characterization of Polymers and Fibres*, 341–358. doi:10.1016/B978-0-12-823986-5.00009-9

Singh, U. P., Roy, B., Jha, S., & Bhattacharyya, S. K. (2001). Microstructure and mechanical properties of as rolled high strength bainitic rail steels. *Materials Science and Technology*, *17*(1), 33–38. doi:10.1179/026708301101509098

Siroky, G., Kraker, E., Rosc, J., Kieslinger, D., Brunner, R., van der Zwaag, S., Kozeschnik, E., & Ecker, W. (2020). Analysis of Sn-Bi Solders: X-Ray Micro Computed Tomography Imaging and Microstructure Characterization in Relation to Properties and Liquid Phase Healing Potential. *Materials (Basel)*, *14*(1), 153. doi:10.3390/ma14010153 PMID:33396361

Smith, A., Stender, M., Pankewitz, A., & Behrens, C. (2014). The quality of ultrasonic extinction based on-line particle size distribution analysers for mineral slurries. *XXVII International Mineral Processing Congress*, 1–22.

Smith, D. S., & Naït-Ali, B. (2021). Thermal properties of ceramic materials. *Encycl. Mater. Tech. Ceram. Glas.*, *1–3*, 855–866. doi:10.1016/B978-0-12-818542-1.00010-2

Snopiński, P. (2023, April 4). Electron Microscopy Study of Structural Defects Formed in Additively Manufactured AlSi10Mg Alloy Processed by Equal Channel Angular Pressing. *Symmetry*, *15*(4), 860. doi:10.3390/sym15040860

Soares, E., Bouchonneau, N., Alves, E., Alves, K., Araújo Filho, O., Mesguich, D., Chevallier, G., Laurent, C., & Estournès, C. (2021, January 16). Microstructure and mechanical properties of AA7075 aluminum alloy fabricated by spark plasma sintering (SPS). *Materials (Basel)*, *14*(2), 430. doi:10.3390/ma14020430 PMID:33467156

Šolcová, P., Nižňanský, M., Schulz, J., Brázda, P., Ecorchard, P., Vilémová, M., & Tyrpekl, V. (2021). Preparation of High-Entropy (Ti, Zr, Hf, Ta, Nb) Carbide Powder via Solution Chemistry. *Inorganic Chemistry*, *60*(11), 7617–7621. doi:10.1021/acs.inorgchem.1c00776 PMID:34028253

Somashekharappa, K. K., & Halappa, R. B. Sr. (2023). Green Chemistry Applications in Electrochemical Sensors. In *Recent Developments in Green Electrochemical Sensors* (pp. 23–37). Design, Performance, and Applications.

Song, Y., Kim, D., Nam, S., Lee, K. A., & Choi, H. (2021). Effect of milling duration on oxide-formation behavior of oxide-dispersion-strengthened high-entropy alloys. *Archives of Metallurgy and Materials*, *66*, 735–740.

Soni, P. R. (2000). *Mechanical alloying: fundamentals and applications*. Cambridge Int Science Publishing.

Srinivas, A., Pavan, D., Venkatesha, B. K., Rao, R. R., & Mohith, L. (2022). Study on mechanical properties of AZ91 magnesium alloy. *Materials Today: Proceedings*, *54*, 291–294. doi:10.1016/j.matpr.2021.09.171

Standish, N. (1985). The kinetics of batch sieving. *Powder Technology*, *41*(1), 57–67. doi:10.1016/0032-5910(85)85075-0

Stanfield, A. D., Manara, D., Robba, D., Hilmas, G. E., & Fahrenholtz, W. G. (2021). Measurement of the melting temperature of ZrB2 as determined by laser heating and spectrometric analysis. *Journal of the American Ceramic Society, 104*(6), 2780–2787. doi:10.1111/jace.17634

Stock, R. F. (2012). *Influencing rolling contact fatigue and wear by different rail grades and contact conditions* [Doctoral dissertation]. University of Leoben, Austria.

Storch, S., Nellessen, D., Schaefer, G., & Reiter, R. (2003, October 1). Selective laser sintering: Qualifying analysis of metal based powder systems for automotive applications. *Rapid Prototyping Journal, 9*(4), 240–251. doi:10.1108/13552540310489622

Storr, B., Moore, L., Chakrabarty, K., Mohammed, Z., Rangari, V., Chen, C.-C., & Catledge, S. A. (2022). Properties of high entropy borides synthesized via microwave-induced plasma. *APL Materials, 10*(6), 061109. doi:10.1063/5.0098276

Strey, N. F., Rezende, A. B., da Silva Miranda, R., da Fonseca, S. T., Mei, P. R., & Scandian, C. (2021). Comparison of rolling contact fatigue damage between railway wheels and twin-disc test specimens. *Tribology International, 160*, 107037. doi:10.1016/j.triboint.2021.107037

Sukanya, R., Sakthivel, M., Chen, S. M., & Chen, T. W. (2018). A new type of terbium diselenide nano octagon integrated oxidized carbon nanofiber: An efficient electrode material for electrochemical detection of morin in the food sample. *Sensors and Actuators. B, Chemical, 269*, 354–367. Advance online publication. doi:10.1016/j.snb.2018.05.013

Sullivan, K. T., Piekiel, N. W., Wu, C., Chowdhury, S., Kelly, S. T., Hufnagel, T. C., Fezzaa, K., & Zachariah, M. R. (2012, January 1). Reactive sintering: An important component in the combustion of nanocomposite thermites. *Combustion and Flame, 159*(1), 2–15. doi:10.1016/j.combustflame.2011.07.015

Suñol, J. J., & Escoda, L. (2022). Novel Materials Synthesis by Mechanical Alloying/Milling. In Materials (Vol. 15, Issue 19). doi:10.3390/ma15196973

Sun, P., Fang, Z. Z., Zhang, Y., & Xia, Y. (2017). Review of the Methods for Production of Spherical Ti and Ti Alloy Powder. *Journal of the Minerals Metals & Materials Society, 69*(10), 1853–1860. doi:10.1007/s11837-017-2513-5

Sun, X., Zhang, H., Li, W., Ding, X., Wang, Y., & Vitos, L. (2019). Generalized Stacking Fault Energy of Al-Doped CrMnFeCoNi High-Entropy Alloy. *Nanomaterials (Basel, Switzerland), 10*(1), 59. doi:10.3390/nano10010059 PMID:31887990

Suryanarayana, C. (2004). Mechanical Alloying and Milling (Issue 1). Marcel Dekker.

Suryanarayana, C., Al Joubori Ahmed A.& Wang Zhi. (2022). Nanostructured Materials and nanocomposites by Mechanical Alloying: An Overview. *Metals and Materials International, 28*, 41–53. doi:10.1007/s12540-021-00998-5

Suryanarayana, C. (1996, August). Recent advances in the synthesis of alloy phases by mechanical alloying/milling. *Metals and Materials, 2*(4), 195–209. doi:10.1007/BF03026094

Suryanarayana, C. (2001). Mechanical Alloying and Milling. *Progress in Materials Science, 46*(1–2), 1–184. doi:10.1016/S0079-6425(99)00010-9

Suryanarayana, C. (2011, June 1). Synthesis of nanocomposites by mechanical alloying. *Journal of Alloys and Compounds, 509*, S229–S234. doi:10.1016/j.jallcom.2010.09.063

Suryanarayana, C. (2019). Mechanical Alloying: A Novel Technique to Synthesize Advanced Materials. *Research, 2019*, 2019/4219812. Advance online publication. doi:10.34133/2019/4219812 PMID:31549063

Suryanarayana, C., & Al-Aqeeli, N. (2013). Mechanically Alloyed Nanocomposites. *Progress in Materials Science, 58*(4), 383–502. doi:10.1016/j.pmatsci.2012.10.001

Suryanarayana, C., & Froes, F. H. (1990). Nanocrystalline Titanium-Magnesium Alloys through Mechanical Alloying. *Journal of Materials Research, 5*(9), 1880–1886. doi:10.1557/JMR.1990.1880

Suryanarayana, C., Ivanov, E., & Boldyrev, V. V. (2001). The science and technology of mechanical alloying. *Materials Science and Engineering A, 304–306*(1–2), 151–158. Advance online publication. doi:10.1016/S0921-5093(00)01465-9

Su, X., & Clayton, P. (1996). Surface-initiated rolling contact fatigue of pearlitic and low carbon bainitic steels. *Wear, 197*(1-2), 137–144. doi:10.1016/0043-1648(95)06888-0

Su, Y., Du, H., Huo, Y., Xu, Y., Wang, J., Wang, L., Zhao, S., & Xiong, S. (2016, June 1). Characterization of cationic starch flocculants synthesized by dry process with ball milling activating method. *International Journal of Biological Macromolecules, 87*, 34–40. doi:10.1016/j.ijbiomac.2015.11.093 PMID:26905465

Tabie, V. M., Li, C., Saifu, W., Li, J., & Xu, X. (2020). Mechanical properties of near alpha titanium alloys for high-temperature applications - a review. *Aircraft Engineering and Aerospace Technology, 92*(4), 521–540. doi:10.1108/AEAT-04-2019-0086

Takahashi, M., & Bhadeshia, H. K. D. H. (1991). A model for the microstructure of some advanced bainitic steels. *Materials Transactions. JIM, 32*(8), 689–696.

Tallarita, G., Licheri, R., Garroni, S., Barbarossa, S., Orrù, R., & Cao, G. (2020). High-entropy transition metal diborides by reactive and non-reactive spark plasma sintering: A comparative investigation. *Journal of the European Ceramic Society, 40*(4), 942–952. doi:10.1016/j.jeurceramsoc.2019.10.031

Tallarita, G., Licheri, R., Garroni, S., Orrù, R., & Cao, G. (2019). Novel processing route for the fabrication of bulk high-entropy metal diborides. *Scripta Materialia, 158*, 100–104. doi:10.1016/j.scriptamat.2018.08.039

Tang, L., Qian, H., Li, C., & Wu, X. (2022). Fretting Fatigue Test and Simulation Analysis of Steam Generator Heat Transfer Tube. *Metals, 13*(1), 67. doi:10.3390/met13010067

Tan, J. L., & Liew, K. W. (2023, March 10). Tribological Behaviour and Microstructure of an Aluminium Alloy-Based g-SiC Hybrid Surface Composite Produced by FSP. *Lubricants (Basel, Switzerland)*, *11*(3), 124. doi:10.3390/lubricants11030124

Tan, Y., Xia, Y., Teng, Z., Chen, C., Zhou, X., & Zhang, H. (2021). Synthesis and enhanced mechanical properties of compositionally complex MAX phases. *Journal of the European Ceramic Society*, *41*(8), 4658–4665. doi:10.1016/j.jeurceramsoc.2021.03.027

Tasak, E., Ziewiec, A., & Tuz, L. (2014). Resistance Welding of High-Manganese Cast Steel with Unalloyed Steel Rails. *Archives of Foundry Engineering*, *14*, 115–120.

Tawade, A. K., Kamble, B. B., Sharma, K. K. K., & Tayade, S. N. (2020). Simultaneous electrochemical investigations of dopamine and uric acid by in situ amino functionalized reduced grahene oxide. *SN Applied Sciences*, *2*(6), 1082. Advance online publication. doi:10.1007/s42452-020-2806-0

Teber, A., Schoenstein, F., Têtard, F., Abdellaoui, M., & Jouini, N. (2012, January 1). Effect of SPS process sintering on the microstructure and mechanical properties of nanocrystalline TiC for tools application. *International Journal of Refractory & Hard Metals*, *30*(1), 64–70. doi:10.1016/j.ijrmhm.2011.06.013

Technology and Challenges in Additive Manufacturing of Duplex Stainless Steels. (2021). *Biointerface Research in Applied Chemistry*, *12*(1), 1110–1119. Advance online publication. doi:10.33263/BRIAC121.11101119

Tenaglia, N. E., Massone, J. M., Boeri, R. E., & Speer, J. G. (2020). Effect of microsegregation on carbide-free bainitic transformation in a high-silicon cast steel. *Materials Science and Technology*, *36*(6), 690–698. doi:10.1080/02670836.2020.1732076

Teng, Z., Tan, Y., Zeng, S., Meng, Y., Chen, C., Han, X., & Zhang, H. (2021). Preparation and phase evolution of high-entropy oxides A2B2O7 with multiple elements at A and B sites. *Journal of the European Ceramic Society*, *41*(6), 3614–3620. doi:10.1016/j.jeurceramsoc.2021.01.013

Terbouche, A., Boulahia, S., Mecerli, S., Ait-Ramdane-Terbouche, C., Belkhalfa, H., Guerniche, D., Sehailia, M., Bachari, K., Mezaoui, D., & Hauchard, D. (2022). A novel hybrid carbon materials-modified electrochemical sensor used for detection of gallic acid. *Measurement*, *187*, 110369. Advance online publication. doi:10.1016/j.measurement.2021.110369

Terzi, M. (2018). Particle Size Distribution Analysis in Aggregate Processing Plants Using Digital Image Processing Methods. *Revista Română de Materiale / Romanian. Journal of Materials*, *48*(4), 514–521.

Tian, Y., Lu, C., Shen, Y., & Feng, X. (2019). Microstructure and corrosion property of CrMnFeCoNi high entropy alloy coating on Q235 substrate via mechanical alloying method. *Surfaces and Interfaces*, *15*, 135–140. Advance online publication. doi:10.1016/j.surfin.2019.02.004

Transparent Market Research. (2022). Particle Size Analyzer Market. *Outlook*, 2022–2031. https://transparencymarketresearch.com/particle-size-analyzer-market.html

Tressia, G., Alves, L. H., Sinatora, A., Goldenstein, H., & Masoumi, M. (2020). Effect of bainitic transformation on the microstructure and wear resistance of pearlitic rail steel. *Industrial Lubrication and Tribology*, *72*(9), 1095–1102. doi:10.1108/ILT-07-2019-0282

Trzepieciński, T., Slota, J., Kaščák, U., Gajdoš, I., & Vojtko, M. (2023, March 14). Friction Behaviour of 6082-T6 Aluminium Alloy Sheets in a Strip Draw Tribological Test. *Materials (Basel)*, *16*(6), 2338. doi:10.3390/ma16062338 PMID:36984219

Tu, J., & Reeves, N. (2019). Feasibility study of microneedle fabrication from a thin nitinol wire using a CW single-mode fiber laser. *Open Engineering*, *9*(1), 167–177. doi:10.1515/eng-2019-0023

Tun, K. S., Kumar, A., & Gupta, M. (2019). Introducing a High Performance Mg-Based Multicomponent Alloy as an Alternative to Al-Alloys. *Frontiers in Materials*, *6*, 1–6. doi:10.3389/fmats.2019.00215

Tunna, J., Sinclair, J., & Perez, J. (2007). A review of wheel wear and rolling contact fatigue. *Proceedings of the Institution of Mechanical Engineers. Part F, Journal of Rail and Rapid Transit*, *221*(2), 271–289. doi:10.1243/0954409JRRT72

Uysal, A. (2016, June 13). Investigation of flank wear in MQL milling of ferritic stainless steel by using nano graphene reinforced vegetable cutting fluid. *Industrial Lubrication and Tribology*, *68*(4), 446–451. doi:10.1108/ILT-10-2015-0141

Vadnere, A. P., & Kalpande, S. D. (2023, May 9). A review on effects of process parameters in milling challenging materials under nanofluid-based MQL conditions. *Industrial Lubrication and Tribology*, *75*(4), 361–371. doi:10.1108/ILT-01-2023-0010

Vega, H., Qin, M., & Luo, J. (2023). Thermodynamics of dual-phase compositionally complex ceramics: A case study of ultrahigh-entropy fluorite-bixbyite refractory oxides. *Journal of the European Ceramic Society*, *43*(5), 2104–2114. doi:10.1016/j.jeurceramsoc.2022.12.033

Velvaluri, P., Soor, A., Plucinsky, P., de Miranda, R. L., James, R. D., & Quandt, E. (2021). Origami-inspired thin-film shape memory alloy devices. *Scientific Reports*, *11*(1), 10988. doi:10.1038/s41598-021-90217-3 PMID:34040022

Verma, L., Andrew, J., Sivakumar, S. M., Balaganesan, G., Vedantam, S., & Dhakal, H. N. (2020). Ballistic Impact Behaviour of Glass/Epoxy Composite Laminates Embedded with Shape Memory Alloy (SMA) Wires. *Molecules (Basel, Switzerland)*, *26*(1), 138. doi:10.3390/molecules26010138 PMID:33396838

Virto. (2022). *Why Sieving?* https://www.virtogroup.com/why-sieving/

Vishwakarma, V., Kaliaraj, G. S., & Amirtharaj Mosas, K. K. (2022). Multifunctional Coatings on Implant Materials—A Systematic Review of the Current Scenario. *Coatings*, *13*(1), 69. doi:10.3390/coatings13010069

Vo, P., Goldbaum, D., & Wong, W. (2015). *Cold-spray processing of titanium and titanium alloys*. Elsevier Inc. doi:10.1016/B978-0-12-800054-0.00022-8

Walunj, G., Bearden, A., Patil, A., Larimian, T., Christudasjustus, J., Gupta, R. K., & Borkar, T. (2020, November 23). Mechanical and tribological behavior of mechanically alloyed Ni-TiC composites processed via spark plasma sintering. *Materials (Basel)*, *13*(22), 5306. doi:10.3390/ma13225306 PMID:33238641

Wang, W. (2016). Introductory Chapter: What is Chemical Sensor? In Progresses in Chemical Sensor. doi:10.5772/64626

Wang, Y.M., & Forssberg, E. (2001). *New milling technology*. Technical report.

Wang, J., Yang, B., Zhong, J., Yan, B., Zhang, K., Zhai, C., Shiraishi, Y., Du, Y., & Yang, P. (2017). Dopamine and uric acid electrochemical sensor based on a glassy carbon electrode modified with cubic Pd and reduced graphene oxide nanocomposite. *Journal of Colloid and Interface Science*, *497*, 172–180. Advance online publication. doi:10.1016/j.jcis.2017.03.011 PMID:28284071

Wang, K., Chen, L., Xu, C., Zhang, W., Liu, Z., Wang, Y., Ouyang, J., Zhang, X., Fu, Y., & Zhou, Y. (2020). Microstructure and mechanical properties of (TiZrNbTaMo)C high-entropy ceramic. *Journal of Materials Science and Technology*, *39*, 99–105. doi:10.1016/j.jmst.2019.07.056

Wang, Q., Sarkar, A., Wang, D., Velasco, L., Azmi, R., Bhattacharya, S. S., Bergfeldt, T., Düvel, A., Heitjans, P., Brezesinski, T., Hahn, H., & Breitung, B. (2019). Multi-anionic and -cationic compounds: New high entropy materials for advanced Li-ion batteries. *Energy & Environmental Science*, *12*(8), 2433–2442. doi:10.1039/C9EE00368A

Wang, W., & Ahn, S. H. (2017). Shape Memory Alloy-Based Soft Gripper with Variable Stiffness for Compliant and Effective Grasping. *Soft Robotics*, *4*(4), 379–389. doi:10.1089/soro.2016.0081 PMID:29251571

Wang, X. L., Wang, G. F., & Zhang, K. F. (2010, May 25). Effect of mechanical alloying on microstructure and mechanical properties of hot-pressed Nb–16Si alloys. *Materials Science and Engineering A*, *527*(13-14), 3253–3258. doi:10.1016/j.msea.2010.02.031

Wang, X., Pan, F., Sun, X., Li, Y., Zhou, J., Wang, Z., & Qin, C. (2022). Performance and Mechanism of Nanoporous Ni@NiO Composites for RhB Ultrahigh Electro-Catalytic Degradation. *Metals*, *13*(1), 38. doi:10.3390/met13010038

Wang, Z., Luo, J., Kuang, W., Jin, M., Liu, G., Jin, X., & Shen, Y. (2022). Strain Rate Effect on the Thermomechanical Behavior of NiTi Shape Memory Alloys: A Literature Review. *Metals*, *13*(1), 58. doi:10.3390/met13010058

Wedd, M., Ward-Smith, S., & Rawle, A. (2019). Particle size analysis. Encyclopedia of Analytical Science, 144–157. doi:10.1016/B978-0-12-409547-2.14522-6

Weiss, I., & Semiatin, S. L. (1998). Thermomechanical processing of beta titanium alloys — An overview. *Materials Science and Engineering A*, *243*(1-2), 46–65. doi:10.1016/S0921-5093(97)00783-1

Weiss, I., & Semiatin, S. L. (1999). Thermomechanical processing of alpha titanium alloys — An overview. *Materials Science and Engineering A*, *263*(2), 243–256. doi:10.1016/S0921-5093(98)01155-1

Wei, X. F., Liu, J.-X., Li, F., Qin, Y., Liang, Y.-C., & Zhang, G.-J. (2019). High entropy carbide ceramics from different starting materials. *Journal of the European Ceramic Society*, *39*(10), 2989–2994. doi:10.1016/j.jeurceramsoc.2019.04.006

Wei, Y., Niu, B., Liu, Q., Liu, Z., & Jiang, C. (2022). First-Principles Calculations of Structural and Mechanical Properties of Cu–Ni Alloys. *Crystals*, *13*(1), 43. doi:10.3390/cryst13010043

Williams, J. C., & Boyer, R. R. (2020). Opportunities and Issues in the Application of Titanium Alloys for Aerospace Components. *Metals*, *10*(6), 705. doi:10.3390/met10060705

Williams, N. (2015). The Magazine for the Metal Additive Manufacturing Industry. *Additive Manufacturing*, *1*(3), 84.

Wills, B. A., & Finch, J. A. (2016b). *Wills' mineral processing technology : an introduction to the practical aspects of ore treatment and mineral recovery*. Academic Press.

Wills, B. A., & Finch, J. A. (2016a). Particle Size Analysis. In *Wills' Mineral Processing Technology* (pp. 91–107). Elsevier., doi:10.1016/B978-0-08-097053-0.00004-2

Winiarski, J. P., Tavares, B. F., de Fátima Ulbrich, K., de Campos, C. E. M., Souza, A. A. U., Souza, S. M. A. G. U., & Jost, C. L. (2022). Development of a multianalyte electrochemical sensor for depression biomarkers based on a waste of the steel industry for a sustainable and one-step electrode modification. *Microchemical Journal*, *175*, 107141. Advance online publication. doi:10.1016/j.microc.2021.107141

Wołosiewicz-Głąb, M., Foszcz, D., & Ogonowski, S. (2017, July 1). Design of the electromagnetic mill and the air stream ratio model. *IFAC-PapersOnLine*, *50*(1), 14964–14969. doi:10.1016/j.ifacol.2017.08.2554

WResearch. (2022). *South Africa Particle Size Analysis Market Outlook*. https://www.6wresearch.com/industry-report/nigeria-denature-alcohol-market-outlook

Wright, A. J., & Luo, J. (2020). A step forward from high-entropy ceramics to compositionally complex ceramics: A new perspective. *Journal of Materials Science*, *55*(23), 9812–9827. doi:10.1007/s10853-020-04583-w

Wu, J., Lin, X., Qiao, H., Zhao, J., Ding, W., & Zhu, R. (2022). Microstructural Evolution and Surface Mechanical Properties of the Titanium Alloy Ti-13Nb-13Zr Subjected to Laser Shock Processing. *Materials (Basel)*, *16*(1), 238. doi:10.3390/ma16010238 PMID:36614576

Wu, Z., Liu, B., Wei, J., Yang, Y., Zhang, X., & Deng, J. (2022). Design of Porous Shape Memory Alloys with Small Mechanical Hysteresis. *Crystals*, *13*(1), 34. doi:10.3390/cryst13010034

Xiang, S., Du, C., Cao, X. & Chen, L. (2022). High-Entropy Ceramics Diborides for Hypersonic Applications. *Lect. Notes Electr. Eng.*, *821*, 247–256.

Xie, H. (2022). Research Progress of High-entropy Ceramic Materials. *Cailiao Daobao. Cailiao Daobao*, *36*, 20070201–20070208.

Xiong, B., Long, W., Chen, Z., Xia, C., Wan, H., & Yan, Y. (2009, March 5). Effects of element proportions on microstructures of Nb/Nb5Si3 in situ composites by spark plasma sintering. *Journal of Alloys and Compounds*, *471*(1-2), 404–407. doi:10.1016/j.jallcom.2008.03.095

Xu, Z. (2019). Study on microwave sintering process and surface texture characteristics of ceramic materials. In *IOP Conference Series: Materials Science and Engineering* (Vol. 677, No. 2, p. 022078). IOP Publishing. 10.1088/1757-899X/677/2/022078

Xue, D., Yuan, R., Zhou, Y., Xue, D., Lookman, T., Zhang, G., Ding, X., & Sun, J. (2016). Design of High Temperature Ti-Pd-Cr Shape Memory Alloys with Small Thermal Hysteresis. *Scientific Reports*, *6*(1), 28244. doi:10.1038/srep28244 PMID:27328764

Yadav, S., Zhang, Q., Behera, A., Haridas, R. S., Agrawal, P., Gong, J., & Mishra, R. S. (2021, October 5). Role of binder phase on the microstructure and mechanical properties of a mechanically alloyed and spark plasma sintered WC-FCC HEA composites. *Journal of Alloys and Compounds*, *877*, 160265. doi:10.1016/j.jallcom.2021.160265

Yahya, N. A., Rashid, T. N. I. T. A., & Hasmaliza, M. (2022). Effect of sintering temperature on physical properties of Trong clay and prefired material in low clay whiteware ceramic fabrication. *Materials Today: Proceedings*, *66*, 2716–2719. doi:10.1016/j.matpr.2022.06.501

Yamanaka, K., Mori, M., Onuki, Y., Sato, S., & Chiba, A. (2022). Dislocation Density of Electron Beam Powder Bed Fusion Ti–6Al–4V Alloys Determined via Time-Of-Flight Neutron Diffraction Line-Profile Analysis. *Metals*, *13*(1), 86. doi:10.3390/met13010086

Yamauchi, K., Ohkata, I., Tsuchiya, K., & Miyazaki, S. (Eds.). (2011). *Shape memory and superelastic alloys: Applications and technologies*. Elsevier. doi:10.1533/9780857092625

Yamazaki, K., & Haga, T. (2022). High-Speed and Low-Load Twin-Roll Casting of Al–5%Mg Strip. *Metals*, *13*(1), 72. doi:10.3390/met13010072

Yang, R., Hao, Y., & Li, S. (2014). Development and Application of Low-Modulus Biomedical Titanium Alloy Ti2448. Biomedical Engineering: Trends in Materials Science, 1–25.

Yang, G., Xue, B., Li, Z., Zhou, G., Zhang, S., Lu, N., Wen, L., & Zhang, D. (2022). Influence of Stress Field and Temperature Field on Residual Stress of 2A14 Aluminum Alloy Based on In Situ SAXS Method. *Materials (Basel)*, *16*(1), 170. doi:10.3390/ma16010170 PMID:36614509

Yang, L., Dai, Q., Liu, L., Shao, D., Luo, K., Jamil, S., Liu, H., Luo, Z., Chang, B., & Wang, X. (2020, June 1). Rapid sintering method for highly conductive Li7La3Zr2O12 ceramic electrolyte. *Ceramics International*, *46*(8), 10917–10924. doi:10.1016/j.ceramint.2020.01.106

Yang, S., Liang, P., Gao, F., Song, D., Jiang, P., Zhao, M., & Kong, N. (2022). The Comparation of Arrhenius-Type and Modified Johnson–Cook Constitutive Models at Elevated Temperature for Annealed TA31 Titanium Alloy. *Materials (Basel)*, *16*(1), 280. doi:10.3390/ma16010280 PMID:36614619

Yang, X., Zhang, X., Chen, J., Wang, Z., Li, X., Zhang, X., & Gao, L. (2022). Study of Hydrogen-Induced Plastic Damage Response of 7085-T7651 High-Strength Aluminum Alloy. *Metals*, *13*(1), 40. doi:10.3390/met13010040

Yang, Y., Song, L., Gao, T., & Nagy, Z. K. (2015, December 2). Integrated upstream and downstream application of wet milling with continuous mixed suspension mixed product removal crystallization. *Crystal Growth & Design*, *15*(12), 5879–5885. doi:10.1021/acs.cgd.5b01290

Yang, Z., & Zhang, F. (2020). Nanostructured bainitic bearing steel. *ISIJ International*, *60*(1), 18–30. doi:10.2355/isijinternational.ISIJINT-2018-693

Yan, X., Hu, J., Wang, L., Chai, Z., Sun, W., & Xu, W. (2021). The Coupled Effect of Thermal and Mechanical Stabilities of Austenite on the Wear Resistance in a 0.2C–5Mn-1.6Si Steel down to Cryogenic Temperatures. *Wear*, *486–487*, 204116. doi:10.1016/j.wear.2021.204116

Yan, X., Hu, J., Yu, H., Wang, C., & Xu, W. (2021). Unraveling the Significant Role of Retained Austenite on the Dry Sliding Wear Behavior of Medium Manganese Steel. *Wear*, *476*, 203745. doi:10.1016/j.wear.2021.203745

Yan, Y., Lu, C., Chen, Z., Zhuo, Y., Wang, C., & Jia, Q. (2023, April 17). Microstructures and High Temperature Tensile Behaviours of Laser Powder Bed Fusion Fabricated Al-Mn-Sc Alloy. *Metals*, *13*(4), 788. doi:10.3390/met13040788

Yasui, T., Hayashi, K., & Fukumoto, M. (2022). Behaviors of Micro-Arcs, Bubbles, and Coating Growth during Plasma Electrolytic Oxidation of β-Titanium Alloy. *Materials (Basel)*, *16*(1), 360. doi:10.3390/ma16010360 PMID:36614699

Ye, B., Wen, T., Huang, K., Wang, C. Z., & Chu, Y. (2019). First-principles study, fabrication, and characterization of (Hf 0.2 Zr 0.2 Ta 0.2 Nb 0.2 Ti 0.2)C high-entropy ceramic. *Journal of the American Ceramic Society*, *102*(7), 4344–4352. doi:10.1111/jace.16295

Yeh, J. W., Yeh, A. C. & Chang, S. Y. (2016). Potential applications and prospects. *High-Entropy Alloy. Fundam. Appl.*, 493–512. . doi:10.1007/978-3-319-27013-5_15

Yeh, J. W. (2006). Recent progress in high-entropy alloys. *Ann. Chim. Sci. des Mater.*, *31*, 633–648.

Yola, M. L., Gupta, V. K., Eren, T., Şen, A. E., & Atar, N. (2014). A novel electro analytical nanosensor based on graphene oxide/silver nanoparticles for simultaneous determination of quercetin and morin. *Electrochimica Acta*, *120*, 204–211. Advance online publication. doi:10.1016/j.electacta.2013.12.086

Yolton, C. F., & Froes, F. H. S. (2015). *Conventional titanium powder production*. Elsevier Inc. doi:10.1016/B978-0-12-800054-0.00002-2

Yoon, B. K., Lee, B. A., & Kang, S. J. (2005, October 1). Growth behavior of rounded (Ti, W) C and faceted WC grains in a Co matrix during liquid phase sintering. *Acta Materialia*, *53*(17), 4677–4685. doi:10.1016/j.actamat.2005.06.021

Yuan, K., Tu, T., Shen, C., Zhou, L., Liu, J., Li, J., Xie, K., & Zhang, G. (2022). Self-ball milling strategy to construct high-entropy oxide coated LiNi0.8Co0.1Mn0.1O2 with enhanced electrochemical performance. *J. Adv. Ceram.*, *11*(6), 882–892. doi:10.1007/s40145-022-0582-6

Yu, K., Su, J., Liu, M., Li, C., Lin, Y., Zhao, M., & Hao, C. (2023, March 17). Facile Synthesis of Al-20Si@Al2O3 Nanosheets Composite Powders and Its Refinement Performance on Primary Silicon in Al-20Si Alloy. *Crystals*, *13*(3), 514. doi:10.3390/cryst13030514

Yu, T., & Breslin, C. B. (2020). Review—Two-Dimensional Titanium Carbide MXenes and Their Emerging Applications as Electrochemical Sensors. *Journal of the Electrochemical Society*, *167*(3), 037514. Advance online publication. doi:10.1149/2.0142003JES

Yu, X., Wu, H., Zhang, Y., Yuan, R., Gu, Y., & Feng, Y. (2021). Influence of niobium microalloying on the thermal stability of medium-carbon carbide-free bainitic steel. *Journal of Materials Research and Technology*, *12*, 487–499. doi:10.1016/j.jmrt.2021.03.033

Zaara, K., Daza, J., Ben Mbarek, W., & Suñol, J.-J. (2022). Structural, Thermal and Magnetic Analysis of Two Fe-X-B (X = Nb, NiZr) Nanocrystalline Alloy. *Materials (Basel)*, *16*(1), 155. doi:10.3390/ma16010155 PMID:36614494

Zapata, D., Jaramillo, J., & Toro, A. (2011). Rolling contact and adhesive wear of bainitic and pearlitic steels in low load regime. *Wear*, *271*(1-2), 393–399. doi:10.1016/j.wear.2010.10.009

Zhang, C., Fu, H., Lin, J., & Lei, Y. (2019). The effect of silicon on microstructure and wear resistance in bainitic steel. *Transactions of the Indian Institute of Metals*, *72*(5), 1231–1244. doi:10.1007/s12666-019-01611-5

Zhang, C., Li, S., Fu, H., & Lin, Y. (2020). Microstructure evolution and wear resistance of high silicon bainitic steel after austempering. *Journal of Materials Research and Technology*, *9*(3), 4826–4839. doi:10.1016/j.jmrt.2020.03.003

Zhang, C., Mu, A., Wang, Y., & Zhang, H. (2019). Study on Dynamic Mechanical Properties and Constitutive Model Construction of TC18 Titanium Alloy. *Metals*, *10*(1), 44. doi:10.3390/met10010044

Zhang, J., Xu, X., & Chen, L. (2018). An ultrasensitive electrochemical bisphenol A sensor based on hierarchical Ce-metal-organic framework modified with cetyltrimethylammonium bromide. *Sensors and Actuators. B, Chemical*, *261*, 425–433. Advance online publication. doi:10.1016/j.snb.2018.01.170

Zhang, L., & Chen, L. (2019). A Review on Biomedical Titanium Alloys : Recent Progress and Prospect. *Advanced Engineering Materials*, *21*(4), 1–29. doi:10.1002/adem.201801215

Zhang, P., Jia, C., Zhao, Y., Luo, H., Tan, X., Ma, X., & Wang, Y. (2019). Detection of tiopronin in body fluids and pharmaceutical products using red-emissive DNA-stabilized silver nanoclusters as a fluorescent probe. *Mikrochimica Acta*, *186*(9), 609. Advance online publication. doi:10.1007/s00604-019-3730-0 PMID:31392427

Zhang, P., Jia, D., Yang, Z., Duan, X., & Zhou, Y. (2013). Influence of ball milling parameters on the structure of the mechanically alloyed SiBCN powder. *Ceramics International, 39*(2), 1963–1969. Advance online publication. doi:10.1016/j.ceramint.2012.08.047

Zhang, R., Zheng, C., Chen, C., Lv, B., Gao, G., Yang, Z., Yang, Y., & Zhang, F. (2021). Study on fatigue wear competition mechanism and microstructure evolution on the surface of a bainitic steel rail. *Wear, 482-483*, 203978. doi:10.1016/j.wear.2021.203978

Zhang, W., Wang, D., Chen, J., Ghassemi-Armaki, H., Carlson, B., & Feng, Z. (2022). A New Perspective of Post-Weld Baking Effect on Al-Steel Resistance Spot Weld Properties through Machine Learning and Finite Element Modeling. *Journal of Manufacturing and Materials Processing*, *7*(1), 6. doi:10.3390/jmmp7010006

Zhang, Y., Guo, W.-M., Jiang, Z.-B., Zhu, Q.-Q., Sun, S.-K., You, Y., Plucknett, K., & Lin, H.-T. (2019). Dense high-entropy boride ceramics with ultra-high hardness. *Scripta Materialia, 164*, 135–139. doi:10.1016/j.scriptamat.2019.01.021

Zhang, Y., Guo, Y., Chen, Y., Cao, Y., Qi, H., & Yang, S. (2019). Microstructure and Mechanical Properties of Al-12Si Alloys Fabricated by Ultrasonic-Assisted Laser Metal Deposition. *Materials (Basel), 13*(1), 126. doi:10.3390/ma13010126 PMID:31888072

Zhang, Y., Jiang, Z.-B., Sun, S.-K., Guo, W.-M., Chen, Q.-S., Qiu, J.-X., Plucknett, K., & Lin, H.-T. (2019). Microstructure and mechanical properties of high-entropy borides derived from boro/carbothermal reduction. *Journal of the European Ceramic Society*, *39*(13), 3920–3924. doi:10.1016/j.jeurceramsoc.2019.05.017

Zhang, Y., Liu, W., Liu, Y., Zhang, M., Tian, Y., & Chen, L. (2022). Research Progress on Corrosion Behaviors and Improvement Methods of Medical Degradable Mg–Based Alloys. *Metals*, *13*(1), 71. doi:10.3390/met13010071

Zhang, Y., Sun, S.-K., Guo, W.-M., Xu, L., Zhang, W., & Lin, H.-T. (2021). Optimal preparation of high-entropy boride-silicon carbide ceramics. *J. Adv. Ceram.*, *10*(1), 173–180. doi:10.1007/s40145-020-0418-1

Zhang, Y., Wang, P., Xia, Y., Zhou, P., Xie, Y., Xu, S., & Zhang, C. (2017). Application of a Shape-Memory Alloy Concentrator in Displaced Patella Fractures: Technique and Long-Term Results. *The Journal of Knee Surgery*, *30*(2), 166–173. doi:10.1055/s-0036-1584187 PMID:27206070

Zhang, Y., Zuo, T. T., Tang, Z., Gao, M. C., Dahmen, K. A., Liaw, P. K., & Lu, Z. P. (2014). Microstructures and properties of high-entropy alloys. *Progress in Materials Science*, *61*, 1–93. doi:10.1016/j.pmatsci.2013.10.001

Zhang, Z., Zhang, J., Zhao, X., Cheng, X., Jiang, S., & Zhang, Q. (2022). Effects of Al-Mg on the Microstructure and Phase Distribution of Zn-Al-Mg Coatings. *Metals*, *13*(1), 46. doi:10.3390/met13010046

Zhao, D., Vdonin, N., Bezgans, Y., Radionova, L., & Glebov, L. (2022). Correlating Electrode Degradation with Weldability of Galvanized BH 220 Steel during the Electrode Failure Process of Resistance Spot Welding. *Crystals*, *13*(1), 39. doi:10.3390/cryst13010039

Zhao, N., Nash, P., & Yang, X. (2005, December 30). The effect of mechanical alloying on SiC distribution and the properties of 6061 aluminum composite. *Journal of Materials Processing Technology*, *170*(3), 586–592. doi:10.1016/j.jmatprotec.2005.06.037

Zhao, P., Zhu, J., Li, M., Shao, G., Wang, H., & He, J. (2023). Influence of molten-salt-synthesised high-entropy (V0.2Cr0.2Mo0.2W0.2Ni0.2)B powder characteristics on superhard mechanism during spark plasma sintering. *Journal of Alloys and Compounds*, *947*, 169618. doi:10.1016/j.jallcom.2023.169618

Zhao, W., Liu, L., Zhang, F., Leng, J., & Liu, Y. (2019). Shape memory polymers and their composites in biomedical applications. *Materials Science and Engineering C*, *97*, 864–883. doi:10.1016/j.msec.2018.12.054 PMID:30678978

Zhao, X., Zhong, Q., Zhai, P., Fan, P., Wu, R., Fang, J., Xiao, Y., Jiang, Y., Liu, S., & Li, W. (2022). Microstructure and Wear Resistance of Laser-Clad Ni–Cu–Mo–W–Si Coatings on a Cu–Cr–Zr Alloy. *Materials (Basel)*, *16*(1), 284. doi:10.3390/ma16010284 PMID:36614623

Zheng, C., Dan, R., Zhang, F., Lv, B., Yan, Z., Shan, J., & Long, X. (2014). Effects of retained austenite and hydrogen on the rolling contact fatigue behaviours of carbide-free bainitic steel. *Materials Science and Engineering A*, *594*, 364–371. doi:10.1016/j.msea.2013.11.072

Zheng, C., Lv, B., Zhang, F., Yang, Z., Kang, J., She, L., & Wang, T. (2016). A novel microstructure of carbide-free bainitic medium carbon steel observed during rolling contact fatigue. *Scripta Materialia*, *114*, 13–16. doi:10.1016/j.scriptamat.2015.11.023

Zheng, X., Zhuang, X., Lei, Y., Chu, Z., Xu, J., Gao, L., & Sun, X. (2019). Corrosion Behavior of the Ti–6Al–4V Alloy in Sulfate-Reducing Bacteria Solution. *Coatings*, *10*(1), 24. doi:10.3390/coatings10010024

Zhong, Y., Sabarou, H., Yan, X., Yang, M., Gao, M. C., Liu, X., & Sisson, R. D. Jr. (2019). Exploration of high entropy ceramics (HECs) with computational thermodynamics - A case study with LaMnO3±δ. *Materials & Design*, *182*, 108060. doi:10.1016/j.matdes.2019.108060

Zhou, M., Xu, G., Tian, J., Hu, H., & Yuan, Q. (2017). Bainitic transformation and properties of low carbon carbide-free bainitic steels with Cr addition. *Metals, 7*(7), 1-13.

Zhou, H., Hu, L., Sun, Y., Zhang, H., Duan, C., & Yu, H. (2016). Synthesis of nanocrystalline AZ31 magnesium alloy with titanium addition by mechanical milling. *Materials Characterization*, *113*, 108–116. doi:10.1016/j.matchar.2016.01.014

Zhou, Y. L., Niinomi, M., & Akahori, T. (2004). Effects of Ta content on Young's modulus and tensile properties of binary Ti – Ta alloys for biomedical applications. *Materials Science and Engineering A*, *371*(1-2), 283–290. doi:10.1016/j.msea.2003.12.011

Zhou, Y., Niinomi, M., Akahori, T., Nakai, M., & Fukui, H. (2007). Comparison of Various Properties between Titanium-Tantalum Alloy and Pure Titanium for Biomedical Applications. *Materials Transactions*, *48*(3), 380–384. doi:10.2320/matertrans.48.380

Zhu, G., & Liu, Z. (2005). Article. *Sheng wu yi xue gong cheng xue za zhi = Journal of biomedical engineering = Shengwu yixue gongchengxue zazhi, 22*(5), 1082–1084.

Zhu, J., Tan, Z., Tian, Y., Gao, B., Zhang, M., Wang, J., & Weng, Y. (2020). Effect of tempering temperature on microstructure and mechanical properties of bainitic railway wheel steel with thermal damage resistance by alloy design. *Metals, 10*(9), 1-13.

Zhu, S-S., Kan, H-M., & Feng, X-J. (2016). *The Preparation Method of Aluminum-magnesium Alloys*. doi:10.2991/ame-16.2016.50

Zupanič, F., Žist, S., Albu, M., Letofsky-Papst, I., Burja, J., Vončina, M., & Bončina, T. (2023, April 7). Dispersoids in Al-Mg-Si Alloy AA 6086 Modified by Sc and Y. *Materials (Basel), 16*(8), 2949. doi:10.3390/ma16082949 PMID:37109786

About the Contributors

Santhy A currently serves as Assistant Professor (Senior grade) at the Department of Chemistry, Amrita School of Physical Sciences, Amritapuri Campus, Amrita Vishwa Vidyapeetham. She has joined Amrita School of Physical Sciences in 2014. She pursued her MSc Chemistry from Sree Narayana College, Kollam in 2011 and qualified CSIR-NET in chemical sciences in 2013. Santhy received her Ph.D in 2023 in the area of electrochemical sensors for biological, pharmaceutical and environmental applications under the guidance of Dr. Beena S, Assistant Professor (Senior grade) at the Department of Chemistry, Amrita School of Physical Sciences, Amritapuri Campus.

Gaurav Kumar Bansal, Scientist, Materials Engineering Division, CSIR – National Metallurgical Laboratory, Jamshedpur, India, is a Scientist working on the research and development of advanced steel grades for the benefit of the environment and society. His research interests mainly include Thermodynamic and Kinetic simulation for alloy and process design, Heat treatment and Thermo-mechanical processing, Microstructure-property correlation, etc. He has 10+ years of experience, and completed over 20 nos. of Scientific and Industry-oriented R&D projects. Mr. Bansal has more than 10 Patents and 20+ peer-reviewed scientific publications to his credit. He is also the recipient of AWSAR Award 2019 for best science story, B.R. Nijhawan award 2020 and 2021 for the top SCI publications, S.P. Mehrotra award 2020 for best Colloquium speaker, etc.

Sateeshkumar Kanakannavar is working as an Assistant Professor in the Department of Mechatronics and Automation Engineering, Indian Institute of Information Technology Bhagalpur, Bihar. He did his PhD from National Institute of Technology Karnataka, Surathkal. His current interest is in the performing sector required for 3D braided polymer composites, involving conceptualization, infrastructure development and engineering the preform to meet the performance requirements of the end product.

Mohammad Mohsin Khan is an accomplished Assistant Professor in the Mechanical Engineering Department at the esteemed National Institute of Technology, Srinagar. He has garnered a reputation as a prolific researcher and educator with extensive experience in teaching and research. Dr. Khan's research contributions are highly regarded in the academic community. He has published over 40 papers in national and international journals of repute and has presented his research findings at various international conferences. Dr. Khan is dedicated to advancing the engineering field as a member of several prestigious research societies, including the Material Research Society of India, the Tribology Society of India, and the Institution of Engineers. In addition to his research pursuits, Dr. Khan has distinguished himself as an exceptional teacher and project supervisor. He teaches at undergraduate and postgraduate levels and has guided over 25 projects for B.Tech and M.Tech students. Dr. Khan's dedication to education and research has been recognized with numerous awards and accolades. Dr. Khan's research interests are focused on ultra-high-temperature ceramic composites, light metal alloys, and bio-composites. His work in these areas has the potential to contribute to materials science and engineering significantly.

Deogratius Titus Maiga is a post-doctoral research fellow at Mintek, working under the Division of Measurement and Control, the Section of Instrumentation, and the subsection of Research and Development. He holds a Ph.D. (Chem) degree from Unisa and an MSc (Chem) degree from the University of Johannesburg. With 16 years of experience in research, mainly in analytical chemistry, water quality, environmental chemistry, nanomaterials, and nanotechnology, he possesses a diverse expertise in various scientific domains. Currently, Dr. Maiga is involved in several R&D projects at Mintek, including the investigation of the robustness of the Cynoprobe Cyanide Analyzer, carbon management (C2 Meter), and the utilization of mixed substrates in the passive treatment of AMD/CMD. Additionally, he is actively engaged in co-supervising and mentoring master's students, further contributing to the development of young researchers in his field. Dr. Maiga's primary objective is to leverage his skills and experience to enhance research, development, and innovation (RDI) outputs at Mintek, thereby fostering advancements and breakthroughs in relevant scientific areas.

Gurudas Mandal is serving as Assistant Professor in the department of Metallurgical Engineering at Kazi Nazrul University, Asansol-713340, West Bengal, India. His research areas are structure-process-properties correlations of metals and materials viz., Steel, Composite, and magnetic materials etc. He has taught courses on Physical Metallurgy, Phase Transformation, Corrosion Science, Heat Treatment Technology, and Metal Processing for more than three years. He is an active editorial board member and reviewer in several reputable journals. His collaborative and interdisciplinary research influences and enhances cutting-edge technologies on an international scale.

Khuthadzo Mudzanani is a research engineer at Mintek-Measurement and Control Division (MAC). She holds a PhD in Chemical Engineering from the University of the Witwatersrand. Her doctorate study was focused on the evaluation of the feasibility and optimization of beneficiating municipal wastewater sediments for the synthesis of methane as a pathway for renewable energy integrated with combined heat and power systems. Other research fields includes functionalization of the adsorption materials for reclamation of municipal and industrial wastewater. She is currently a chief investigator of the project titled “Development of a robust online instrument that measures particle size with the target applications in mining operations”. Dr Khuthadzo is also steer guiding the carbon management project which aims to address the challenges of activated carbon applications in mineral processing.

Harveer Singh Pali is working as an Assistant Professor in Mechanical Engineering Department at National Institute of Technology, Srinagar. NIT Srinagar is premier institute of National importance under aegis Ministry of Education, Government of India. Dr. Pali has a sustained track record of quality and interesting teaching experience of 15 years. He always enjoyed teaching Engineering Thermodynamics, Applied Thermodynamics, RAC, Thermal Energy Systems and Renewable Energy. Dr. Pali has excellent research contribution in terms of research publication in journal of repute (SCI, SCOPUS etc.). He has published more than 30 papers international journals of repute and more than 20 in proceedings of International conferences. He also published 2 patents 04 book chapters. He has guided 06 M. Tech thesis and 20+ B. Tech Projects. He presently guiding 04 Ph.D. students.

Tabrez Qureshi holds an M.Tech degree in Mechanical Engineering and is currently pursuing his Ph.D. in the same field at the National Institute of Technology, Srinagar, Jammu and Kashmir, India. He is specifically focused on designing and synthesizing advanced high-entropy ceramics for ultra-high-temperature applications. As a senior research fellow, he has worked extensively on the various “Fabrication techniques of high-entropy borides” and “High-entropy hydrides for hydrogen storage”.

Rejithamol R. received her M. Sc. degree in Organic Chemistry from M. G. University campus, Kottayam in 2009. She then worked as a Project Assistant at the Regional Research Laboratory (now NIIST), Trivandrum in the area of organic synthesis for two years and joined the Department of Chemistry at Amrita School of Physical Sciences, Amritapuri Campus in 2013. She received her Ph. D. in the area of electrochemical sensors in 2020 from Amrita Vishwa Vidyapeetham. Her research interests include fabrication of electrochemical sensors for biomolecules like amino acids, steroids.

Beena S. currently serves as Assistant Professor (Sl. Gd.) in the Department of Chemistry, School of Physical Sciences, Amrita Vishwa Vidyapeetham, Amritapuri Campus. She received her Ph. D. in 2011 from Cochin University of Science and Technology in the area of electrochemical sensors for the determination of metal ions. She also worked at the Polish Academy Sciences, Poland as part of DST INDO-POLISH project. She is currently involved in the fabrication of disposable electrochemical sensors for point of care diagnosis, food analysis and electrode materials for energy devices.

Index

M

N

P

R

S

T

V

W

www.ingramcontent.com/pod-product-compliance
Lightning Source LLC
LaVergne TN
LVHW061545290926
849162LV00004B/918